Lecture Notes in Mathematics

Edited by A. Dold and B. Eckmann

1209

Differential Geometry Peñíscola 1985

Proceedings of the 2nd International Symposium held at Peñíscola, Spain, June 2–9, 1985

Edited by A. M. Naveira, A. Ferrández and F. Mascaró

Springer-Verlag

Berlin Heidelberg New York London Paris Tokyo

Editors

Antonio M. Naveira
Angel Ferrández
Francisca Mascaró
Departamento de Geometría y Topología, Facultad de Matemáticas
Burjasot, Valencia, Spain

Mathematics Subject Classification (1980): 22EXX, 53AXX, 53BXX, 53CXX, 57DXX, 58AXX, 58EXX, 58GXX

ISBN 3-540-16801-X Springer-Verlag Berlin Heidelberg New York
ISBN 0-387-16801-X Springer-Verlag New York Berlin Heidelberg

Printing and binding: Druckhaus Beltz, Hemsbach/Bergstr.
2146/3140-543210

PREFACE

The 1st International Symposium on Differential Geometry was held in Peñíscola in October, 1982, and the Proceedings were published in the Lecture Notes in Mathematics , No. 1045. Due to the interest raised in the mathematical community we get encouraged to continue with periodical meetings on Differential Geometry and related topics. Following this line we have decided to organize the 2nd International Symposium on Differential Geometry also held in Peñíscola, June 2-9, 1985.

This Symposium was attended by approximately seventy mathematicians from all over the world. The present volume includes the texts of most of the contributions presented covering several areas of Differential Geometry and related topics; like Riemannian manifolds and submanifolds, Hermitian and Kaehlerian manifolds, symplectic and contact structures, foliations and analysis on manifolds. The editors regret that due to a general editorial requirement of homogeneity in a Lecture Notes volume, it was not possible to include other interesting contributions. All papers have been examined by referees and we want to thank them for their valuable task.

We would like to thank to all the following institutions which have contributed to the success of the meeting with generous financial support: Ministerio de Educación y Ciencia (C.A.I.C.Y.T.), Consellería de Cultura, Educación y Ciencia de la Comunidad Autónoma Valenciana, Facultad de Ciencias Matemáticas de la Universidad de Valencia, Diputaciones Provinciales of Valencia and Castellón and Cultural Services of the French Embassy in Madrid. We thank also to the Instituto de Estudios de Administración Local de Peñíscola for allowing us to use their premises during this Symposium, and Peñíscola's Council for their kind hospitality with the organizing committee and with all the participants and J.M. Yturralde, author of the poster of the meeting. Finally, we would like to acknowdledge the colaboration of all the members of the "Departamento de Geometría y Topología" of the Universidad de Valencia.

April, 1986
The Editors.

LIST OF PARTICIPANTS

E. Abbena
U. Torino, Italy

V. Aldaya
U. Valencia, Spain

S.I. Andersson
U. Göteborg, Sweden

T. Aubin
U. Paris VI, France

A. Asada
U. Sinsyu, Japan

R.L. Bishop
U. Illinois, U.S.A.

R.A. Blumenthal
U. St. Louis, U.S.A.

N. Bokan
U. Beograd, Yugoslavia

E. Calabi
U. Pennsylvania, U.S.A.

F.J. Carreras
U. Valencia, Spain

B.Y. Chen
U. Michigan, U.S.A.

D. Chinea
U. La Laguna, Spain

Ph. Delanoë
U. Nice, France

M. de León
U. Santiago, Spain

H. Donnelly
Purdue U., U.S.A.

F.J. Echarte
U. Sevilla, Spain

M. Falcitelli
U. Bari, Italy

M. Fernández
U. Santiago, Spain

A. Ferrández
U. Valencia, Spain

E. Gallego
U.A. Barcelona,Spain

S. Garbiero
U. Torino, Italy

O. Gil-Medrano
U. Valencia, Spain

S. Gillot
U. Paris VI, France

V.V. Goldberg
New Jersey Inst. of Tech., U.S.A.

J.C. González
U. La Laguna, Spain

A. Gray
U. Maryland, U.S.A.

J. Grifone
U. Toulouse, France

K. Grove
U. Maryland, U.S.A.

L. Gualandri
U. Bolonia, Italy

G. Guasp
U.A. Barcelona, Spain

M.T. Iglesias
U. Santiago, Spain

I. Kashiwada
U. Ochanomizu, Japan

R.S. Kulkarni
U. Indiana, U.S.A.

A. Lichnerowicz
Collège de France, France

M. Fernández Andrés
U. Sevilla, Spain

L.M. Fernández Fernández
U. Sevilla, Spain

A. Machado
U. Lisboa, Portugal

Y. Maeda
U. Keio, Japan

J.M. Margalef
C.S.I.C. Madrid, Spain

F. Marhuenda
U. Valencia, Spain

V. Marino
U. Calabria, Italy

R.A. Marinosci
U. Lecce, Italy

F. Mascaró
U. Valencia, Spain

R.S. Millman
N.S.F., Washington, U.S.A.

V. Miquel
U. Valencia, Spain

J. Monterde
U. Valencia, Spain

A. Montesinos
U. Valencia, Spain

S. Montiel
U. Granada, Spain

A.M. Naveira
U. Valencia, Spain

A. Pastor
U. Valencia, Spain

A.M. Pastore
U. Bari, Italy

P.Piccinni
U. Roma, Italy

G. Lupacciolu
U. Roma, Italy

M. Llabres
U.A. Barcelona, Spain

J.F. Pommaret
Ec. Nat. Ponts et Ch., Paris, France

A.H. Rocamora
U. Valencia, Spain

B. Rodríguez
U. Santiago, Spain

A. Ros
U. Granada, Spain

C. Ruiz
U. Granada, Spain

M.R. Salgado
U. Santiago, Spain

S. Segura
U. Valencia, Spain

K. Sekigawa
U. Niigata, Japan

M. Sekizawa
U. Tokyo Gakugei, Japan

R. Sulanke
Humboldt U., Berlin, East Germany

Ph. Tondeur
U. Illinois, U.S.A.

J.F. Torres Lopera
U. Santiago, Spain

F. Urbano
U. Granada, Spain

L. Vanhecke
U. Leuven, Belgium

F. Varela
U. Murcia, Spain

H.E. Winkelnkemper
U. Maryland, U.S.A.

TABLE OF CONTENTS

CAUCHY UNIQUENESS IN THE RIEMANNIAN OBSTACLE PROBLEM

Stephanie B. Alexander
I. David Berg
Richard L. Bishop

Department of Mathematics
University of Illinois
1409 West Green Street
Urbana, Illinois 61801

§1. Introduction. In a Riemannian manifold-with-boundary it is not generally true that geodesics (locally shortest paths) have the Cauchy uniqueness property. For example, whenever there is a boundary direction in which the boundary bends away from the interior, there will be an obvious one-parameter family of distinct geodesics of a given sufficiently small length which start in that direction. Two geodesics of such a family coincide on an initial segment, after which one of them is a geodesic of the interior. The following theorem states that this is the only manner in which Cauchy uniqueness fails. By an involute of a geodesic β is meant a geodesic which has the same initial point, initial tangent vector, and length as β, and which consists of a maximal initial segment in common with β followed by a nontrivial geodesic segment of the interior.

Theorem 1 (Cauchy uniqueness for manifolds-with-boundary). Every boundary point has a neighborhood in which: if two geodesic segments with the same initial point, initial tangent vector and length do not coincide, then one of them has its right endpoint in the interior and is an involute of the other.

In studying the geodesics of Riemannian manifolds-with-boundary, we are studying, for example, the geometry of wavefront propagation around an obstacle in an isotropic medium (since the orthogonal trajectories of the wavefronts are geodesics in the appropriate Riemannian manifold-with-boundary). In another approach to the analysis of bifurcation of geodesics, Arnol'd has studied the singularities of wave fronts generated by obstacles in general position [A]; for an extensive bibliography, see [ABB]. Here we are interested in what controls the tendency of geodesics emanating from a boundary point, sometimes bearing on the boundary and sometimes travelling through the interior, to pull apart from one another. We deal throughout with a C^∞ Riemannian manifold-with-boundary M, with C^∞ boundary B. In this setting, the geodesics cannot be described by differential equations with Lipschitz continuous coefficients. Our problem, that of analyzing how the geodesics of M are controlled by the interaction between the geodesic equations of the boundary and the interior, is one to which routine techniques do not apply.

In [ABB], it was shown that M cannot have "positively infinite curvature" at the boundary. Specifically, an estimate was found for the least distance at which two geodesics emanating from a point p can rejoin (i.e., the cut radius of M), and the rate at which geodesics from p can pull together. The estimate is in terms of the tubular radius of M, namely the supremum of all R for which M can be imbedded in some Euclidean space so that every point at distance R or less from M is the center of a closed ball which meets M at a single point. This single invariant reflects upper bounds of curvature and lower bounds of cut radius for both the interior and the boundary of M.

On the other hand, M can certainly have "negatively infinite curvature" at the boundary, in the above sense of having a family of geodesics with the same initial tangent. However, our proof of Theorem 1 has the following consequence. For any $C > 1$ there is a $\rho > 0$ depending only on C and the tubular radius of M such that: if the endpoint of a geodesic from any point p is moved along the boundary at a fixed distance $s \leq \rho$ from p, then the ratio of the endpoint separation to the initial tangent separation is bounded above by Cs. In this sense, we have established integral bounds for the tangential curvature at a boundary point.

No assumption is made on the boundary except smoothness. If we were to assume that every geodesic segment had finitely many switchpoints (points at which it switches from nontrivial boundary segment to nontrivial interior segment), then Cauchy uniqueness would be straightforward. However, besides boundary segments, interior segments and switchpoints, a geodesic may contain accumulation points of switch points, which we call intermittent points. Indeed, it is easy to construct a geodesic segment containing a set of positive measure of intermittent points. When intermittent points are allowed, the proof of Cauchy uniqueness becomes quite delicate. Intermittent points are not rare, in the sense that, for example, arbitrarily close in the C^2 norm to any negatively curved surface in E^3 is a surface for which the manifold-with-boundary lying to one side has geodesics with intermittent points. On the other hand, manifolds-with-boundary whose geodesics have intermittent points are apparently not generic in any reasonable sense. An important motivation for studying the general case is that this seems to be the most direct way to obtain curvature bounds, independent of the number and behavior of switchpoints, even in the generic case.

It is shown in [ABB] that every point of B has a neighborhood in which no involute of any geodesic segment can end on B. (This is because locally an involute of a geodesic β lies above β with respect to the inward normals to B.) Therefore Theorem 1 is an immediate consequence of the following statement:

(*) two geodesics with the same initial tangent vector at p must coincide on an interval if each geodesic touches B arbitrarily close to p.

The proof of claim (*) will use the following regularity theorem. By a collar neighborhood for B is meant a neighborhood of some boundary point which is foliated by interior geodesics normal to B.

Theorem 2. [ABB] Any geodesic β of M is C^1. The acceleration of β exists except at the (countably many) switchpoints, and vanishes at the intermittent points. If β lies in a collar neighborhood for B, then its normal projection to B is C^2 with locally Lipschitz second derivative.

§2. Proof of claim (*). On a collar neighborhood for B at p, consider coordinates $x_1,\dots,x_n$ adapted to B: that is, let x_n be the distance from B, and let the x_k for $k < n$ be arbitrary coordinates on B which are extended to be constant on the geodesics normal to B. Then the equation of a geodesic β of M has the following form ([ABB]):

(1) $$x_k'' = -\sum_{i,j} x_i' \, x_j' \, \Gamma_{ijk}$$

(2) $$x_n'' = -\kappa_\beta - \sum_{i,j<n} x_i' \, x_j' \, \Gamma_{ijn}.$$

Here the Γ_{ijk} are the connection coefficients of the interior of M, and κ_β is defined on the boundary segments of β to be the normal curvature of B in the direction of β', and at all other points of β to be 0. (1) holds everywhere, and (2) holds except at switchpoints. (For instance, on boundary segments the right-hand side of (2) vanishes, as required, because $-\Gamma_{ijn}$ is the second fundamental form of B.) Since the Γ_{nnk} vanish, we may rewrite (1) as follows:

(3) $$x_n'' = F_k(x_1,\dots,x_n,x_1',\dots,x_{n-1}') + x_n' \, G_k(x_1,\dots,x_n,x_1',\dots,x_{n-1}')$$

where the F_k and G_k are C^∞ in their $2n-1$ arguments.

Suppose β and γ are geodesics with the same initial tangent vector at p. Set $\beta_i = x_i \circ \beta$, $\gamma_i = x_i \circ \gamma$, $X = (\gamma_1 - \beta_1,\dots,\gamma_{n-1} - \beta_{n-1})$, $Y = X'$, $z = \gamma_n - \beta_n$, $F = (F_1,\dots,F_{n-1})$, and $G = (G_1,\dots,G_{n-1})$. Then (3) gives

$$Y' = F \circ \gamma - F \circ \beta + \gamma_n' \, G \circ \gamma - \beta_n' \, G \circ \beta$$
$$= F \circ \gamma - F \circ \beta + \gamma_n'(G \circ \gamma - G \circ \beta) + (zG \circ \beta)' - z(G \circ \beta)' .$$

Therefore

$$\|Y(t)\| = \left\|\int_0^t Y'\right\|$$
$$= \left\|(z\, G \circ \beta)(t) + \int_0^t [-z(G \circ \beta)' + (F \circ \gamma - F \circ \beta) + \gamma_n'(G \circ \gamma - G \circ \beta)]\right\|.$$

Now we apply the regularity theorem for geodesics and the fact that F and G are functions of $x_1,\dots,x_n,x_1',\dots,x_{n-1}'$ only. It follows that γ_k' and β_k' for $k < n$ are C^1, and hence so are $F \circ \gamma$, $F \circ \beta$, $G \circ \gamma$ and $G \circ \beta$; furthermore, γ_n' and β_n' are continuous. Therefore

(4) $$\|Y(t)\| < A_1|z(t)| + \int_0^t [A_2|z| + A_3\|X\| + A_4\|Y\|]$$

for t sufficiently small, where

(5) $$\|X(t)\| = \left\|\int_0^t Y\right\| < t \max_{[0,t]}\|Y\|.$$

Now suppose that the following bounded slope condition were known to hold for t sufficiently small:

(a) $$|z| < C\|X\|.$$

Then (4), (5) and **(a)** imply:

$$\|Y(t)\| < D\, t \max_{[0,t]}\|Y\|.$$

From this it follows that Y vanishes for t sufficiently small, and hence X and z vanish also, so that β and γ coincide.

Thus the proof of claim **(*)** reduces to proving that **(a)** holds for geodesics β and γ having the same initial tangent vector, each touching the boundary arbitrarily close to p. **(a)** may be viewed as saying that β and γ do not behave even occasionally as though one were an involute of the other. **(a)** is obviously true if t is restricted to those values at which both β and γ are on B. For other values of t, however, showing that the horizontal separation $\|X\|$ does not become small with respect to the vertical separation $|z|$ seems to be delicate. (Recall that since β and γ have the same initial tangent, we have no control on how quickly $\|X\|$ approaches 0.) In the next section we prove a general lemma from which **(a)** follows.

§3. Proof of inequality **(a)**. The following lemma states that if p is a boundary point, then for any $C > 0$ there is a neighborhood in M of p in which **(a)** holds for any two geodesics from p which end on the boundary. We take N to be a Riemannian extension of M, without boundary and of the same dimension as M.

Lemma. For any θ, $0 < \theta < \pi/2$, there is a neighborhood U of p in which: if β and γ are two geodesics from p which end on B, then any vector which looks from a point of β to a point of γ in the exponential map of N makes an angle greater than θ with the inward normal field Z of B.

Proof. First we show that p has a neighborhood U such that the N-distance $d(t) = d(\beta(t), \gamma(t))$ is increasing for any two geodesics in U from p. (Note that when β and γ have the same initial tangent, it is nontrivial to claim that d is increasing on any interval at all.) A differential inequality is proved in [ABB] for the Euclidean distance $f(t)$ between two geodesics of speed no more than one, where N has been isometrically imbedded in a Euclidean space (of any codimension). Specifically, except at the countably many points where f'' fails to exist,

$$f'' \geq -K^2 f$$

where $R = 1/K$ is the tubular radius of the imbedding. That is, f is never more concave than an appropriate sinusoid of period $2\pi R$. It is immediate that the cut radius of M at p is no less than πR. It follows also that, if the geodesics have the same initial point,

$$f'(t) > A\, f(t)/t$$

for $A > 0$ and t sufficiently small. Since it can be verified by comparing the first variation formulas in N and in the ambient Euclidean space that $|d' - f'| < Bf$ for $B > 0$, we obtain $d'(t) > 0$ for $0 < t < c$.

We also need a fact about geodesics in the Riemannian manifold N without boundary. If q is in a tube over an N-geodesic segment S, that is, in the diffeomorphic image under $\exp_N$ of all tangent vectors normal to S whose length does not exceed some constant, then the sight vector from S to q is the unit vector normal to S in the direction of q. By a <u>θ-wedge</u> $(0 < \theta < \pi/2)$ over S is meant all points in a tube over S whose sight vector makes an angle of at most θ with some fixed parallel vector field along S. Given p and θ, there are θ^* $(\theta < \theta^* < \pi/2)$ and a neighborhood U of p such that in U, for any θ-wedge W and θ^*-wedge W^* over any N-geodesic segment S, we have:

(b) if both endpoints of an N-geodesic segment T in W^* take the maximum sight angle θ^*, then any points of T which lie in the interior of W lie closer to S than does either endpoint of T.

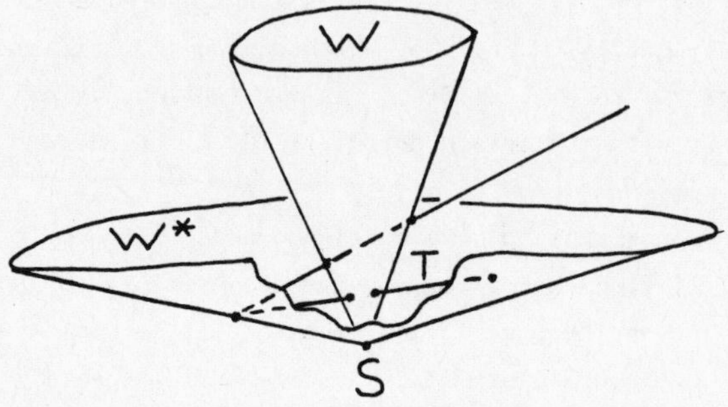

The figure illustrates this claim in E^4 by showing the projection into a 3-plane normal to S. In general, θ^* and U certainly can be chosen so that curves lying in an exponentiated hyperplane normal to S and having sufficiently small normal curvature have the property described in **(b)**. But parallel translation along S defines an identification of each exponentiated hyperplane with the initial one, which preserves W, W^* and distance from S. The claim **(b)** follows from the fact that if U is sufficiently small then the corresponding projection map carries the N-geodesics T into curves of arbitrarily small normal curvature. This in turn may be seen from the geodesic equation in Fermi coordinates on the tube over S.

Given θ, we now show that there is a neighborhood U of p in which: if β and γ are two geodesics from p which start at angle less that $\pi/4$ from each other and end on B, and if γ lies in a tube over β, then the sight vector from β to $\gamma(t)$, for any t, makes angle greater than θ with the inward normal field Z. This suffices to prove the lemma. (Restricting to vectors normal to β is permissible because the tangent vectors to β and γ may be assumed to be arbitrarily close to their initial values, and in particular, arbitrarily close to normal to Z.)

Given θ, choose U and θ^* as in the first two paragraphs. Suppose for the sake of contradiction that for β and γ as above, the sight vector $X(t_0)$ from β to $\gamma(t_0)$ makes angle no greater than θ with Z. Reparametrize β so that $X(t_0)$ is based at $\beta(t_0)$. Let S be a two-sided involute of β: namely, the N-geodesic segment defined on the same parameter interval as β and having the same velocity vector at t_0 as β. Consider the θ^*-wedge W^* along S centered on the parallel extension of $Z(\beta(t_0))$. If U has been chosen sufficiently small, we may assume that for any such wedge: S remains in M; B does not intersect the interior of W^*; and β remains in an inverted wedge along S of sufficiently small angle to ensure that

$$d(q,S) \le d(q,\beta) \tag{6}$$

for all q in W^*. Since $\gamma(t_0)$ is in the interior of W^*, it lies on some maximal nontrivial N-geodesic segment T of γ. Since γ begins and ends on B, this segment T enters W^* at some $\gamma(t)$ in ∂W^*, where $t < t_0$, and leaves W^* at $\gamma(u)$, $u > t_0$. Therefore by **(b)**,

$$d(\gamma(t),S) \ge d(\gamma(t_0),S).$$

However, from (6) and the fact that the distance between β and γ is increasing, we have

$$d((\gamma(t),S) \le d(\gamma(t),\beta) \le d(\gamma(t),\beta(t)) < d(\gamma(t_0),\beta(t_0)) = d(\gamma(t_0),S).$$

This contradiction completes the proof of the lemma, and of Theorem 1.

REFERENCES

[A] V. I. Arnol'd, Singularities in the calculus of variations, J. of Soviet Math. 27, 2679-2712 (1984).

[ABB] Stephanie B. Alexander, I. D. Berg, R. L. Bishop, The Riemannian obstacle problem, Illinois J. Math., to appear.

NON-ABELIAN HODGE THEORY VIA HEAT FLOW*

Stig I. Andersson
Research Group for Global Analysis and Applications

Chalmers University of Technology and
University of Göteborg
S-412 96 Göteborg, Sweden

Synopsis

The very recent efforts in non-abelian cohomology theory have resulted in a reasonable structure in dimensions < 3.

This paper treats the natural continuation of this theory, in giving a formulation of the associated Hodge theory, exhibiting the harmonic element in each given cohomology class.

In the abelian category, the (heat propagation) semigroup generated by the Laplace-Beltrami operator on p-forms, preserves cohomology (Milgram/Rosenbloom) and can be used to construct the Hodge element.

The appearance of natural differential operators also in the non-abelian case makes it possible to employ a similar construction, yielding a non-abelian Hodge theory.

This construction of the Hodge element differs from the one proposed by B. Gaveau.

Table of contents

* Research supported by STU under contract 85-3370 and by NFR under contracts 3610-115, 5342-100.

0. Introduction

An extension of sheaf cohomology to the non-abelian category is desirable for many reasons. Besides general reasons of mathematical completeness, there are also a number of important situations where a formulation in terms of non-abelian cohomology is natural and/or necessary.

Such situations are for instance;

- the classification of fiber spaces, where isomorphism classes of vector bundles (of rank n) on the manifold X $\Big\} \cong H^1(X, GL(n))$
- extension theory for groups, c.f. [At];
- Galois theory, c.f. [Se] ;
- the Riemann-Hilbert problem, c.f. [On 2] and [R];
- Bäcklund theory and the Yang-Mills problem in mathematical physics, where a central problem is to obtain information about vector bundle connections from prescribed curvature data. Indeed, in a certain sense, Yang-Mills theory **is** just a 2-dimensional non-abelian Hodge theory.

Whereas de Rham theory is concerned with describing (singular) cycles in terms of cohomology classes, i.e. roughly the isomorphism

$$\underset{\text{(p-forms)}}{H^p(M)} \times \underset{\text{(p-cycles)}}{H_p(M)} \longrightarrow \mathbf{C}$$

given by

$$([\omega], [c]) \longrightarrow \int_c \omega \in \mathbf{C},$$

Hodge theory is the natural continuation and selects in each cohomology class the unique *harmonic* element.

Somewhat independent of the precise definitions chosen in the formulation of the non-abelian cohomology theory (NAC) itself, Hodge theory can be compactly formulated using the analogue of heat flow in the abelian situation [MR].

This paper is concerned with such a formulation of Hodge theory and to a lesser extent with the underlying NAC, which we shall only briefly review in the next section.

There exists already one formulation of non-abelian Hodge theory, given by Gaveau [G], which we shall sketch in Section III.1. for reasons of comparison and completeness. Our approach is disjoint from the one of Gaveau.

Acknowledgements: I am grateful to Professor Antonio Naveira for his invitations to lecture in Valencia and to this conference, thereby also allowing me to enjoy the hospitality and enthusiasm of his research group.

A kind invitation by Professor Rolf Sulanke (Berlin) to Reinhardsbrunn (Thüringen) enabled me to meet Professor Arkadij Oniščik (Jaroslavl) to whom I am deeply indebted for generously sharing with me his insight into non-abelian cohomology.

As usual, discussions with Professor Akira Asada (Matsumoto) contributed much of the general inspiration and motivation for this work.

I. NON-ABELIAN COHOMOLOGY THEORY (NAÇ)

I.0. General Status

There is by now a reasonable version (or even two versions) for dimensions < 3, including Hodge theory, non-abelian Poincaré Lemma (proven in two different ways, c.f. Asada [As 6] and Gross [Gr]) and an associated theory of characteristic classes.

As for dimension ≥ 3 there is a tentative Poincaré 3-Lemma as well as some tentative formulations of specific aspects of the general theory. It is worthwhile observing though, that the formulation of Hodge theory proposed here and the attached theory of characteristic classes are not restricted to low dimensions.

Generally speaking, there exists three lines of development,

- the "abstract" approach, working entirely inside the frame of homological algebra,
- the 1- and 2-dimensional cohomologies by Oniščik and Tolpygo, with a stronger geometric flavour and formulated as a Čech theory [On 1-4], [To 1-3]
- the strongly geometric approach by Asada [As 1-6], [An 1-2], utilizing tools from global analysis (partial differential operators on vector bundles) and obstruction theory.

The abstract approach has its roots in the works in the 1950's by Grothendiek and this line of development was pursued by among others P. Dedecker, J. Frenkel, T.A. Springer and J. Giraud.
Since this development has found a very elaborate form in the book of Giraud [Gir] and since this formulation will play little rôle for us, we shall leave it out of our brief characterization.

We shall however need some concepts and ideas from the other two approaches, to a brief description of which we shall therefore devote the next two subsections.

I.1. The 1- and 2-cohomologies of Oniščik and Tolpygo.

Initiated by A.L. Oniščik in a series of papers [On 1-4] and motivated by various classification problems, this approach was elaborated by A.K. Tolpygo [To 1-3] and B. Gaveau [G].

The basic notion is the one of a *non-abelian cochain complex* (NÇÇ).

Def. 1.1. A NÇÇ is a triplet $K = \{K^o, K^1, K^2\}$ such that

- K^o, K^1 are groups and K^2 just a set with a base point;
- K^o acts on K^1 and K^2 by automorphisms, σ_1 and σ_2 respectively;
- there exists a twisted homomorphism $\delta_o : K^o \longrightarrow K^1$

i.e $\delta_o (ab) = \delta_o (a) \sigma_1 (a) (\delta_o(b)), \qquad a, b \in K^o$

and a homomorphism $\delta_1 : K^1 \longrightarrow K^2$ with the property

$$\delta_1 \rho (a) (\gamma) = \sigma_2 (a) \delta_1 (\gamma), \tag{1.1}$$

$\gamma \in K^1$, $a \in K^o$, with $\rho (a) (\gamma) := \delta_o (a) \sigma_1 (a) (\gamma) \in K^2$.

Clearly (1.1) implies $\delta_1 \delta_o (a) = e \in K^o$.

Def. 1.2. For a NÇÇ $K = \{K^o, K^1, K^2\}$,

K^j, $j = 0, 1, 2$ are the *j-cochains* and

$Z^j (K) := \mathrm{Ker}\, \delta_j$, $j = 0, 1$ the *j-cocycles* . For the cohomologies we define

$$H^0 (K) := Z^0 (K)$$
$$H^1 (K) := Z^1 (K) / \rho (K^o).$$

Note: $\rho(K^o)(\gamma) = \{\rho(a)(\gamma) \mid a \in K^o\}$ is the orbit of $\gamma \in K^1$, $\rho(K^o) = \{\rho(K^o)(\gamma) \mid \gamma \in K^1\}$ and in particular $\rho(K^o)(e) = \text{Ran}\ \delta_o$.

We view $H^1(K)$ as a set with base point Ran δ_o.

Let now, K be a NÇÇ and L a subcomplex with inclusion homomorphism

$$i : L \longrightarrow K.$$

Def. 1.3. Let $\xi_L 1 := \{a \in K^o \mid \delta_o(a) \in L^1\}$ and define

$$H^o(K/L) := \{[a] \in K^o / L^o \mid a \in \xi_L 1\}$$

Theorem 1.4. Let K and L be as above, then

$$e \longrightarrow H^o(L) \xrightarrow{i^*} H^o(K) \xrightarrow{p} H^o(K^o/L^o) \xrightarrow{\delta^*} H^1(L) \xrightarrow{i^*} H^1(K)$$

is *exact* , $p : K^o \longrightarrow K^o / L^o$ being the canonical projection and δ^* a Bockstein operator.

In a very natural way, based on the standard Čech cohomology of a topological space with coefficients in a sheaf of groups, one has

Def. 1.5. Let X be a topological space. A *non-abelian sheaf (cochain) complex* (NSÇ) is a triplet $K = \{K^o, K^1, K^2\}$, such that

- K^o, K^1 are sheaves of groups over X, K^2 is a sheaf of sets with base section ;
- K^o acts on K^1 and K^2 by automorphisms, σ_1 and σ_2 resp.;
- there exist a twisted homomorphism $\delta_o : K^o \longrightarrow K^1$ and a homomorphism $\delta_1 : K^1 \longrightarrow K^2$, such that $K_x = \{K_x{}^o, K_x{}^1, K_x{}^2\}$ is a NÇÇ relative to them for each $x \in X$.

Remark 1.6. To any NSÇ K there is associated the NÇÇ $C^\infty(X,K) := \{C^\infty(X,K^o), C^\infty(X,K^1), C^\infty(X,K^2)\}$

Def 1.7. Let K be a NSÇ, then we introduce the *sheaf cohomology sets* as

$$\zeta^o(K) = \text{Ker}\,\delta_o \quad ; \quad \zeta^1(K) := \text{Ker}\,\delta_1/\rho\,(K^o)$$

(where δ_o, δ_1, ρ now denote the corresponding sheaf maps). Furthermore, let F be a sheaf of groups over X such that $\zeta^o(K) = F$ and $\zeta^1(K) = \{e\}$, i.e.

$$e \longrightarrow F \overset{i}{\longrightarrow} K^o \overset{\delta^o}{\longrightarrow} K^1 \overset{\delta^1}{\longrightarrow} K^2$$

is exact, then the NSÇ is said to be a *resolution* of the sheaf F.

For a resolution one has naturally induced exact cohomology sequences (c.f. [On 3], p 63).

The model for a NSÇ is provided by the *differential of a Lie-group valued map* :

Let G be a Lie group with Lie algebra $\mathfrak{g}$, X a C^∞-manifold.
Consider

$C^\infty(X, G)$: = sheaf of germs of smooth maps $X \longrightarrow G$;

$\Lambda^p(X, \mathfrak{g}) := C^\infty(X, \Lambda^p T^*(X) \otimes \mathfrak{g})$ ($\mathfrak{g}$-valued differential p-forms)

We define a map $d_o : C^\infty(X, G) \longrightarrow \Lambda^1(X, \mathfrak{g})$ by

$$d_o(f)_x := T_{f(x)}\left(\gamma\left(f(x)^{-1}\right)\right) \cdot T_x(f) : T_x(X) \longrightarrow \mathfrak{g} \qquad (1.2)$$

where $\gamma(s): G \longrightarrow G$ is the left translation diffeomorphism $x \longrightarrow sx$.
The differential (1.2) has the following properties;

- for $f_1, f_2 \in C^\infty(X, G)$, $d_o(f_1 f_2)_x = \text{Ad}\left(f_2(x)^{-1}\right) d_o(f_1)_x + d_o(f_2)_x$;
- $d_o(f^{-1})_x = -\,\text{Ad}\left(f(x)\right) d_o(f)_x$;
- $f_1 f_2^{-1}$ locally constant $\Leftrightarrow d_o(f_1) = d_o(f_2)$;
- for the exponential map exp: $\mathfrak{g} \longrightarrow G$ one has in any point $\xi \in \mathfrak{g}$

$$d_o\,(\exp)_\xi = \sum_{p=0}^{\infty} \frac{1}{(1+p)!} \left(\mathrm{ad}\,(-\xi)\right)^p$$

Remark 1.8. In case $G = \mathbf{R}^n$, $d_o\,(f) = df$ (ordinary differential) and in case $G \subset GL\,(\mathbf{R},n)$ $d_o(f) = df \cdot f^{-1}$ (logarithmic differential).

Def. 1.9. Let $d_1 : \Lambda^1\,(X, \mathfrak{g}) \longrightarrow \Lambda^2\,(X, \mathfrak{g})$ be given by

$$d_1\,(\alpha) = d\,\alpha + \frac{1}{2}\,[\alpha,\alpha] = d\,\alpha + \alpha \wedge \alpha$$

Clearly $d_1\, d_o = 0$ and one has the

Non-abelian Poincaré 1-Lemma:: $d_1\,(\alpha) = 0 \Leftrightarrow \alpha = d_o\,(f)$ locally.

Let $C_t^\infty\,(X,G)$ be the sheaf of germs of constant G-valued maps. We shall later on be interested in the properties of the sequence

$$0 \longrightarrow C_t^\infty\,(X,G) \longrightarrow C^\infty\,(X,G) \xrightarrow{d_o} \Lambda^1(x,\mathfrak{g}) \xrightarrow{d_1} \Lambda^2\,(x,\mathfrak{g}) \qquad (1.3)$$

and various extensions thereof, but here we just mention the trivially verified

Lemma 1.10. $\{C^\infty\,(X,G), \Lambda^1\,(X, \mathfrak{g}), \Lambda^2\,(X, \mathfrak{g})\}$ with the crossed homomorphism d_o: $C^\infty\,(X,G) \longrightarrow \Lambda^1\,(X, \mathfrak{g})$ and the homomorphism d_1: $\Lambda^1\,(X, \mathfrak{g}) \longrightarrow \Lambda^2\,(X, \mathfrak{g})$ is a NSÇ. Here the representation of $C^\infty\,(X,G)$ in $\Lambda^1\,(X, \mathfrak{g})$ and $\Lambda^2\,(X, \mathfrak{g})$ is the one induced by Ad.

This scheme, has been further elaborated by Tolpygo, to yield a definition of 2-cohomology sets (c.f. [To 1, To 3]).

I.2. The PDO Approach. Asada Connections.

The basic observation in this approach is that not only are vector bundle connections a special kind of partial differential operators (PDO), but also - and conversely - can connections be naturally attached to a given PDO (or even pseudodifferential operator, c.f. [An 2]) on the section of a vector bundle.

Initiated and pursued in a series of studies by A. Asada [As 1-6], additional aspects were treated by S.I. Andersson [An 1-2].
We shall here just indicate the basic notions and leave the refinements for the interested reader, since only some notions will appear in the formulation of Hodge theory.

Let $\mathrm{Diff}_k(E_1, E_2)$ be the differential operators of order k between the section of the vector bundles $E_1 \longrightarrow X$, $E_2 \longrightarrow X$, i.e. $P \in \mathrm{Diff}_k(E_1, E_2)$ implies

$$P : C_1 \longrightarrow C_2, \quad \left(C_j = C^\infty(X, E_j)\right)$$

and on $U_i \cap U_j$; $P_i\, t^1_{ij} = t^2_{ij}\, P_j$ (locality)
(1.4)

for the local restriction $P_i := P \,|\, C^\infty(v_i, E_1 \,|\, U_i)$ and where t^1_{ij}, t^2_{ij} are the transition functions for the bundles E_1 and E_2 respectively;

$$t^1_{ij} : v_i \cap v_j \longrightarrow GL(n_1) \qquad (n_1 = \mathrm{rank}\, E_1).$$

Here $\{v_i\}$ denotes just open sets, $v_i \subset X$.

Given now a third vector bundle $E \longrightarrow X$, we consider the bundles $E_1 \otimes E$, $E_2 \otimes E$ with transition functions T^1_{ij}, T^2_{ij} respectively. We define the (index-preserving) lifting $\wp \in \mathrm{Diff}_k(E_1 \otimes E, E_2 \otimes E)$ of $P \in \mathrm{Diff}_k(E_1, E_2)$ such that $\sigma_k(\wp) = \sigma_k(P) \otimes \mathrm{Id}_E$ holds for the symbols. As a rule, locality is violated for $\wp$ i.e. (1.4) no longer holds.

This motivates

Def. 1.11. $Q = \{Q_i\}$, $Q_i \in \operatorname{Diff}_r(E_1 \otimes E \,|\, \nu_j, E_2 \otimes E \,|\, \nu_i)$, $r < k$

is an *Asada E-connection of P* iff $\wp + Q \in \operatorname{Diff}_k(E_1 \otimes E, E_2 \otimes E)$, i.e.

$$(\wp_i + Q_i)\, T^1{}_{ij} = T^2{}_{ij}\, (\wp_j + Q_j) \quad \text{on } \nu_i \cap \nu_j \tag{1.5}$$

Using the local picture, one easily describes the obstruction to (1.5);

$$W_{ij} := Q_i\, T^1{}_{ij} - T^2{}_{ij}\, Q_j \tag{1.6}$$

as well as its symbol $\sigma_{k-1}(W_{ij})$ (obviously to highest order the symbol of (1.6) vanishes). Let $\sigma_k(W) = \{\sigma_k(W_{ij})\}$, then assuming $0 = \sigma_{k-1}(W) = \ldots\ldots = \sigma_{k-j}(W)$; there exists an Asada E-connection of P of order $\leq k - (j+2)$ iff $\sigma_{k-(j+1)}(W) = 0$.

We call $\sigma_{k-(j+1)}(W)$ the *obstruction of order* $k-(j+2)$.

The Asada construction now proceeds by the following series of observations:

- $\sigma_{k-s}(W_{ij}) \in C^\infty(U_i \cap U_j, \zeta_{k-s} \,|\, U_i \cap U_j)$ where ζ_{k-s} is the vector bundle
  $\zeta_{k-s} := \operatorname{Hom}(E_1, E_2) \otimes S^{k-s}(T^*(X)) \otimes \operatorname{Hom}(E,E)$:
  ($S^p(E)$: p:th symmetric product of the bundle E)
- $\sigma_{k-s}(W_{ij})$ is independent of Q for order (Q) $\leq k-s$;
- $\sigma_{k-s}(W_{ij}) = 0$ in $C^\infty(X, \zeta_{k-s}) \Leftrightarrow$ there exists an E-connection of P of order $k-(s+1)$;
- $\sigma_{k-s}(W_{ij})$ is a twisted 1-cocycle in the sense that

$$\sigma_{k-s}(W_{ij})\, T^1{}_{jr} + T^2{}_{ij}\, \sigma_{k-s}(W_{jr}) = \sigma_{k-s}(W_{ir}) \text{ on } U_i \cap U_j \cap U_r, \text{ and}$$

$$T^2{}_{ji}\, \sigma_{k-s}(W_{ij})\, T^1{}_{ji} = -\, \sigma_{k-s}(W_{ji}) \text{ on } U_i \cap U_j \;;$$

$\sigma_{k-s}(W_{ij})$ thus determines an *obstruction class* $\Sigma_{k-s} := \{[\sigma_{k-s}(W_{ij})]\}$ in $H^1(X, N_s)$ (N_s is the range of a certain easily computable differential operator of order s).
In other words; there exists an E-connection of p of order k-(s+1) iff $\Sigma_{k-s} = 0$ in $H^1(X, N_s)$ i.e. iff Σ_{k-s} is a 1-coboundary.

From this, it is easy to formulate exact sheaf sequences, the corresponding cohomology sequences as well as the analogues of Chern classes etc (c.f. As [1], An [2]).

The extension to complexes of PDO is equally simple and it is not difficult to see that we have in fact generated another example of NSÇ.

II. HEAT FLOW APPROACH TO HODGE THEORY FOR ELLIPTIC COMPLEXES

The version of Hodge theory which we shall generalize to the non-abelian situation, is the heat flow approach. First formulated in 1951 by Milgram and Rosenbloom [MR] for closed Riemannian manifolds, this approach was later elaborated and generalized by Spencer (complete manifolds with restrictions on the curvature tensor [Sp]), Yosida (open manifolds [Y]) and - still more completely - Gaffney [Ga].
Basically one is here concerned with the global behaviour of the heat kernel for the Laplacian.
As a starting point, we shall in this section give a slight generalization of this theory to the context of *elliptic complexes*.

Remark 2.1. For the complex variable situation and the connection to spectral geometry c.f. [St].

Remark 2.2 We could as well have formulated the theory for general Fredholm complexes, a degree of generality which we shall not need however in dealing just with elliptic (pseudo-) differential operators on a closed manifold.

Let X be a closed n-dimensional C^∞-manifold and $E_j \longrightarrow X$, $j = 0,\ldots,N$ a sequence of C^∞-vector bundles with C^∞-sections $C_j = C^\infty(X,E_j)$.

Let $L_j: C_j \longrightarrow C_{j+1}$ be differential operators of fixed order k and consider

$$(E): 0 \longrightarrow C_o \xrightarrow{L_o} C_1 \longrightarrow \dots\dots \xrightarrow{L_{N-1}} C_N \longrightarrow 0$$

which we assume to be an elliptic complex, i.e.

- $L_j L_{j-1} = 0, \ \forall j$ and
- the associated symbol sequence ;

$$0 \longrightarrow \pi^*(E_o) \xrightarrow{\sigma(L_o)} \pi^*(E_1) \dots\dots \xrightarrow{\sigma(L_{N-1})} \pi^*(E_N) \longrightarrow 0$$

is exact. Here $\sigma(L_j)$ = (global) k-symbol of L_j.

$\{L_j\}$ are the differentials of degree +1 of the cochain complex (E) with j-cochains C_j. As usual we define the cohomology for (E) as

$$H^j(E): = Z^j(E)/B^j(E)$$

with the j-cocycles $Z^j(E): = \mathrm{Ker}\,(L_j: C_j \longrightarrow C_{j+1})$ and
the j-coboundaries $B^j(E): = \mathrm{Ran}\,(L_{j-1}: C_{j-1} \longrightarrow C_j)$.

To (E) one can associate various conjugated complexes and the one of interest to us here is the *adjoint complex* (obtained by algebraic conjugation), (E*). This is defined by equipping each vector bundle $E_j \longrightarrow X$ with a Hermitean structure (arising e.g. from a Riemannian structure on X), inducing a pre-Hilbert structure on the smooth sections C_j which we denote by $(.,.)_j$.

Let L^*_j be the adjoint with respect to $(.,.)_j$, i.e.

$$(L^*_j u, \upsilon)_j = (u, L_j \upsilon)_{j+1}, \qquad \upsilon \in C_j, \quad u \in C_{j+1}$$

$L_j^*: C_{j+1} \longrightarrow C_j$ are the differentials of degree -1 of the adjoint chain complex

$$(E^*): 0 \longleftarrow C_o \overset{L_o^*}{\longleftarrow} C_1 \longleftarrow \ldots \overset{L^*_{N-1}}{\longleftarrow} C_N \longleftarrow 0$$

($C^*_j \cong C_j$ via $(.,.)_j$), for which we define the homology by

$$H_j\,(E^*) := Z_j(E^*)/B_j(E^*)$$

with the j-cycles $Z_j(E^*) := \mathrm{Ker}\,(L^*_{j-1}: C_j \longrightarrow C_{j-1})$ and
the j-boundaries $B_j\,(E^*) := \mathrm{Ran}\,(L^*_j: C_{j+1} \longrightarrow C_j)$.

Remark 2.3 We could obtain further interesting situations by considering other conjugated complexes, like *the dual complex* (topological conjugation), the *transposed complex* (geometric conjugation), *the dual of the transposed* etc. For their (co-) homologies we could easily formulate the analogues of e.g. Serre and Poincaré duality and the de Rham theorem.

Remark 2.4 From $\sigma\,(L^*_j) = (-1)^k\,\sigma\,(L_j)^*$, $\sigma\,(L_{j+1}\,L_j) = \sigma\,(L_{j+1}) \cdot \sigma\,(L_j)$ and $L^*_{j-1}\,L^*_j = (L_j \cdot L_{j-1})^* = 0$ we have that (E*) is also an elliptic complex.

Remark 2.5 Because of elliptic regularity, all the properties formulated for (E) could as well have been formulated for the complex obtained by taking closures in Sobolev norms

$$L_j: H^s\,(X, E_j) \longrightarrow H^{s-k}\,(X, E_{j+1}).$$

Def. 2.6 A complex

$$(P): 0 \longleftarrow C_o \overset{P_o}{\longleftarrow} C_1 \overset{P_1}{\longleftarrow} \ldots\ldots \overset{P_{N-1}}{\longleftarrow} C_N \longleftarrow 0$$

is called a parametrix for (E) if

$$P_j\,L_j + L_{j-1}\,P_{j-1} \equiv \mathrm{Id}_j\ (\mathrm{mod}\ C^\infty) \qquad (2.1)$$

with $P_j \in L^{-k}\,(X; E_{j+1}, E_j)$ (the pseudodifferential operators $C_{j+1} \longrightarrow C_j$ of order -k and type (1,0)) and Id_j = identity map $C_j \longrightarrow C_j$. Relation (2.1) means that there exists a $K_j \in L^{-\infty}(X; E_{j+1}, E_j)$ (the smoothing operators) for each $j = 0,\ldots, N$ such that

$$P_j L_j + L_{j-1} P_{j-1} = Id_j - K_j$$

Note, that since X is compact, the K_j are *compact* operators.

Remark 2.7 A parametric is thus, in the language of homological algebra just a cochain homotopy (of pseudodifferential operators)

$$Id_j \sim K_j$$

between the identity and a particular endomorphism $K = \{K_j\}$ of the complex, consisting of compact operators.

That $K = \{K_j\}$ is an endomorphism of the complex follows from a trivial calculation, $L_j K_j = K_{j+1} L_j$.
Hence, the induced maps on the cohomologies $[K_j] : H^j(E) \longrightarrow H^j(E)$ are just the identities

$$[K_j]\,[f] = [f] \quad ([f] = \text{cohomology class of } f \in Z^j(E))$$

i.e.

$$K_j f \equiv f \ (\mathrm{mod}\ B^j(E)),\ \forall\, f \in Z^j(E) \qquad (2.2)$$

Def. 2.8 To the complex (E), we associate the Laplacians

$$\Delta_j := L^*_j L_j + L_{j-1} L^*_{j-1} : C_j \longrightarrow C_j,\ j = 0,\ldots,N$$

Trivially we have that $L_j \Delta_j = \Delta_{j+1} L_j$ and the Laplacians are thus endomorphisms of (E).

Lemma 2.9 The Δ_j are elliptic of order 2k and have self-adjoint extensions of which we shall fix one, denoted by D_j, to work with henceforth.

Proof: This follows from the symbol calculus

$$\sigma(\Delta_j) = (-1)^k\,(\sigma(L_j^*)\,\sigma(L_j) + \sigma(L_{j-1})\,\sigma(L^*_{j-1}))$$

and the exactness of the symbol sequence. That Δ_j is essentially self-adjoint is a consequence of the compactness of X. Since namely Δ_j is elliptic and $2k > 0$, $\Delta_j \pm iI$ is also elliptic.Suppose u solves $\Delta_j u = \pm iu$, $u \in L^2(X, E_j)$. By ellipticity $u \in C_j$ and since $\Delta_j^* = \Delta_j$ (formally s.a.) we have that $(\Delta_j u,u)_j = \pm i\,(u, u)_j$ i.e. $(u, u)_j = 0$ so $u = 0$ and Δ_j is hence e.s.a.

Def. 2.10 A section $f \in C_j$ is called *harmonic* if $f \in \operatorname{Ker} D_j$. We denote by

$$\Pi_j: C_j \longrightarrow \operatorname{Ker} D_j$$

the orthogonal projection onto the space of harmonic sections.

Lemma 2.11 $f \in C_j$ is harmonic iff $L_j f = 0$ and $L^*_{j-1} f = 0$.

Proof: Let f be harmonic, then $0 = (L^*_j L_j f + L_{j-1} L^*_{j-1} f, f)_j = (L_j f, L_j f)_{j+1} + (L^*_{j-1} f, L^*_{j-1} f)_{j-1} = \|L_j f\|_{j+1} + \| L^*_{j-1} f\|_{j-1}$, implying $L_j f = 0$, $L^*_{j-1} f = 0$. The converse statement is obvious.

One has now the fundamental

Theorem 2.12 To each elliptic complex (E) on a closed manifold X, there exists a parametrix (P).

Proof: Cf. [RS], Section 3.2.3.1.

Furthermore, one has the central result in Hodge theory for elliptic complexes.

Theorem 2.13 (Hodge-Kodaira) Let (E) be an elliptic complex over the closed manifold X. Then there exists a parametrix

$$(Q): \longleftarrow C_o \xleftarrow{Q_o} C_1 \xleftarrow{Q_1} \cdots\cdots \xleftarrow{Q_{N-1}} C_N \longleftarrow 0$$

$Q_j \in L^{-k}(X; E_{j+1}, E_j)$ and with associated smoothing operators $K_j = \Pi_j \in L^{-\infty}(X; E_j, E_j)$ i.e.

$$Id_j = \Pi_j + Q_j L_j + L_{j-1} Q_{j-1} \qquad j = 0,...., N.$$

Proof. Cf. Kotake [Ko].

Corollary 2.14. For an elliptic complex (E) over a closed manifold X, we have that:

- $L_j C_j$ is closed in C_{j+1}, Ker L_j has a topological complement in C_j and there is a topological decomposition

$$C_j = \text{Ran } L^*_j \oplus \text{Ran } L_{j-1} \oplus \text{Ker } D_j.$$

- the map Ker $D_j \ni f \longrightarrow [f] \in H^j$ (E) is an isomorphism

$$\text{Ker } D_j \cong H^j (E)$$

- dim H^j (E) $< \infty$, $\forall j$.

Proof: Let (P) be a parametrix

$$Id_j - K_j = P_j L_j + L_{j-1} P_{j-1}$$

i.e. $(Id_j - K_j) : Z^j (E) \longrightarrow B^j (E) \longrightarrow Z^j (E)$ and since $(Id_j - K_j) \mid Z^j (E)$ is Fredholm, codim Ran $(Id_j - K_j) < \infty$ in $Z^j(E)$ so codim $B^j (E) < \infty$ i.e. dim $H^j (E) < \infty$. By the same argument; $L_j C_j = B^{j+1}$ (E) is closed in Z^{j+1} (E) and hence in C_{j+1}.
The topological decomposition is immediate from the existence of the special parametrix in Thm 2.13 with $K_j = \Pi_j$.
To show that $Z^j (E) \equiv \text{Ker } L_j$ has a topological complement in C_j, define

$$F_j : Z^j (E) \oplus B^{j+1} (E) \longrightarrow C_j \quad (B^{j+1} \equiv \text{Ran } L_j)$$

by $\quad F_j (a,b) : = a + P_j (b).$

F_j is Fredholm: $\forall\, u \in C_j$ we have

$$(Id_j - K_j)u = L_{j-1} P_{j-1} u + P_j L_j u = F_j (L_{j-} P_{j-1}u, L_j u)$$

$(Id_j - K_j)$ Fredholm $\Rightarrow$ Ran F_j closed and dim Coker $F_j < \infty$.
On the other hand, for $(a,b) \in Z^j(E) \oplus B^{j+1}(E)$ we have;

$$L_j F_j (a,b) = L_j P_j (b) = (L_j P_j + P_{j+1} L_{j+1}) (b) = (Id_{j+1} - K_{j+1}) (b)$$

so

$$\text{Ker } F_j \subset \text{Ker } L_j F_j = \text{Ker } (Id_{j+1} - K_{j+1}) \Big| B^{j+1} (E),$$

hence

$$\dim \text{Ker } F_j \leq \dim \text{Ker } (Id_{j+1} - K_{j+1}) \Big| B^{j+1} (E) < \infty.$$

Let $(a,b) \in$ Ker F_j i.e. $a = - P_j (b)$ or Ker $F_j = \{0\} \oplus P_j^{-1} (Z^j(E)) \cap B^{j+1} (E)$.

F_j induces an isomorphism

$$(\text{Ker } F_j)^{\perp} \equiv Z^j(E) \oplus \left(B^{j+1} (E)/P_j^{-1} (Z^j(E)) \cap B^{j+1} (E) \right) \longrightarrow \text{Ran } F_j \subset C_j$$

Since F_j is Fredholm, codim Ran $F_j < \infty$, Ran F_j closed.

Now,

$$\text{Ran } F_j = Z^j(E) \oplus P_j \left(B^{j+1} (E)/P_j^{-1} (Z^j(E)) \cap B^{j+1} (E) \right) =$$
$$= Z^j(E) \oplus \left(P_j (B^{j+1}) / Z^j (E) \cap P_j (B^{j+1} (E)) \right)$$

and if H_j is a topological complement to Ran F_j in C_j;

$$C_j = H_j \oplus \text{Ran } F_j ,$$

a topological complement to $Z^j(E)$ in C_j is given simply by

$$H_j \oplus \left(P_j (B^{j+1}(E)) / Z^j (E) \cap P_j (B^{j+1} (E)) \right).$$

The isomorphism Ker $D_j \cong H^j (E)$ is clear from Theorem 2.13 and Remark 2.7.

Remark 2.15 An abstract complex (E) with dim H^j (E) < ∞ $\forall_j$ is called a *Fredholm complex*. Such complexes have a well-defined *index*;

$$\text{index (E)} := \sum_{i=0}^{N} (-1)^i \dim H^i (E)$$

Next, we shall formulate in this context the alternative approach by Milgram and Rosenbloom.

Given the Laplacians $D_j : C_j \longrightarrow C_j$, we consider the associated Cauchy problem for the heat operator in each C_j of the elliptic complex;

$$\begin{aligned} &\frac{\partial u}{\partial t} - D_j = 0, && (t, x) \in \mathbf{R}_+ \times X \\ &u(o,x) = u_o(x) \in C_j, && x \in X \end{aligned} \qquad (2.3)$$

Here $u(t,x) \in C^\infty(\mathbf{R}_+ \times X, E_j)$, i.e. letting $\{e_i\}$ be a local frame at $x \in X$ for $E_j \longrightarrow X$ so that $f \in C_j(\nu) \equiv C^\infty(\nu, E_j \mid \nu)$, ν some neighbourhood $\nu \ni x$, has the form

$$f = \sum a^r(e)\, e_r, \qquad a^r(e) \in C^\infty(\nu),$$

we extend the coefficients $a^r(e)$ to have a C^∞-dependence on a parameter $t \in \mathbf{R}_+$.

Clearly D_j generates a strongly continuous semigroup $T_j(t) := e^{tD_j}$ which in fact is holomorphic. From standard semigroup theory one has that

- $f \in L^2(E_j) \Rightarrow T_j(t) f \in C_j \ \forall\, t > 0$,
- $u(t,x) := T_j(t)\, u_o \in C^\infty(\mathbf{R}_+ \times X, E_j)$,
- $T_j(t)$ is a strongly continuous semigroup on each $H^s(E_j)$, $s \in \mathbf{R}$.

Since X is closed, D_j is Fredholm and being self-adjoint has the spectral representation

$$D_j = \sum_{i=0}^{\infty} \lambda_i E_i(\lambda_i) \ \text{ i.e. } \ T_j(t) = \sum_{i=0}^{\infty} e^{\lambda_i t} E_i(\lambda_i) \qquad (2.4)$$

The solution of the Cauchy problem (2.3) is thus given by $u(t,x) = T_j(t)\, u_o(x)$.

By strong continuity of the semigroup

$$f_\infty : = \underset{t \to \infty}{\text{s-lim}} \; T_j (t) f$$

exists for any $f \in C_j$.

Obviously;

- $T_j (t) f_\infty = f_\infty, \forall f \in C_j$ by the semigroup property , and
- $f_\infty \in \text{Ker } D_j , \forall f \in C_j$, i.e. the harmonic projection is given by

$$\Pi_j = \underset{t \to \infty}{\text{s-lim}} \; T_j (t) : C_j \longrightarrow \text{Ker } D_j \tag{2.5}$$

This follows since $T_{\infty, j} : = \underset{t \to \infty}{\text{s-lim}} \; T_j (t) = E (0)$, the projection onto Ker D_j, or - alternatively - since

$$\left(\frac{\partial}{\partial t} - D_j\right) T_j (t) = 0 \Rightarrow 0 = \left(\frac{\partial}{\partial t} - D_j\right) T_j (t) f_\infty = \frac{\partial f_\infty}{\partial t} - D_j f_\infty \Rightarrow D_j f_\infty = 0.$$

From the uniqueness of the Cauchy problem for parabolic equations we also have that

$$L_j T_j (t) = T_{j+1} (t) L_j$$

so that $\{T_j(t)\}$ is a cochain endomorphism, $T_j (t) Z^j(E) \subset Z^j(E)$ and $T_j (t) B^j(E) \subset B^j(E)$ ("the j-cocycle (j-coboundary) property propagates under the heat flow").

Lemma 2.16. Let $f \in Z^j (E)$ and let $f_t (x) \equiv f (t,x) = T_j (t) f (x) \in Z^j (E)$ be the corresponding solution of (2.3). Then $[f_t] \in H^j (E)$ is independent of t i.e. in particular $f_\infty \equiv f \,(\text{mod } B^j (E))$.

Proof: Define the "Kronecker index"

$$< \cdot , \cdot >_j : H^j (E) \times H_j (E) \longrightarrow \mathbf{C}$$

by $<[u],[\partial] >_j : = (u, \partial)_j$, $(\cdot , \cdot)_j : C_j \times C_j{}^* \longrightarrow \mathbf{C}$ being the pairing between algebraic duals, and $u \in Z^j (E)$, $\partial \in Z_j (E)$.

This index is well-defined, since for $u \in Z^j (E)$, $L_{j-1}\, s \in B^j (E)$, $s \in C_{j-1}$ and

$\partial \in Z_j(E), L^*_j \psi \in B_j(E), \psi \in C_{j+1}$ we have that

$$(u + L_{j-1} s, \partial + L^*_j \psi)_j = (u, \partial)_j + (u, L^*_j \psi)_j + (L_{j-1} s, \partial)_j + (L_{j-1} s, L^*_j \psi)_j =$$
$$= (u, \partial)_j + (L_j u, \psi)_{j+1} + (s, L^*_{j-1} \partial)_{j-1} + (L_j L_{j-1} s, \psi)_{j+1} = (u, \partial)_j.$$

Furthermore, let for $f \in Z^j(E)$, $\partial \in Z_j(E)$

$$\Phi(t) := \langle [f_t], [\partial] \rangle_j.$$

By continuity of the scalar product and since $T_j(t) Z^j(E) \subset Z^j(E)$, we obtain

$$\frac{d\Phi}{dt} = (D_j T_j(t) f, \partial)_j = (L_{j-1} L^*_{j-1} f_t, \partial)_j = (L^*_{j-1} f_t, L^*_{j-1} \partial)_{j-1} = 0.$$

Thus f_∞ obtained in (2.5) is the Hodge element in the class $[f]$, $f \in Z^j(E)$ arbitrary.

III. NON-ABELIAN HODGE THEORY

Apart from the heat-flow characterization, there is another abelian characterization of the Hodge element, which one may try to generalize to the non-abelian situation. This is the property that the Hodge element in a given cohomology class minimizes a certain L^2-norm and has been pursued by B. Gaveau.

III.1. The Gaveau Approach

Consider again the NSÇ $\{C^\infty(X,G), \Lambda^1(X, \mathfrak{g}), \Lambda^2(X, \mathfrak{g})\}$ with the homomorphisms d_o and d_1 and the sequence

$$0 \longrightarrow C_t^\infty(X,G) \longrightarrow C^\infty(X,G) \xrightarrow{d_o} \Lambda^1(X, \mathfrak{g}) \xrightarrow{d_1} \Lambda^2(X, \mathfrak{g})$$

There is a natural action μ of $C^\infty(X,G)$ on $\Lambda^1(X, \mathfrak{g})$;

Def. 3.1. $C^\infty(X,G) \ni f \longrightarrow \mu(f) : \Lambda^1(X, \mathfrak{g}) \longrightarrow \Lambda^1(X, \mathfrak{g})$ by $\mu(f)\,\omega = f\,\omega\,f^{-1} + d_o(f^{-1})$ is an action, i.e. $\mu(f_1 f_2) = \mu(f_1)\,\mu(f_2)$ $\left(d_o(f^{-1})_x = -\operatorname{Ad}(f(x))\; d_o(f)_x\right)$

In this context, it is natural to make the

Def. 3.2. Let $\omega \in M^1(X, \mathfrak{g}) := \left\{ \omega \in \Lambda^1(X, \mathfrak{g}) \mid d\omega + \omega \wedge \omega = 0 \right\}$ (the $\underline{d_1\text{-closed}}$ or <u>flat connections</u>). Then the non-abelian cohomology class of ω is the set

$$[\omega] := \left\{ \mu(f)\,\omega \mid f \in C^\infty(X,G) \right\}.$$

Observation (in the abelian case) : On a compact manifold, the Hodge element in a given class is the unique element minimizing the L^2-norm among all representatives of the class.

Assume henceforth, that the Lie algebra $\mathfrak{g}$ is compact.

Then, $\Lambda^p(X, \mathfrak{g}) \cong \Lambda^p(X) \otimes \mathfrak{g}$ $\left(\Lambda^p(X) = \text{p-forms}\right)$ carries a natural scalar product in that a Riemannian metric on X induces a Riemannian metric on p-forms and for the $\mathfrak{g}$-part we have the Killing form.

Theorem 3.3. (Gaveau) Let $B_p(\cdot,\cdot)$ = (Riemannian metric on p-forms) x (Killing form), then there exists an element $f_o \in C^\infty(X, G)$ realizing an absolute minimum of the Hodge functional

$$\Phi_\omega(f) := \int_X B_p\left(\mu(f)\,\omega, \mu(f)\,\omega\right) dx\,,$$

defined on $C^\infty(X, G)$.

Def 3.4. The harmonic element in $[\omega]$, $\omega \in M^1(X, \mathfrak{g})$ is given by $\mu(f_o)\,\omega$.

As far as Hodge theory is concerned, this is the essence of the Gaveau approach.

III.2. Heat Flow Approach.

The first situation which we shall consider is the one where the forms have coefficients in a bundle.
We shall consider a *compact* Lie group G with Lie algebra $\mathfrak{g}$.

Let p: $P \longrightarrow X$ be a principal bundle with structure group G. Choosing a representation ρ of G in the space V, the principal bundle can alternatively be viewed as the associated vector bundle $\pi := E \equiv P \times_\rho V \longrightarrow X$.

Let $C_j := C^\infty\left(X, \Lambda^j T^*(X) \otimes E\right)$, the E-valued j-forms and let ∇ be a (G-) connection on the bundle, i.e. a linear differential operator

$$\nabla : C_o \longrightarrow C_1 \qquad\qquad (C_o = C^\infty\text{-sections of E})$$

satisfying $\nabla(f\,\phi) = f\,\nabla\phi + df \otimes \phi$, $f \in C^\infty(X)$, $\phi \in C_o$.
∇ has the natural extension to $\nabla_p : C_p \longrightarrow C_{p+1}$ by

$$\nabla_p\,(\omega \otimes \phi) = d\omega \otimes \phi + (-1)^p\,\omega \wedge \nabla(\phi),$$

using the decomposition $C_p \cong \Lambda^p \otimes C_o$ $\left(\Lambda^p = C^\infty(X, \Lambda^p T^*(X)\right)$, the ρ-forms). ∇_p is the covariant derivative associated to the connection ∇.

Since the group G is compact, the Killing form is negative semi-definite and we have the decomposition

$$\mathfrak{g} = \mathfrak{g}_o \oplus \mathfrak{g}^1,$$

$\mathfrak{g}_o$ being the null-space of the Killing form.

We shall equip C_j with the natural scalar product which is induced by a Riemannian structure on X (Riemannian metric on Λ^j) times the Killing form. Denote this scalar product by $B_j(\cdot,\cdot)$.

By ∇^*_p: C_{p+1} we denote the adjoint w.r.t. B_j (the codifferential).

We shall thus be interested in the sequence

$$C_o \xrightarrow{\nabla} C_1 \xrightarrow{\nabla_1} \ldots\ldots \xrightarrow{\nabla_p} C_{p+1} \longrightarrow \ldots \tag{3.1}$$

and its adjoint.

Clearly, we shall say that <u>$\omega \in C_j$ is ∇-harmonic</u> if

$$\begin{aligned} &\nabla_p \, \omega = 0 \\ &\nabla^*_{p-1} \, \omega = 0. \end{aligned} \tag{3.2}$$

The sequence (3.1) is no longer a complex in general and the obstruction is measured by the *curvature*

$$R_p := \nabla_{p+1} \nabla_p : C_p \longrightarrow C_{p+2}.$$

One sees that $R_p \in C^\infty \big(X, \Lambda^2 T^* (X) \otimes \mathrm{Hom}\,(E,E)\big)$, i.e. has locally a representation as a matrix of p-forms.

In the principal bundle picture, $R_p \in C^\infty \big(X, \Lambda^2 T^* (X) \otimes \mathfrak{g}_E\big)$, $\mathfrak{g}_E \cong P\times_{ad}\mathfrak{g}$.

We shall hence be interested in the problem (3.2). Formally, we make for each step in the sequence the

Def. 3.5. The Laplacian associated to (3.1) and its adjoint is given by

$$\Delta_j : \nabla^*_j \nabla_j + \nabla_{j-1} \nabla^*_{j-1} : C_j \longrightarrow C_j.$$

Clearly $\mathrm{Ker}\,\Delta_j = \big\{ \omega \in C_j \mid \omega_j \text{ satisfies } (3.2)\big\}$.

The heat kernel argument used to describe Ker Δ_j has much in common with the ones used to handle the Hodge problem for (1.3). This Hodge problem, together with the one just presented will be our main examples and we shall carry out the arguments for both of them in an analogous way. For this we need some more facts about (1.3), which we now present, before passing over to the heat kernel arguments.

From (1.3) and Def. 3.2. we have the exact sequence

$$0 \longrightarrow C^{\infty}{}_t\,(X,G) \longrightarrow C^{\infty}\,(X,G) \xrightarrow{d_o} M^1(X,g) \longrightarrow 0 \tag{3.3}$$

and we also considered the complex (1.3)

$$0 \longrightarrow C^{\infty}{}_t\,(X,G) \longrightarrow C^{\infty}\,(X,G) \xrightarrow{d_o} \Lambda^1(X,g) \xrightarrow{d_1} \Lambda^2(X,g)\,,$$

d_1 given by Def. 1.9.

Along with (3.3) we also consider the adjoint complex and we shall formally be interested in the

$$\delta_1 := d^*{}_1 d_1 + d_o d^*{}_o : \Lambda^1\,(X,g) \longrightarrow \Lambda^1\,(X,g). \tag{3.4}$$

where the adjoints are formed w.r.t. $B_p\,(\cdot,\cdot)$, as in Theorem 3.3.

The cohomology of the sequence (3.3) is naturally given by (c.f. Def. 1.2)

$$H^o\,(X,G) := \text{Ker}\, d_o\,;\quad H^1\,(X,G) := \text{Ker}\, d_1/\text{Ran}\, d_o$$

The exact sequence (3.3) generates on the cohomology level the exact sequence

$$0 \longrightarrow H^o\big(X,C_t^{\infty}(X,G)\big) \longrightarrow H^o\big(X,C^{\infty}(X,G)\big) \longrightarrow H^o\big(X,M^1,(X,g)\big) \longrightarrow$$
$$H^1\big(X,C_t(X,G)\big) \longrightarrow H^1\big(X,C^{\infty}(X,G)\big) \longrightarrow H^1\big(X,M^1,(X,g)\big),$$

which has been extended by Asada [As 5], to include H^2-sets as well.

Now, from the point of view of the general theory, as developed in II, these two model situations do not quite fit into this scheme.
However, and this indicates the usefulness of the general theory, despite this, there is enough structural similarity in both situations to allow - with obvious modifications - the construction of Hodge elements.
Since we have given the genral theory quite in detail in earlier sections, we shall here just point out the problems in the two situations under consideration and indicate why the construction in II still applies.
We leave the details out of this presentation.

As for the situation (3.1)-(3.2), it is of course *not a complex* and for this reason does not fit into the general scheme. However, although $R_p \in C^\infty(X, \Lambda^2 T^*(X) \otimes \mathrm{Hom}(E,E))$ does not vanish in general, (3.1) is an *elliptic semi-complex* which is sufficiently close to an elliptic complex to enable us to carry through the heat flow argument.
That (3.1) is an elliptic semi-complex means that the associated symbol sequence

$$0 \longrightarrow \Pi^*(E_o) \xrightarrow{\sigma(\nabla)} \Pi^*(E_1) \xrightarrow{\sigma(\nabla_1)} \Pi^*(E_2) \longrightarrow \ldots \xrightarrow{\sigma(\nabla_{n-1})} \Pi^*(E_o) \longrightarrow 0$$

($E_j := \Lambda^j T^*(X) \otimes E$) is exact. This is easily verified by computing the symbols;

$$\sigma_1(\nabla_j)(x, \xi)(\upsilon \otimes e) = (\xi \wedge \upsilon) \otimes e,$$

$(x, \xi) \in T^*_x(X)$, $\upsilon \in \Lambda^j T^*_x(X)$, $e \in E_x$ (the fiber of E at x).

To an elliptic semi-complex, a number of elliptic operators can be associated, among others the Laplacians Def. 2.8 (Def. 3.5). This observation is enough to make possible the construction of parametrices in section II and yields, modulo obvious modifications, a Hodge theory.

As for the complex (1.3) ((3.3)) and the Hodge problem associated with the operator δ_1 (3.4), the complication here is, that strictly speaking the involved opertors are not

Despite this, and in fact more generally in the situation of a NSC being a resolution (Def. 1.7) (making suitable assumptions on K^i, i=0,1,2), one can still carry through the arguments of section II, working with

$$\delta_1 \mid \mathrm{Ker} d_1 = \delta'_1 .$$

Without giving the necessary estimates here, we shall state the main results as follows;

- δ'_1 generates a 1-parameter semigroup, $T_1(t)$ say, which solves the associated Cauchy problem for the heat flow;
- $\omega_o \in \mathrm{Ker} d_1 \subset \Lambda^1(X, g) \Rightarrow T_1(t)\, \omega_o \in \mathrm{Ker} d_1, \forall\, t > 0$ (ω_o being the Cauchy datum) so that the 1-cocycle property is propagataed under the heat flow;
- $\omega_\infty = \underset{t \longrightarrow \infty}{\text{s-lim}}\, T_1(t)\, \omega_o$ exists for $\omega_o \in \mathrm{Ker} d_1$ and $\omega_\infty \in \mathrm{Ker}\, \delta_1$. In fact:

- the operator Π_1 from (2.5)

  $$\Pi_1 : \Lambda^1(X,g) \longrightarrow \mathrm{Ker}\, \delta_1$$

  is surjective.
- the analogue of Lemma 2.16 holds true.

As a final remark, we would like to point out that the formalism developed above has much in common with the situation considered in so-called Yang-Mills theory (c.f. [B-L], [Bou 1,2]).

A central issue in this context being the classification of *harmonic curvatures* (Lie-algebra valued 2-forms satisfying a system of partial differential equations), the construction of Hodge elements in a 2-cohomology setting should be of direct relevance. This topic will be dealt with in a forthcoming publication.

REFERENCES

[An 1]
S.I. Andersson: VECTOR BUNDLE CONNECTIONS AND LIFTINGS OF PARTIAL DIFFERENTIAL OPERATORS.
Lect. Notes in Math. (Springer) 905, 119-132, 1982.

[An 2]
S.I.Andersson: PSEUDODIFFERENTIAL OPERATORS AND CHARACTERISTIC CLASSES FOR NON-ABELIAN COHOMOLOGY.
Lect. Notes in Math. (Springer) 1045, 1-10 (1984).

[As 1]
A. Asada: CONNECTIONS OF DIFFERENTIAL OPERATORS,
J. Fac. Sci. Shinshu Univ. 13, 87-102, 1978.

[As 2]
A. Asada: FLAT CONNECTIONS OF DIFFERENTIAL OPERATORS AND ODD DIMENSIONAL CHARACTERISTIC CLASSES, ibid. 17, 1-30 (1982).

[As 3]
A. Asada: FLAT CONNECTIONS OF DIFFERENTIAL OPERATORS AND RELATED CHARACTERISTIC CLASSES,
Lect. Notes in Math. (to appear).

[As 4]
A. Asada: NON-ABELIAN DE RHAM THEORIES,
to appear in Proc. of Int. Coll. on Differential Geometry, Hajduszoboszló 1984 (Coll. Soc. János Bolyai, North -Holland).

[As 5]
A: Asada: CURVATURE FORMS WITH SINGULARITIES AND NON-INTEGRAL CHARACTERISTIC CLASSES,
Lect. Notes in Math. Vol 1139, 152-168 (1985).

[As 6]
A. Asada: NON-ABELIAN POINCARÉ LEMMA.,
Proc. of the Int. Conf. on Differential Geometry, Peniscola (Spain) 1985. To appear in Lect. Notes in Math.

[At]
M. Atiyah: COMPLEX ANALYTIC CONNECTIONS IN FIBRE BUNDLES,
Trans. Am. Math. Soc 85, 181-207 (1957).

[B-L]
J-P Bourguignon, H.B. Lawson: YANG-MILLS THEORY.
Its Physical Origins and Differential Geometric Aspects. in "Seminar on Differential Geometry" (Ed. S-T, Yau)
Annals of Mathem. Studies 102 (1982) Princeton Univ. Press.

[Bou 1]
J-P Bourguignon: Groupe de gauge élargi et connexions stables. Progr in Math. 7 (Birkhäuser 1980) in "Vector Bundles and Differential Equations" (Ed. A. Hirschowitz).

[Bou 2]
J-P Bourguignon: ANALYTICAL PROBLEMS ARISING IN GEOMETRY: Examples from Yang-Mills Theory.
Jber. d. Dt. Math.-Verein. 87, 67-89 (1985).

[G]
B. Gaveau: Intégrales harmoniques non-abéliennes, Bull. Sci. Math ., 2e série, 106, 113-169 (1982).

[Ga]
M.P. Gaffney: THE HEAT EQUATION METHOD OF MILGRAM AND ROSENBLOOM FOR OPEN RIEMANNIAN MANIFOLDS
Ann. Math. 60, 458-466 (1954).

[Gir]
J. Giraud: Cohomologie non-abélienne, Grundl. d. Math. Wiss.179, (Springer) 1971.

[Gr]
L. Gross: A POINCARÉ LEMMA FOR CONNECTION FORMS
J. Funct. Anal. 63 1-46 (1985).

[Ko]
T.Kotake: THE FIXED POINT THEOREM OF ATIYAH-BOTT VIA PARABOLIC OPERATORS
Comm. Pure Appl. Math. 22 789-806 (1969).

[MR]
A.N.Milgram, P.C. Rosenbloom: HARMONIC FORMS AND HEAT CONDUCTION
I. CLOSED RIEMANNIAN MANIFOLDS.
Proc. Nat. Acad. Sci. (USA). 37, 180-184 (1951).

[On 1]
A.L. Oniščik: ON THE CLASSIFICATION OF FIBER SPACES,
Sov. Math. Dokl. 2, 1561-1564 (1961).

[On 2]
A.L. Oniščik: CONNECTIONS WITH ZERO CURVATURE AND THE DE RHAM THEOREM,
Sov. Math. Dokl. 5, 1654-1657 (1964).

[On 3]
A.L. Oniščik: SOME CONCEPTS AND APPLICATIONS OF NON-ABELIAN COHOMOLOGY THEORY,Trans. Moscow Math. Soc. 17, 49-98 (1967).

[On 4]
A.L. Oniščik: ON COMPLETELY INTEGRABLE EQUATIONS ON HOMOGENOUS SPACES,
Mat. Zametki 9 (14), 365-373 (1970) (russian).

[R]
H. Röhrl, Das Riemann-Hilbertsche Problem der Theorie der linearen Differentialgleichungen. Math. Ann. 133, 1-25 (1957).

[RS]
S. Rempel, B-W Schulze: INDEX THEORY OF ELLIPTIC BOUNDARY PROBLEMS. Akademie-Verlag, Berlin 1982.

[Se]
J-P Serre: Cohomologie Galoisienne; Lect. Notes Math. 5 Springer Verlag, Berlin 1964.

[Sp]
D. Spencer: A GENERALIZATION OF A THEOREM BY HODGE,
Proc, Nat. Acad. Sci. (USA) 38, 533-534 (1952).

[St]
N.K. Stanton: THE HEAT EQUATION IN SEVERAL COMPLEX VARIABLES,
Bull. Am. Mat. Soc. 11, 65-84 (1984).

[To 1]
A.K. Tolpygo: ON TWO-DIMENSIONAL COHOMOLOGY THEORIES,
Usp. Mat. Nauk. 27 (5), 251-252, (1972) (russian).

[To 2]
A.K. Tolpygo: UNIVERSALITY OF NON-ABELIAN COHOMOLOGY THEORIES
Mat. Sbornik 91 (2), 267-278 (1973) (russian).

[To 3]
A.K.Tolpygo: TWO-DIMENSIONAL COHOMOLOGY AND SPECTRAL SEQUENCES IN NON-ABELIAN THEORY.
in "Questions in group theory and homological algebra". Jaroslavl 1977 p. 156-197. (russian).

[Y]
K. Yosida: AN ERGODIC THEOREM ASSOCIATED WITH HARMONIC INTEGRALS,
Proc. Jap. Acad., 27 540-543 (1951).

NON ABELIAN POINCARE LEMMA

Akira Asada
Department of mathematics, Faculty of Science, Sinsyu University
Matsumoto, Nagano pref., Japan

De Rham theory concerns the differential geometry of complex line bundles. From this point of view, de Rham theory is constructed using the group C* and the differential operator d. Similarly, we constructed sets referring to the differential geometry of complex vector bundles ([4], [6]). We call these sets non-abelian de Rham sets. They are constructed from the group G=GL(n,C) and the following differential operators d^e (and d^e_φ).

$$d^e f = e^{-f}d(e^f) = df + \sum_{n=1}^{\infty}\frac{(-1)^n}{(n+1)!}(\mathrm{ad}\ f)^n(df),\quad (\mathrm{ad}\ f)u = fu-uf,$$

$$d^e\theta = d\theta + \theta\wedge\theta = d\theta + \frac{1}{2}[\theta,\theta],$$

$$d^e_\varphi\Theta = d\Theta + [\varphi,\Theta],\quad [\varphi,\Theta] = \varphi\wedge\Theta - \Theta\wedge\varphi.$$

In a sense, the usual de Rham theory is the global theory of d. The local theory of d is the Poincaré lemma. Similarly, non-abelian de Rham sets concern the global study of d^e and the Frobenius theorem shows the local solvability condition of the equation $d^e f=\theta$ is $d^e\theta=0$ (cf. [6]). This is the 1-dimensiomal non-abelian Poincaré lemma and it is one reason why the 1-dimensional non-abelian de Rham set is understood well (cf. [7], [11], [12], [16]). On the other hand, the local integrability condition of the equation

$$d\theta + \theta\wedge\theta = \Theta,$$

seems to be unknown (cf. 9). In this article, we give this condition. It is the problem to state the 2-dimensional non-abelian Poincaré lemma. Constructions of all local solutions are also done (cf.[8]).

Our main tool treating these problems is Peano series or chronological calculus of differential forms. (Right) Peano series $P_{r,\theta}(\varphi) = P_\theta(\varphi)$ of a 1-form θ with the initial data φ is defined as follows:

$$P_\theta(\varphi) = \sum_{n=0}^{\infty} I_\theta{}^n(\varphi), \quad I_\theta{}^0(\varphi) = \varphi, \quad I_\theta{}^n(\varphi) = I(I_\theta{}^{n-1}(\varphi)\wedge\theta), \quad n \geq 1.$$

Here I is the chain homotopy (Green operator) of d ([15]). The left and adjoint Peano series $P_{\ell,\theta}(\varphi)$ and $P_{ad,\theta}(\varphi)$ are defined similarly. For functions (forms on R^1), this series was defined in [13], and named Peano series in [14]. Including operator valued functions, this series considered in[1],[2] under the name chronological calculus. See also [10].But it seems that Peano series of forms were never considered. We show that

$$dP_\theta(\varphi) = P_\theta(\varphi)\wedge\theta + P_{-\theta}(d\varphi - (-1)^p I(P_\theta(\varphi)\wedge d^{e,p}\theta)),$$

where p is the degree of φ and $d^{e,p}\theta = d\theta + (-1)^p\theta\wedge\theta$ (§1, $(8)_r$). To use this formula (and similar formulas for $P_{\ell,\theta}(\varphi)$ and $P_{ad,\theta}(\varphi)$), we solve equations $d\Phi + \Phi\wedge\theta = 0$, $d\Phi + \Phi\wedge\theta = \Psi$, etc., in §2. As a special case, we get a local integrability condition of the equation $d^e_\varphi\Theta = \Psi$ (Lemma 5,Corollary). It is the 3-dimensional (non-abelian) Poincaré lemma. But our integrability condition (14) leaves much to be desired. On the other hand, we get by $(8)_r$ the following chain homotopy formula (§2,(11))

$$\theta = d^e k^e(\theta) + P_\theta(1)^{-1} P_{-\theta}(I(P_\theta(1) d^e\theta)), \quad k^e(\theta) = \log P_\theta(1) .$$

To get the 2-dimensional non-abelian Poincaré lemma, one starts from (11) and sets

$$J_\theta(\varphi) = P_{(-1)^{p-1}\theta}(I(P_\theta(1)\varphi)P_\theta(1)^{-1}, \quad p = \deg\varphi \geqq 1.$$

If Θ is a 2-form, the gauge transformation of $J_\theta(\Theta)$ by $P_\theta(1)$ is denoted $\kappa^e{}_\theta(\Theta)$. Explicitly, we have

$$\kappa^e{}_\theta(\Theta) = \theta + P_\theta(1)^{-1}J\ (\Theta - d^e\theta).$$

Then we obtain (§3,(20))

$$d^e\kappa^e{}_\theta(\Theta) = \Theta - P_\theta(1)^{-1}J_\theta(d^e{}_\theta\Theta) + P_\theta(1)^{-1}J_\theta(P_\theta(1)^{-1}(J_\theta(d^e\theta)\wedge\Theta -$$

$$- J_\theta(\Theta)\wedge d^e\theta)) + P_\theta(1)^{-1}J_\theta(\Theta)\wedge P_\theta(1)^{-1}J_\theta(\Theta - d^e\theta)\ .$$

In this right hand side, the first term is the desired term, the second term is the obstruction term and the third and fourth terms are remainder terms, since these last 2-terms are $O(||x||^{c+2})$ if $(||(\Theta-d^e\theta)(x)||=O(||x||^c)$, $c\geqq 0$. Next, starting from a pair of a 1-form and a 2-form $\{\theta_0,\Theta_0\}$, we define a series of pairs of 1-form and 2-form $\{\theta_n,\Theta_n\}$ as follows:

$$\theta_n = \kappa^e{}_{\theta_{n-1}}(\Theta_{n-1}),\ \Theta_n = \Theta_{n-1} - P_{\theta_{n-1}}(1)^{-1}J_{\theta_{n-1}}(d^e_{\theta_{n-1}}\Theta_{n-1}),\ n\geqq 1.$$

Then we have the following asymptotic chain homotopy formula (cf.[18])

$$||(\Theta_n - d^e\theta_n)(x)|| = O(||x||^{2n}),\ ||x||\to 0.$$

At this stage, we do not know whether $\lim \theta_n$ exists or not. But we can show (§3,Theorem 3);

<u>2-dimensional non-abelian Poincaré lemma</u>. If $d^e{}_{\theta_n}\Theta=0$ for all n, $\lim \theta_n = \theta$ exists on some neighborhood of the origin and we have $d^e\theta = \Theta$.

This θ is smooth if both θ_0 and Θ_0 are smooth. But our estimate of the convergence radius is not sharp. To get a sharp estimate of the convergence radius of θ would be useful to understand the geometric meaning of the 2-dimensional non-abelian de Rham set.

If θ is a local solution of the equation $d^e\theta = \Theta$, another local solution is given by $\theta+\eta$, where η satisfies the equation

$$d\eta + \eta\wedge\eta + [\theta,\eta] = 0,$$

(§4,(34)). We transform this equation to the integral equation $\eta = P_{ad,-\theta}(\Lambda - I(\eta\wedge\eta))$, where Λ is a 1-form. This equation has a unique solution η for given Λ(Lemma 12). The solution is constructed as follows: Set $\eta_1 = P_{ad,-\theta}(\Lambda)$, $\eta_n = -P_{ad,-\theta}(I(\sum_{k=1}^{n-1}\eta_k\wedge\eta_{n-k}))$, $n \geqq 2$, and $q_\theta(\Lambda) = \sum\eta_n$, η is given by $q_\theta(\Lambda)$. Of course, $q_\theta(\Lambda)$ does not satisfy (34) in general. But this construction suggests that the system

$$d\eta_1 + [\theta,\eta_1] = 0, \quad d\eta_n + [\theta,\eta_n] + \sum_{k=1}^{n-1}\eta_k\wedge\eta_{n-k} = 0, \quad n \geqq 2,$$

is in a sense, the "linearization" of (34). We end this article by showing the following formula which gives a relation between $q(\Lambda)$ and the gauge transformation of θ;

$$q_\theta(d(f - I[f,\theta])) = e^{-f}\theta e^{f} - \theta + e^{-f}d(e^{f}), \quad \text{if } [d^e\theta, f] = 0.$$

In this article, we assume functions and forms take the values in the Lie algebra of GL(n,C). But we may take U(n) or SU(n) instead of GL(n,C). We also note that even if d is changed to a (pseudo) differential operator D, we may obtain similar results, if D has a "good" Green operator (cf.[3],[5]).

§1. Peano series or chronological calculus of differential forms

1. Throughout this article, functions and differential forms take values in the complex matrices. They are always defined on some star-like neighborhood of the origin of R^m. m always mean the dimension of the space. If φ is a p-form and ψ is a q-form, we set $[\varphi,\psi]=\varphi\wedge\psi-(-1)^{pq}\psi\wedge\varphi$. We also denote $\mathrm{ad}\,\varphi(\psi)=[\varphi,\psi]$. By definition, we have the following formulas

$$d[\varphi,\psi] = [d\varphi,\psi]+(-1)^p[\varphi,d\psi],$$

$$(-1)^{pr}[\varphi,[\psi,\omega]]+(-1)^{pq}[\psi,[\omega,\varphi]]+(-1)^{qr}[\omega,[\varphi,\psi]] = 0,$$

where $\deg\varphi=p$, $\deg\psi=q$ and $\deg\omega=r$. By this second formula, we have

(1) $$[\varphi,[\varphi,\psi]] = [\varphi\wedge\varphi,\psi], \quad \text{if } \varphi \text{ is an odd degree form.}$$

Definition. We set

$$d^e f = e^{-f}d(e^f) = df + \sum_{n=1}^{\infty}\frac{(-1)^n}{(n+1)!}(\mathrm{ad}\,f)^n(df),$$

$$d^e\theta = d\theta+\theta\wedge\theta = d\theta+\tfrac{1}{2}[\theta,\theta],$$

$$d^e_\varphi\Theta = d\Theta+[\varphi,\Theta].$$

Note. d^e_φ is determined relative to φ.

By definition, we have $d^e(d^e f)=0$ and $d^e_\theta(d^e\theta)=0$. This second equality is the Bianchi identity. We also use following notations.

$$d^{e,p}\theta = d\theta+(-1)^p\theta\wedge\theta, \quad d^e_\chi\theta = d\theta-\theta\wedge\theta.$$

By definition , we have

(2) $$(-1)^p d^{e,p}((-1)^p\theta) = d^e\theta,$$

$$d^{e,p}(-\theta) = -d^e_{\ell}\theta, \text{ p is even}, \quad d^{e,p}(-\theta) = d^e_{\ell}(-\theta) = -d^e\theta, \text{ p is odd}$$

Since the differential form φ takes value in a vector space, we denote the vector norm of $\varphi(x)$ by $||\varphi(x)||$. We also set

$$\bar{\varphi}(x) = \max_{0 \leqq t \leqq 1} ||\varphi(xt)||, \quad xt = (x_1 t, \ldots, x_n t).$$

By definition, we have

(3) $$\overline{(\varphi+\psi)}(x) \leqq \bar{\varphi}(x) + \bar{\psi}(x), \quad \overline{(\varphi\wedge\psi)}(x) \leqq \bar{\varphi}(x)\bar{\psi}(x).$$

2. Let φ be $\sum \varphi_{i_1,\ldots,i_p} dx_{i_1}\wedge\cdots\wedge dx_{i_p}$. Then we set

$$I\varphi = \sum I(\varphi)_{i_1,\ldots,i_{p-1}} dx_{i_1}\wedge\cdots\wedge dx_{i_{p-1}},$$

$$I(\varphi)_{i_1,\ldots,i_{p-1}}(x) = \int_0^1 t^p \sum_{j=1}^m x_j \varphi_{j,i_1,\ldots,i_{p-1}}(xt)dt.$$

I satisfies $Id\varphi+dI\varphi=\varphi$, $\deg\varphi \geqq 1$, $Idf=f-f(0)$ ([15]) and the following estimate if $\bar{\varphi}(x) \leqq ||x||^c r(x)$, $c \geqq 0$ and $r(x)$ is non-decreasing

(4) $$||I(\varphi)(x)|| \leqq \frac{||x||^{c+1}}{p+c} {}_mC_p r(x).$$

<u>Definition</u>. Let θ be a 1-form, φ a p-form ($p \geqq 0$). Then we set

$$I_{r,\theta}(\varphi) = I(\varphi\wedge\theta), \quad I_{\ell,\theta}(\varphi) = I(\theta\wedge\varphi), \quad I_{ad,\theta}(\varphi) = I([\theta,\varphi]).$$

Lemma 1. We have

(5) $$||I_{r,\theta}{}^{n}(\varphi)(x)|| \text{ (and } ||I_{\ell,\theta}{}^{n}(\varphi)(x)||) \leqq \frac{||x||^{n}}{n!}({}_{m}C_{p}\bar{\theta}(x))^{n}\bar{\varphi}(x)$$

$$||I_{ad,\theta}{}^{n}(\varphi)(x)|| \leqq \frac{||x||^{n}}{n!}(2\,{}_{m}C_{p}\bar{\theta}(x))^{n}\,\bar{\varphi}(x).$$

Proof. Since we get $||I_{r,\theta}(\varphi)(x)|| \leqq ||x||/(p+1)\,{}_{m}C_{p}\bar{\theta}(x)\bar{\varphi}(x)$ by (3) and (4), we have $||I_{r,\theta}{}^{n}(\varphi)(x)|| \leqq p!\,||x||^{n}/(p+n)!\,({}_{m}C_{p}\bar{\theta}(x))^{n}\,\bar{\varphi}(x)$ by (4) Hence we obtain the first inequality because $p!/(p+n)! \leqq 1/n!$. The second inequality follows from the inequality $\overline{[\theta,\varphi]}(x) \leqq 2\bar{\theta}(x)\bar{\varphi}(x)$.

Corollary. $\sum_{n=0}^{\infty} I_{r,\theta}{}^{n}(\varphi)$, $\sum_{n=0}^{\infty} I_{\ell,\theta}{}^{n}(\varphi)$ and $\sum_{n=0}^{\infty} I_{ad,\theta}{}^{n}(\varphi)$ converge uniformly and absolutely on any starlike neighborhood of the origin on which both θ and φ are defined. Here $I_{r,\theta}{}^{0}(\varphi)$, $I_{\ell,\theta}{}^{0}(\varphi)$ and $I_{ad,\theta}{}^{0}(\varphi)$ mean φ.

Definition. We define right, left and adjoint Peano series (or chronological calculus) of θ with the initial data φ by

(6) $$P_{r,\theta}(\varphi) = \sum_{n=0}^{\infty} I_{r,\theta}{}^{n}(\varphi), \quad P_{\ell,\theta}(\varphi) = \sum_{n=0}^{\infty} I_{\ell,\theta}{}^{n}(\varphi),$$

$$P_{ad,\theta}(\varphi) = \sum_{n=0}^{\infty} I_{ad,\theta}{}^{n}(\varphi).$$

Note. In the rest, we shall often denote by $P_{\theta}(\varphi)$, $I_{\theta}(\varphi)$ and $I_{\theta}{}^{n}(\varphi)$ the quantities $P_{r,\theta}(\varphi)$, $I_{\ell,\theta}(\varphi)$ and $I_{ad,\theta}{}^{n}(\varphi)$, etc., for simplicity.

Lemma 2. (i). $P_{\theta}(\varphi)$ is a p-form if φ is a p-form. If p=0, $P_{\theta}(f)$ is invertible near the origin, provided f(0) is invertible.

(ii). $P_{\theta}(\varphi)$ is linear in φ.

(iii). The following formula holds

(7) $$P_\theta(\varphi) - I_\theta(P_\theta(\varphi)) = P_\theta(\varphi - I_\theta(\varphi)) = \varphi.$$

(iv). The **following** estimates hold.

$$||P_{r,\theta}(\varphi)(x)|| \text{ (and } ||P_{\ell,\theta}(\varphi)(x)||) \leqq \bar{\varphi}(x)e^{m^{C}p\bar{\theta}(x)||x||},$$

$$||P_{ad,\theta}(\varphi)(x)|| \leqq \bar{\varphi}(x)e^{2_{m}C_{p}\bar{\theta}(x)||x||}.$$

Proof. (i),(ii) and (iii) follow from the definition. (iv) follows from lemma 1.

3. Lemma 3. Let φ be a p-form. Then we have

$(8)_r$ $$dP_{r,\theta}(\varphi) = P_{r,\theta}(\varphi)_\wedge\theta + P_{r,-\theta}(d\varphi-(-1)^p I(P_{r,\theta}(\varphi)_\wedge d^{e,p}\theta)),$$

$(8)_\ell$ $$dP_{\ell,\theta}(\varphi) = \theta_\wedge P_{\ell,\theta}(\varphi) + P_{\ell,\theta}(d\varphi-I(d^e{}_\ell\theta_\wedge P_{\ell,\theta}(\varphi))),$$

$(8)_{ad}$ $$dP_{ad,\theta}(\varphi) = [\theta,P_{ad,\theta}(\varphi)] + P_{ad,\theta}(d\varphi-I[d^e{}_\ell\theta,P_{ad,\theta}(\varphi)]).$$

Proof. Since we have $dI_{r,\theta}{}^0(\varphi)=d\theta$ and $dI_{r,\theta}(\varphi)=\varphi_\wedge\theta-I((-1)^p\varphi_\wedge d\theta+ +d\varphi_\wedge\theta)$, we assume

$(9)_r$ $$dI_{r,\theta}{}^n(\varphi)$$

$$= I_{r,\theta}{}^{n-1}(\varphi)_\wedge\theta - I(\sum_{s=0}^{n-2}((-1)^s I_{r,\theta}{}^s((-1)^p I_{r,\theta}{}^{n-s-1}(\varphi)_\wedge d\theta+$$

$$+I_{r,\theta}{}^{n-s-2}(\varphi)_\wedge(\theta_\wedge\theta))) - (-1)^{n-1}I_{r,\theta}{}^{n-1}((-1)^p\varphi_\wedge d\theta+d\varphi_\wedge\theta)).$$

Then, since $I(I_{r,\theta}{}^s(\varphi)_\wedge\theta)=I_{r,\theta}{}^{s+1}(\varphi)$, we have

$$dI_{r,\theta}{}^{n+1}(\varphi) = dI(I_{r,\theta}{}^n(\varphi)_\wedge\theta) =$$

$$= I_{r,\theta}{}^{n}(\varphi)\wedge\theta - I(dI_{r,\theta}{}^{n}(\varphi)\wedge\theta+(-1)^{p}I_{r,\theta}{}^{n}(\varphi)\wedge d\theta)$$

$$= I_{r,\theta}{}^{n}(\varphi)\wedge\theta - I(I_{r,\theta}{}^{n-1}(\varphi)\wedge\theta\wedge\theta -$$

$$-I((\sum_{j=0}^{n-2}(-1)^{s}I_{r,\theta}{}^{s}((-1)^{p}I_{r,\theta}{}^{n-s-1}(\varphi)\wedge d\theta)\wedge\theta+$$

$$+(I_{r,\theta}{}^{n-s-2}(\varphi)\wedge(\theta\wedge\theta)))\wedge\theta-(-1)^{n-1}I_{r,\theta}{}^{n-1}((-1)^{p}\varphi\wedge d\theta+d\varphi\wedge\theta)\wedge\theta)+$$

$$+(-1)^{p}I_{r,\theta}{}^{n}(\varphi)\wedge d\theta))$$

$$= I_{r,\theta}{}^{n}(\varphi)\wedge\theta - I(\sum_{s=0}^{n-1}((-1)^{s}I_{r,\theta}{}^{s}((-1)^{p}I_{r,\theta}{}^{n-s}(\varphi)\wedge d\theta+$$

$$+I_{r,\theta}{}^{n-s-1}(\varphi)\wedge(\theta\wedge\theta)))-(-1)^{n}I_{r,\theta}{}^{n}((-1)^{p}\varphi\wedge d\theta+d\varphi\wedge\theta)).$$

Hence $(9)_r$ is true for any $n\geqq 2$. Then by (5), we get for some $M>0$

$$||dI_{r,\theta}{}^{n}(\varphi)(x)|| \leqq M^{n}\sum_{k=0}^{n-1}\frac{1}{k!(n-k-1)!} < C(2eM)^{n}\cdot n^{(1-n)/2},\quad ||x||<R$$

where $R>0$ is arbitrary but fixed. Hence we get $dP_{r,\theta}(\varphi)=\sum_{n=0}^{\infty}dI_{r,\theta}{}^{n}(\varphi)$ and we have $(8)_r$ by $(9)_r$.

Since we have $dI_{\ell,\theta}(\varphi)=\theta\wedge\varphi-I(d\theta\wedge\varphi-\theta\wedge d\varphi)$ and $dI_{ad,\theta}(\varphi)=[\theta,\varphi]-$ $-I([d\theta,\varphi]-[\theta,d\varphi])$, we obtain

$(9)_\ell$
$$dI_{\ell,\theta}{}^{n}(\varphi) = \theta\wedge I_{\ell,\theta}{}^{n-1}(\varphi) - I(\sum_{s=0}^{n-2}I_{\ell,\theta}{}^{s}(d\theta\wedge I_{\ell,\theta}{}^{n-s-1}(\varphi)-$$
$$-(\theta\wedge\theta)\wedge I_{\ell,\theta}{}^{n-s-2}(\varphi))-I_{\ell,\theta}{}^{n-1}(d\theta\wedge\varphi-\theta\wedge d\varphi)),$$

$(9)_{ad}$
$$dI_{ad,\theta}{}^{n}(\varphi) = [\theta, I_{ad,\theta}{}^{n-1}(\varphi)]-I(\sum_{s=0}^{n-2}I_{ad,\theta}{}^{s}([d\theta, I_{ad,\theta}{}^{n-s-1}(\varphi)]-$$
$$-[\theta\wedge\theta, I_{ad,\theta}{}^{n-s-2}(\varphi)])-I_{ad,\theta}{}^{n-1}([d\theta,\varphi]-[\varphi,d\theta])),$$

by the same way. Hence we have $(8)_\ell$ and $(8)_{ad}$.

Note. Set $P_\theta(\varphi)=\Phi$, $(8)_r$, $(8)_\ell$ and $(8)_{ad}$ are written as follows:

$(8)_r'$ $$d\Phi = \Phi_\wedge\theta + P_{r,-\theta}(d\varphi - (-1)^p I(\Phi_\wedge d^{e,p}\theta)),$$

$(8)_\ell'$ $$d\Phi = \theta_\wedge\Phi + P_{\ell,\theta}(d\varphi - I(d^e{}_\ell\theta_\wedge\Phi)),$$

$(8)_{ad}'$ $$d\Phi = [\theta,\Phi] + P_{ad,\theta}(d\varphi - I[d^e{}_\ell\theta,\Phi]).$$

§2. Linear differential equations for differential forms

4. By direct calculations, we have

Lemma 4. Let Φ be a p-form. Then we obtain

$(10)_r$ $$\Phi_\wedge d^{e,p}\theta = 0, \quad \text{if } d\Phi = \Phi_\wedge\theta,$$

$(10)_\ell$ $$d^e{}_\ell\theta_\wedge\Phi = 0, \quad \text{if } d\Phi = \theta_\wedge\Phi,$$

$(10)_{ad}$ $$[d^e{}_\ell\theta,\Phi] = 0, \quad \text{if } d\Phi = [\theta,\Phi].$$

Theorem 1. (i). A p-form Φ satisfies the equation $d\Phi=\Phi_\wedge\theta$ if and only if $\Phi=P_{r,\theta}(\Lambda)$, where $d\Lambda=0$ and $\Phi_\wedge d^{e,p}\theta=0$.

(ii). A p-form Φ satisfies the equation $d\Phi=\theta_\wedge\Phi$ if and only if $\Phi=P_{\ell,\theta}(\Lambda)$, where $d\Lambda=0$ and $d^e{}_\ell\theta_\wedge\Phi=0$.

(iii). A p-form Φ satisfies the equation $d\Phi=[\theta,\Phi]$ if and only if $\Phi=P_{ad,\theta}(\Lambda)$, where $d\Lambda=0$ and $[d^e{}_\ell\theta,\Phi]=0$.

Proof. By assumption, we get $\Phi-I_\theta(\Phi)=dI\Phi$. Hence we have the theorem by (7).

Corollary. If $d^{e,p}\theta=0$ (resp. $d^e_\ell\theta=0$), $P_{r,\theta}(\Lambda)$ (resp. $P_{\ell,\theta}(\Lambda)$ or $P_{ad,\theta}(\Lambda)$) is a solution of the equation $d\Phi=\Phi\wedge\theta$ (resp. $d\Phi=\theta\wedge\Phi$ or $d\Phi=[\theta,\Phi]$) if and only if $d\Lambda=0$.

Example. If p=0, the equation $dg=g\theta$ has a solution if and only if $d^e\theta=0$ assuming g to be invertible (Frobenius' theorem).
In general, we get by $(8)_r$

$$(11)' \qquad dP_\theta(1) = P_\theta(1)\theta - P_{-\theta}(I(P_\theta(1)d^e\theta)),$$

where P_θ means $P_{r,\theta}$ and 1 is the identity matrix. Since $P_\theta(1)(0)=1$, define $\log P_\theta(1)$ by $(\log P_\theta(1))(0)=0$ and set $k^e(\theta)=\log P_\theta(1)$, we have

$$(11) \qquad d^e k^e(\theta) + P_\theta(1)^{-1}P_{-\theta}(I(P_\theta(1)d^e\theta)) = \theta.$$

This is the chain homotopy for the non-abelian differential d^e on 1-forms and the 1-dimensional non-abelian Poincaré lemma (Frobenius' theorem) follows from this formula.

5. Similarly as lemma 4, we have by direct calculations

Lemma 5. Let Φ be a p-form and Ψ a (p+1)-form. Then we obtain

$$(12)_r \qquad d\Psi + \Psi\wedge\theta + (-1)^p\Phi\wedge d^{e,p}\theta = 0, \quad \text{if } d\Phi = \Phi\wedge\theta + \Psi,$$

$$(12)_\ell \qquad d\Psi - \theta\wedge\Psi + d^e_\ell\theta\wedge\Phi = 0, \quad \text{if } d\Phi = \theta\wedge\Phi + \Psi,$$

$$(12)_{ad} \qquad d\Psi - [\theta,\Psi] + [d^e_\ell\theta,\Phi] = 0, \quad \text{if } d\Phi = [\theta,\Phi] + \Psi.$$

Theorem 2. (i). The equation $d\Phi-\Phi\wedge\theta=\Psi$, $\deg\Psi=p+1$, has a solution if and only if there exists a closed form Λ such that

$$(13)_r \qquad P_{r,\theta}(\Lambda)\wedge d^{e,p}\theta = d\Psi + \Psi\wedge\theta + (-1)^p P_{r,\theta}(I\Psi)\wedge d^{e,p}\theta .$$

(ii). The equation $d\Phi - \theta\wedge\Phi = \Psi$ has a solution if and only if there exists a closed form Λ such that

$$(13)_\ell \qquad d^e{}_\ell\theta\wedge P_{\ell,\theta}(\Lambda) = d\Psi - \theta\wedge\Psi + d^e{}_\ell\theta\wedge P_{\ell,\theta}(I\Psi).$$

(iii). The equation $d\Phi - [\theta,\Phi] = \Psi$ has a solution if and only if there exists a closed form Λ such that

$$(13)_{ad} \qquad [d^e{}_\ell\theta, P_{ad,\theta}(\Lambda)] = d\Psi - [\theta,\Psi] + [d^e{}_\ell\theta, P_{\ell,\theta}(I\Psi)].$$

Proof. If a solution Φ exists, it satisfies $\Phi - I_\theta(\Phi) = I\Psi + dI\Phi$. Hence by (7), we have $\Phi = P_\theta(I\Psi + dI\Phi)$. This shows the theorem.

Corollary (3-dimensional (non-abelian) Poincaré lemma). Let Ψ be a 3-form. Then the equation $d^e_\theta\Theta = \Psi$ has a local solution if and only if there exists a closed 2-form Λ such that

$$(14) \qquad [d^e\theta, P_{ad,-\theta}(\Lambda)] = d\Psi + [\theta,\Psi] - [d^e\theta, P_{ad,-\theta}(I\Psi)].$$

Proof. Since $d^e_\theta\Theta = d\Theta - [-\theta,\Theta]$, we have the corollary by (2) and $(13)_{ad}$.

Note. Theorem 2 and theorem 3 suggest following successive method for solving the equations $d\Phi = \Phi\wedge\theta$, $d\Phi = \Phi\wedge\theta + \Psi$, etc.. For example, consider the equation $d^e_\theta\Theta = \Psi$, we define a sequence of 2-forms $\Theta_0, \Theta_1, \ldots$ as follows:

$$\Theta_0 = I\Psi + \Lambda\ , \quad d\Lambda = 0, \quad \Theta_n = I[\Theta_{n-1}, \theta],\ n \geq 1.$$

By definition, each Θ_n $(n \geq 1)$ satisfies the following differential equation

$$(15) \qquad d\Theta_n = [\Theta_{n-1}, \theta] + \Psi_n, \quad \sum_{n=1}^{\infty} \Psi_n = Id\Psi.$$

One of the integrability condition of (15) is

$$(16) \qquad [\Theta_{n-1}, d\theta] + d\Psi_n = -([\Theta_{n-2}, \theta\wedge\theta] + [\Psi_{n-1}, \theta]).$$

Similarly, we derive the systems of equations $d\Phi_n = \Phi_{n-1}\wedge\theta$, $d\Phi_n = \theta\wedge\Phi_{n-1}$ and $d\Phi_n = [\theta, \Phi_{n-1}]$ from the equations $d\Phi = \Phi\wedge\theta$, $d\Phi = \theta\wedge\Phi$ and $d\Phi = [\theta, \Phi]$. Their integrability conditions are given by $\Phi_n\wedge d\theta + (-1)^p \Phi_{n-1}\wedge(\theta\wedge\theta) = 0$, $d\theta\wedge\Phi_n - -(\theta\wedge\theta)\wedge\Phi_{n-1} = 0$ and $[d\theta, \Phi_n] - [\theta\wedge\theta, \Phi_{n-1}] = 0$.

<u>Example</u>. If each Φ_n satisfies $d\Phi_n = \Phi_{n-1}\wedge\theta$ and $d\Phi_0 = 0$, $\Phi_z = \sum z^n \Phi_n$ satisfies the equation $d\Phi_z = \Phi_z\wedge(z\theta)$. If θ is $h^{-1}dh = df$, that is, $h = e^f$ and $[f, df] = 0$, then $\Phi_n = f^n/n!\Lambda$ and $\Phi_z = \Lambda e^{zf}$ (cf. [17], [18]).

§3. 2-dimensional non-abelian Poincaré lemma

6. In this section, P_θ means $P_{r,\theta}$.

<u>Definition</u>. Let θ be a 1-form and Θ a p-form $(p \geq 1)$. Then we set

$$(17) \qquad J_\theta(\Theta) = P_{(-1)^{p-1}\theta}(I(P_\theta(1)\Theta)).$$

By definition, $J_\theta(\Theta)$ is a (p-1)-form and linear in Θ.

<u>Lemma 6</u>. We have

$$(18) \qquad dJ_\theta(\Theta) =$$

$$= (-1)^{p-1}J_\theta(\Theta)\wedge\theta + P_\theta(1)\Theta - J_\theta(d^e{}_\theta\Theta) +$$

$$+ J_\theta(P_\theta(1)^{-1}(J_\theta(d^e\theta)\wedge\Theta - J_\theta(\Theta)\wedge d^e\theta)).$$

<u>Proof</u>. By (2) and $(8)_r$, we get

$$dJ_\theta(\Theta) = (-1)^{p-1}J_\theta(\Theta)\wedge\theta + P_{(-1)^p\theta}(dI(P_\theta(1)\Theta)-I(J_\theta(\Theta)\wedge d^e\theta)).$$

Again by $(8)_r$, we get

$$dI(P_\theta(1)\Theta) = P_\theta(1)\Theta - Id(P_\theta(1)\Theta) =$$

$$= P_\theta(1)\Theta - I(P_\theta(1)\theta\wedge\Theta + P_\theta(1)d\Theta - P_{-\theta}(I(P_\theta(1)d^e\theta)\wedge\Theta).$$

On the other hand, by (7), we get

$$P_{(-1)^p\theta}(P_\theta(1)\Theta) = P_\theta(1)\Theta + (-1)^p P_{(-1)^p\theta}(I(P_\theta(1)\Theta\wedge\theta).$$

Hence we have

$$P_{(-1)^p\theta}(dI(P_\theta(1)\Theta)) = P_\theta(1)\Theta - P_{(-1)^p\theta}(I(P_\theta(1)(d\Theta+\theta\wedge\Theta-(-1)^p\Theta\wedge\theta))+$$

$$+ P_{(-1)^p\theta}(P_{-\theta}(I(P_\theta(1)d^e\theta)\wedge\Theta)).$$

Since $d\Theta+\theta\wedge\Theta-(-1)^p\Theta\wedge\theta$ is $d^e{}_\theta\Theta$, we obtain (18).

<u>Corollary</u>. Let Θ be a 2-form. Then set

(19) $$\kappa^e{}_\theta(\Theta) = \Theta + P_\theta(1)^{-1}J_\theta(\Theta - d^e\theta),$$

we have

(20) $$d^e(\kappa^e_\theta(\Theta))$$

$$= \Theta - P_\theta(1)^{-1}J_\theta(d^e_\theta\Theta) + P_\theta(1)^{-1}J_\theta(P_\theta(1)^{-1}(J_\theta(d^e\theta)_\wedge\Theta-$$

$$-J_\theta(\Theta)_\wedge d^e\theta)) + P_\theta(1)^{-1}J_\theta(\Theta)_\wedge P_\theta(1)^{-1}J_\theta(\Theta-d^e\theta).$$

<u>Proof</u>. Since $J_\theta(\Theta)$ is a 1-form, we have by $(8)_r$ and (18)

$$d(J_\theta(\Theta)P_\theta(1)^{-1}) = dJ_\theta(\Theta)P_\theta(1)^{-1} + J_\theta(\Theta)P_\theta(1)^{-1}{}_\wedge dP_\theta(1)P_\theta(1)^{-1} =$$

$$= -J_\theta(\Theta)_\wedge\theta P_\theta(1)^{-1} + P_\theta(1)\Theta P_\theta(1)^{-1} - J_\theta(d^e_\theta\Theta)P_\theta(1)^{-1} +$$

$$+ J_\theta(P_\theta(1)^{-1}(J_\theta(d^e\theta)_\wedge\Theta-J_\theta(\Theta)_\wedge d^e\theta))P_\theta(1)^{-1} + J_\theta(\Theta)_\wedge\theta P_\theta(1)^{-1} +$$

$$+ J_\theta(\Theta)P_\theta(1)^{-1}P_{-\theta}(I(P_\theta(1)d^e\theta)P_\theta(1)^{-1}$$

$$= P_\theta(1)\Theta P_\theta(1)^{-1} - J_\theta(d^e_\theta\Theta)P_\theta(1)^{-1} + J_\theta(P_\theta(1)^{-1}(J_\theta(d^e\theta)_\wedge\Theta-$$

$$-J_\theta(\Theta)_\wedge d^e\theta))P_\theta(1)^{-1} + J_\theta(\Theta)P_\theta(1)^{-1}{}_\wedge J_\theta(d^e\theta)P_\theta(1)^{-1}.$$

Hence we obtain

(20)' $$d^e(J_\theta(\Theta)P_\theta(1)^{-1})$$

$$= P_\theta(1)\Theta P_\theta(1)^{-1} - J_\theta(d^e_\theta\Theta)P_\theta(1)^{-1} + J_\theta(P_\theta(1)^{-1}(J_\theta(d^e\theta)_\wedge\Theta-$$

$$-J_\theta(\Theta)_\wedge d^e\theta))P_\theta(1)^{-1} + J_\theta(\Theta)P_\theta(1)^{-1}{}_\wedge J_\theta(\Theta-d^e\theta)P_\theta(1)^{-1}.$$

On the other hand, by $(8)_r$, we get

$$P_\theta(1)^{-1}(J_\theta(\Theta)P_\theta(1)^{-1})P_\theta(1) + P_\theta(1)^{-1}dP_\theta(1) = \Theta + P_\theta(1)^{-1}J_\theta(\Theta-d^e\theta)$$

Since this right hand side is $\kappa^e_\theta(\Theta)$, we have (20) by (20)'.

7. Definition. Let $\theta=\theta_0$ and $\Theta=\Theta_0$ be a pair of a 1-form and a 2-form. Then we define a series of pairs of 1-forms $\theta_n=\theta_n(\theta,\Theta)$ and 2-forms $\Theta_n = \Theta_n(\theta,\Theta)$, n=1,2,..., as follows:

$$\theta_n = \kappa^e_{\theta_{n-1}}(\Theta_{n-1}),$$

$$\Theta_n = \Theta_{n-1} - P_{\theta_{n-1}}(1)^{-1}J_{\theta_{n-1}}(d^e_{\theta_{n-1}}\Theta_{n-1}).$$

Example. If $\theta=0$ and $d\Theta=0$, we have $\theta_1=I(\Theta)$ and $\Theta_1=\Theta-I[I(\Theta),\Theta]$. If $d^e_{\theta_1}\Theta=0$, that is $[\theta,I(\Theta)]=0$ in this case, θ_2 is given by $I(\Theta)-$ $-P_{I(\Theta)}(1)^{-1}P_{I(\Theta)}(I(P_{I(\Theta)}(1)I(\Theta)\wedge I(\Theta)))$.

By definition, we have

Lemma 7. (i). θ_n and Θ_n are defined on the starlike neighborhood of the origin on which θ and Θ are both defined.

(ii). For any $n\geqq 0$ and $i\geqq 0$, we have

$$\theta_{n+i} = \theta_n(\theta_i,\Theta_i), \quad \Theta_{n+i} = \Theta_n(\theta_i,\Theta_i).$$

(iii). Θ_n is equal to Θ_{n-1} if Θ_{n-1} satisfies the Bianchi identity for θ_{n-1}. Especially, $\Theta_n=\Theta$ and $\theta_n=\kappa^e_{\theta_{n-1}}(\Theta)$ for all $n\geqq 0$ if Θ satisfies Bianchi identity for all θ_n, that is, if we have

(21) $\quad d^e_{\theta_n}\Theta = 0, \quad$ for all $n \geqq 0$.

(iv). If $\Theta_n=d^e\theta_n$ for some n, $n\geqq 0$, we have $\theta_{n+i}=\theta_n$ and $\Theta_{n+i}=\Theta_n$ for all $i\geqq 0$.

Note. By definition , θ_n and Θ_n have the following expressions.

(22) $$\theta_n = \theta_0 + \sum_{i=0}^{n-1} P_{\theta_i}(1)^{-1} J_{\theta_i}(\Theta_i - d^e\theta_i).$$

(23) $$\Theta_n = \Theta - \sum_{j=0}^{n-1} k^e{}_{\theta_j,\Theta_j}{}^{n-j-1}(d^e{}_{\theta_j}\Theta).$$

Here $k^e{}_{\theta,\Theta}{}^i(\Phi)$ is given by

$$k^e{}_{\theta,\Theta}{}^i(\Phi) = P_\theta(1)^{-1}J_\theta(\Phi) + \sum_{i \geqq i_1 > \cdots > i_k > 0} (-1)^k P_{\theta_{i_1}}(1)^{-1} J_{\theta_{i_1}}(d^e{}_{\theta_{i_1}}($$

$$(\cdots(P_{\theta_{i_k}}(1)^{-1}J_{\theta_{i_k}}(P_\theta(1)^{-1}J_\theta(\Phi)))\cdots)).$$

By (22) and (23), we obtain

(24) $$d^e\theta_n + \sum_{j=0}^{n-1} k^e{}_{\theta_j,\Theta_j}{}^{n-j-1}(d^e{}_{\theta_j}\Theta) -$$

$$- P_{\theta_{n-1}}(1)^{-1}J_{\theta_{n-1}}(P_{\theta_{n-1}}(1)^{-1}(J_{\theta_{n-1}}(d^e\theta_{n-1})\wedge\Theta_n -$$

$$-J_{\theta_{n-1}}(\Theta_{n-1})\wedge d^e\theta_{n-1})) -$$

$$- P_{\theta_{n-1}}(1)^{-1}J_{\theta_{n-1}}(\Theta_{n-1})\wedge P_{\theta_{n-1}}(1)^{-1}J_{\theta_{n-1}}(\Theta_{n-1} - d^e\theta_{n-1})$$

$$= \Theta.$$

8. Lemma 8. If $\bar{\Theta}(x) \leqq C(x)||x||^s$ for some $s \geqq 0$, where $C(x)$ is non decreasing in $||x||$, then we have

(25) $$||J_\theta(\Theta)(x)|| \leqq \frac{||x||^{s+1}}{s+1} C(x) e^{2m\bar{\theta}(x)||x||} .$$

Proof. Since we get $||I(P_\theta(1)\Theta)(x)|| \leqq ||x||^{s+1}/(s+1)C(x) \cdot$ $\cdot e^{m\bar{\theta}(x)||x||}$ by (4) and lemma 2,(iv), we have (25) by lemma 2, (iv).

Lemma 9. If $me^{m\bar{\theta}(x)||x||}||x||<1$, we have

$$(26) \qquad ||P_\theta(1)^{-1}(x)|| \leqq (1 - me^{m\bar{\theta}(x)||x||}||x||)^{-1} .$$

Proof. Since $P\ (1)=1+I_\theta(P_\theta(1))$, we have

$$||P_\theta(1)^{-1}(x)|| \leqq \sum_{n=0}^{\infty} ||I_\theta(P_\theta(1))(x)||^n ,$$

if $||I_\theta(P_\theta(1))(x)||<1$. Since $||I_\theta(P_\theta(1))(x)||\leqq me^{m\bar{\theta}(x)||x||}||x||$ by (4) and lemma 2, (iv), we obtain (26).

Note. It seems that the equality $P_\theta(1)^{-1}=P_{\gamma,-\theta}(1)$ holds (If $d^e\theta=0$, this is true). If it is true, (26) is improved to $||P_\theta(1)^{-1}(x)||\leqq \leqq e^{m\bar{\theta}(x)||x||}$.

Lemma 10. If the inequality

$$||\Theta(x) - d^e\theta(x)|| \leqq C||x||^s$$

holds for some $C>0$ and $s\geqq 0$ on $\{x|\ ||x||<r\}$, there exist constants $c=c(\bar{\theta})>0$ and $a=a(\bar{\theta})>6m||\bar{\theta}(\rho)||$, $\rho>0$ is fixed, such that the inequalities

$$(27) \qquad ||\theta(x) - K^e_\theta(\Theta)(x)|| \leqq \frac{||x||^{s+1}}{s+1}Ce^{a||x||},$$

$$(28) \qquad ||\Theta(x) - (d^e(K^e_\theta(\Theta))(x)+P_\theta(1)^{-1}J_\theta(d^e_\theta\Theta)(x))||$$

$$\leqq \frac{||x||^{s+2}}{s+1}2(A(x))^2e^{a||x||}, \qquad A(x) = \max(1, C\bar{\theta}(x)),$$

hold on $\{x|\ ||x||<c\}$. Here c depends on $\bar{\theta}$ and a. If a is fixed and

$\bar{\theta}_1 \geqq \bar{\theta}_2$, then $c(\bar{\theta}_1) \leqq c(\bar{\theta}_2)$.

Proof. We choose $c'>0$ and $a>0$ as follows: If $||x||<c'$, the inequality

$$(29) \qquad (1 - me^{m\bar{\theta}(x)||x||}||x||)^{-2} e^{4m\bar{\theta}(x)||x||} \leqq e^{a||x||}$$

holds. Then, setting $c= \min(c',r)$, we have (27) by (25). Because we have

$$(1 - me^{m\bar{\theta}(x)||x||}||x||)^{-1} e^{2m\bar{\theta}(x)||x||}$$

$$\leqq (1 - me^{m\bar{\theta}(x)||x||}||x||)^{-2} e^{4m\bar{\theta}(x)||x||},$$

on $||x||<c$.

Since $J_\theta(\Theta)$ is linear in Θ, to evaluate $J_\theta(d^e\theta)\wedge\Theta - J_\theta(\Theta)\wedge d^e\theta$, it is sufficient to evaluate $C(J_\theta(\beta)\Theta + J_\theta(\Theta)\beta)$, where $||\beta(x)|| \leqq ||x||^s$, under our assumption. Then by (25) and (26), we have

$$||P_\theta(1)^{-1} J_\theta(P_\theta(1)^{-1} J_\theta(d^e\theta)\wedge\Theta - J_\theta(\Theta)\wedge d^e\theta))(x)||$$

$$\leqq \frac{||x||^{s+2}}{s+1} C\bar{\theta}(x)(1-me^{m\bar{\theta}(x)||x||}||x||)^{-2} e^{4m\bar{\theta}(x)||x||}.$$

On the other hand, by (25) and (26), we get

$$||(P_\theta(1)^{-1} J_\theta(\Theta)\wedge P_\theta(1)^{-1} J_\theta(\Theta - d^e\theta))(x)||$$

$$\leqq \frac{||x||^{s+2}}{s+1} (C\bar{\theta}(x))^2 (1-me^{m\bar{\theta}(x)||x||}||x||)^{-2} e^{4m\bar{\theta}(x)||x||}.$$

Hence we obtain (28) by the definition of $A(x)$ and (29).

The second assertion of the lemma follows from the definitions of the constants a and c.

Corollary (Asymptotic chain homotopy formula for d^e at 2-forms). We have the following asymptotic formula at $||x||\to 0$.

$$(30)\qquad \Theta(x) = d^e\theta_n(x) + \sum_{j=0}^{n-1} k^e_{\theta_j,\Theta_j}{}^{n-j-1}(d^e_{\theta_j}\Theta)(x) + O(||x||^{2n}).$$

Proof. This follows from induction on n and (24), (28).

Note. More simply, (30) is written as follows:

$$(30)'\qquad ||\Theta_n(x) - d^e\theta_n(x)|| = O(||x||^{2n}),\quad ||x||\to 0.$$

9. Theorem 3 (2-dimensional non-abelian Poincaré lemma). Let Θ be a 2-form, θ_0 a 1-form such that the equalities (21) hold. Then there exists $c>0$ such that $\{\theta_n\}$ and $\{d^e\theta_n\}$ both converge uniformly and absolutely on $\{x|\ ||x||<c\}$. Moreover, setting $\lim_{n\to\infty}\theta_n=\theta$, we have

$$d^e\theta = \Theta,\quad \text{on } \{x|\ ||x|| < c\}.$$

Proof. We fix $R>0$ and $a>6m\cdot\max(\bar\theta(R),1)$. Then, depending on R and a, we take a positive integer s and positive numbers A, B such that

$$s > e^{aR},\quad A > 2\cdot\max(1, \overline{(\Theta-d^e\theta)}(R)\cdot\bar\theta(R)),\quad \frac{a}{6m} > B > \max(\bar\theta(R),1).$$

Using these constants, we define a function r(x) by

$$r(x) = B + s!\left(\frac{1}{(1-A^2||x||)^{s+1}} - 1\right).$$

By definition, we have $(1-me^{mr(x)||x||}||x||)^{-2}e^{4mr(x)||x||}=1+6mB||x||+ +O(||x||^2)$. Hence we can choose $c=c(r,a)>0$ such that $c<\min(R,1)$ and the inequality

$$(1 - me^{mr(x)||x||}||x||)^{-2}e^{4mr(x)||x||} \leqq e^{a||x||}$$

holds if $||x||<c$. Then by (21), (27) and (28), we get

$$(31)_1 \qquad ||\theta_1(x) - \theta_0(x)|| \leqq A^2 s||x||, \quad ||x|| < c,$$

$$(32)_1 \qquad ||\Theta(x) - d^e\theta_1(x)|| \leqq A^2 s||x||^2, \quad ||x|| < c.$$

Next, we assume that we have the inequalities

$$(31)_i \qquad ||\theta_i(x) - \theta_{i-1}(x)|| \leqq A^{2i}\,\frac{(i+s)!}{i!}||x||^i, \quad ||x|| < c,$$

$$(32)_i \qquad ||\Theta(x) - d^e\theta_i(x)|| \leqq A^{2i}\,\frac{(i+s)!}{i!}||x||^{i+1}, \quad ||x|| < c,$$

if $i \leqq n$. Then by $(31)_i$, we obtain on $||x||<c$

$$(33)_n \qquad ||\theta_n(x)|| \leqq ||\theta_0(x)|| + \sum_{i=1}^{n}||\theta_i(x) - \theta_{i-1}(x)|| \leqq$$

$$\leqq B + \sum_{i=1}^{n} A^{2i}\,\frac{(i+s)!}{i!}||x||^i \leqq B + s!\left(\frac{1}{(1-A^2||x||)^{s+1}} - 1\right) = r(x).$$

By $(33)_n$, we get

$$(1 - me^{m\bar{\theta}_n(x)||x||}||x||)^{-2}e^{4m\bar{\theta}_n(x)||x||}$$

$$\leqq (1 - me^{r(x)||x||}||x||)^{-2}e^{4mr(x)||x||},$$

on $||x||<c$. Hence we have by inductive assumption, lemma 10, (21) and the choice of s

$$||\theta_{n+1}(x) - \theta_n(x)|| \leqq A^{2n} \frac{(n+s)!}{(n+1)!} ||x||^{n+1} e^{a||x||},$$

$$||\Theta(x) - d^e\theta_n(x)|| \leqq A^{2(n+1)} \frac{(n+s)!}{(n+1)!} ||x||^{n+2} e^{a||x||},$$

on $||x||<c$. On the other hand, since $A \geqq 2$ and $||x||<c<1$, we obtain by (21) and $(33)_n$

$$A^{2n} \frac{(n+s)!}{(n+1)!} ||x||^{n+1} e^{a||x||} \leqq A^{2(n+1)} \frac{(n+s+1)!}{(n+1)!} ||x||^{n+1},$$

$$A^{2(n+1)} \frac{(n+s)!}{(n+1)!} ||x||^{n+2} e^{a||x||} \leqq A^{2(n+1)} \frac{(n+s+1)!}{(n+1)!} ||x||^{n+2},$$

on $||x||<c$. Hence we have $(31)_{n+1}$, $(32)_{n+1}$ and $(33)_{n+1}$. Therefore we obtain $(31)_n$, $(32)_n$ and $(33)_n$ for all n. Hence $\{\theta_n\}$ and $\{d^e\theta_n\}$ both converge uniformly and absolutely on $\{x|\ ||x||<c\}$. Then by $(32)_n$, set $\lim_{n\to\infty}\theta_n=\theta$, we have $d^e\theta=\Theta$ on $||x||<c$.

Corollary. The equation $d^e\theta=\Theta$ has a local solution if and only if there exists a 1-form θ_0 such that the equalities (21) hold for Θ.

Proof. We need only to show necessity. But it follows from lemma 7, (iii) and (iv).

Note. Let $I=(i_1,\ldots,i_n)$ be an integer vector. Then setting

$$\bar{\varphi}_I(x) = \max_{j_1\leqq i_1,\ldots,j_n\leqq i_n} \max_{0\leqq t\leqq 1} \left|\left|\frac{\partial^{j_1+\cdots+j_n}\varphi}{\partial x_1^{j_1}\cdots\partial x_n^{j_n}}(xt)\right|\right|,$$

we have

$$\left|\left|\frac{\partial^{|I|}P_\theta(\varphi)}{\partial x_1^{i_1}\cdots\partial x_n^{i_n}}(x)\right|\right| \leqq 2^{|I|}\,\bar{\varphi}_I(x) e^{2^{|I|}\,{}_mC_p\,\bar{\theta}_I(x)||x||}, \quad |I|=i_1+\cdots+i_n$$

Hence if Θ and θ_0 are both smooth, the above convergence of $\{\theta_n\}$ is C - convergence. Therefore θ is smooth in this case.

§4. Construction of all local solutions of the equation $d^e\theta = \Theta$

10. If the 1-forms and ' both satisfy the equation $d^e\theta = d^e\theta' = \Theta$, we obtain

$$(34) \qquad d\eta + \eta\wedge\eta + [\theta,\eta] = 0, \quad \theta' = \theta + \eta.$$

Because we have $d^e(\theta+\eta) = d^e\theta + d^e\eta + [\theta,\eta]$.

Lemma 11. If η is a solution of the equation (34), then η satisfies following integral equation for suitable 1-form Λ.

$$(35) \qquad \eta = P_{ad,-\theta}(\Lambda - I(\eta\wedge\eta)).$$

Proof. If η satisfies (34), we have $\eta + I[\theta,\eta] = dI\eta - I(\eta\wedge\eta)$. Hence taking $\Lambda = dI\eta$, η satisfies (35).

Lemma 12. The equation (35) has unique solution near the origin for any Λ.

Proof. We define a series of 1-forms $\eta_1, \eta_2, \ldots,$ by

$$(36)_1 \qquad \eta_1 = P_{ad,-\theta}(\Lambda),$$

$$(36)_n \qquad \eta_n = -P_{ad,-\theta}(I(\sum_{k=1}^{n-1} \eta_k\wedge\eta_{n-k})), \quad n \geqq 2.$$

By lemma 2,(iv), set $M(x) = 2m\bar{\theta}(x)$, we have

$$||\eta_1(x)|| \leqq \bar{\Lambda}(x)e^{M(x)||x||}.$$

We assume that if $1 \leqq k \leqq n-1$, we have the inequalities

$$||\eta_k(x)|| \leqq m^{2(k-1)}(\bar{\Lambda}(x))^k e^{(2k-1)M(x)||x||}||x||^{k-1}. \tag{37}$$

Then we have

$$||\eta_k(x)||\cdot||\eta_{n-k}(x)|| \leqq m^{2(n-2)}(\bar{\Lambda}(x))^n e^{(2n-2)M(x)||x||}||x||^{n-2}.$$

Therefore by (4) and lemma 2, (iv), we obtain

$$||\eta_n(x)|| \leqq \frac{m^2}{n-1}e^{M(x)||x||}\sum_{k=1}^{n-1}||\eta_k(x)||\cdot||\eta_{n-k}(x)||$$

$$\leqq m^2 e^{M(x)||x||}m^{2(n-2)}(\bar{\Lambda}(x))^n e^{(2n-2)M(x)||x||}||x||^{n-1}$$

$$\leqq m^{2(n-1)}(\bar{\Lambda}(x))^n e^{(2n-1)M(x)||x||}||x||^{n-1}.$$

Hence (37) hold for all k. Therefore, taking $c>0$ to satisfy $m^2\bar{\Lambda}(y)e^{2M(y)c}c<1$, $||y||=c$, $\eta = \sum_{n=1}^{\infty}\eta_n$ converges uniformly and absolutely on $\{x|\ ||x||<c\}$. Then by the definitions of $\eta_1, \eta_2, \ldots$, this η satisfies (35). Hence (35) has at least 1 solution.

To show the uniqueness, we assume η and η' both satisfy (35). Then we have

$$\eta - \eta' = P_{ad,-\theta}(I(\eta'\wedge\eta' - \eta\wedge\eta)).$$

We assume $||\eta(x)|| \leqq K$ on $||x||<r$. Then, if $||(\eta-\eta')(x)|| \leqq C||x||^s$, $s \geqq 0$, on $||x||<r$, we have $||(\eta'\wedge\eta'-\eta\wedge\eta)(x)|| \leqq (2CK+C^2r^2)||x||^s$ on $||x||<r$. Hence assuming $r<1$ and taking a suitable $A>0$, we obtain

$$\|(\eta - \eta')(x)\| \leqq \frac{A^n}{n!}\|x\|^n, \quad \|x\| < r,$$

for all n by induction and lemma 2, (iv). Hence we have the uniqueness if $\|x\|<1$.

<u>Corollary</u>. If η is a solution of (34), then η is given by the series $\sum_{n=1}^{\infty}\eta_n$, where each η_n is given by (36) with $\Lambda = dI\eta$.

11. <u>Definition</u>. Let $\eta_1, \eta_2, \ldots,$ be determined by (36). Then we set

$$(38) \qquad q_\theta(\Lambda) = \sum_{n=1}^{\infty}\eta_n.$$

<u>Lemma 13</u>. We have

$$(39) \qquad dq_\theta(\Lambda) + [\theta, q_\theta(\Lambda)] + q_\theta(\Lambda)\wedge q_\theta(\Lambda)$$

$$= d\Lambda + Id\{[\theta, q_\theta(\Lambda)] + q_\theta(\Lambda)\wedge q_\theta(\Lambda)\}.$$

<u>Corollary</u>. $q_\theta(\Lambda)$ satisfies (34) if and only if $d\Lambda=0$ and $d\{[\theta, q_\theta(\Lambda)]+q_\theta(\Lambda)\wedge q_\theta(\Lambda)\}=0$.

<u>Proof</u>. (39) shows sufficiency. If $q_\theta(\Lambda)$ satisfies (34), $[\theta, q_\theta(\Lambda) + q_\theta(\Lambda)\wedge q_\theta(\Lambda)]$ is equal to $-dq_\theta(\Lambda)$. On the other hand, Λ is equal to $dIq_\theta(\Lambda)$ by lemma 11. Hence we have the necessity.

If $\eta_1, \eta_2, \ldots,$ are given by (36), we have

$$(36)' \qquad d\eta_1 + [\theta, \eta_1] = d\Lambda + Id[\theta, \eta_1],$$

$$d\eta_n + [\theta,\eta_n] + \sum_{k=1}^{n-1}\eta_k\wedge\eta_{n-k} = I\{d[\theta,\eta_n] + d(\sum_{k=1}^{n-1}\eta_k\wedge\eta_{n-k}), \quad n \geqq 2.$$

(36)' suggests that the system of equations

$$(40)_1 \qquad d\eta_1 + [\theta,\eta_1] = 0,$$

$$(40)_n \qquad d\eta_n + [\theta,\eta_n] + \sum_{k=1}^{n-1}\eta_k\wedge\eta_{n-k} = 0, \quad n \geqq 2,$$

is the "linearized" system of the non-linear equation (34).

Lemma 14. The integrability condition of the system (40) is given by

$$(41) \qquad [\eta_n, \Theta] = 0, \quad n = 1,2,\ldots, \qquad \Theta = d^e\theta.$$

Proof. By lemma 4, the integrability condition of $(40)_1$ is given by $[\eta_1,\Theta]=0$. By lemma 5, the integrability condition of $(40)_n$, $n\geqq 2$, is given by

$$[\eta_n,\Theta] = -d(\sum_{k=1}^{n-1}\eta_k\wedge\eta_{n-k}) + \left[\sum_{k=1}^{n-1}\eta_k\wedge\eta_{n-k},\theta\right].$$

Applying $(40)_n$ to this right hand side, we get

$$d(\sum_{k=1}^{n-1}\eta_k\wedge\eta_{n-k}) = \sum_{k=1}^{n-1}d\eta_k\wedge\eta_{n-k} - \sum_{k=1}^{n-1}\eta_k\wedge d\eta_{n-k} =$$

$$= -\sum_{k=1}^{n-1}[\theta,\eta_k]\wedge\eta_{n-k} - \sum_{k=1}^{n-1}(\sum_{j=1}^{k-1}\eta_j\wedge\eta_{k-j})\wedge\eta_{n-k} +$$

$$+ \sum_{k=1}^{n-1}\eta_k\wedge[\theta,\eta_{n-k}] + \sum_{k=1}^{n-1}\eta_k\wedge(\sum_{\ell=1}^{n-k-1}\eta_\ell\wedge\eta_{n-k-\ell})$$

$$= \left[\sum_{k=1}^{n-1}\eta_k\wedge\eta_{n-k},\theta\right].$$

Hence we have the lemma.

12. Lemma 15. To set

$$(42)_1 \qquad \eta_1 = df - [f,\theta],$$

$$(42)_n \qquad \eta_n = \frac{(-1)^n}{n!}(\mathrm{ad}\ f)^{n-1}([f,\theta] - df), \quad n \geqq 2,$$

we have

$$(43)_1 \qquad d\eta_1 + [\theta,\eta_1] = [\Theta,f], \quad \Theta = d^e\theta,$$

$$(43)_n \qquad d\eta_n + [\theta,\eta_n] + \sum_{k=1}^{n-1}\eta_k \wedge \eta_{n-k} = \frac{(-1)^n}{n!}(\mathrm{ad}\ f)^n(\Theta), \quad n \geqq 2.$$

Proof. We set $\eta_z = e^{-zf}\theta e^{zf} - \theta + e^{-zf}d(e^{zf})$, where z is a complex parameter (cf. [17]). Then we have $d\eta_z + [\theta,\eta_z] + \eta_z \wedge \eta_z = e^{-zf}\Theta e^{zf} - \Theta$. On the other hand, we know

$$e^{-zf}\theta e^{zf} - \theta + e^{-zf}d(e^{zf}) = \sum_{n=1}^{\infty}\frac{(-1)^n}{n!}z^n(\mathrm{ad}\ f)^n([f,\theta] - df),$$

$$e^{-zf}\Theta e^{zf} - \Theta = \sum_{n=1}^{\infty}\frac{(-1)^n}{n!}z^n(\mathrm{ad}\ f)^n(\Theta).$$

Since z is an indeterminant, we have the lemma by these equalities.

Theorem 4. If $[\Theta,f]$ is equal to 0, we have

$$(44) \qquad q_\theta(d(f - I[f,\theta])) = e^{-f}\theta e^f - \theta + e^{-f}d(e^f).$$

Proof. Since $[f,\Theta]=0$, the system (43) becomes the system (40). Hence $\eta_1, \eta_2, \ldots,$ are given by (36) for some Λ. To determine Λ, we set $df - [f,\theta] = P_{\mathrm{ad},-\theta}(\Lambda)$. Then we have

$$\Lambda = df - [f,\theta] + I[\theta, df-[f,\theta]] =$$

$$= df - [f,\theta] + Id[f,\theta] = d(f - I[f,\theta]).$$

Hence we have the theorem.

Note. In general, if $P_{ad,-\theta}(\Lambda)=df-[f,\theta]$, we get

(45) $$\Lambda = d(f - I[f,\theta]) + I[\Theta, f].$$

Hence it seems that the equality

(44)' $$q_\theta(d(f-I[f,\theta])+I[\Theta,f]) = e^{-f}\theta e^{f} - \theta + e^{-f}d(e^{f})$$

holds for any f. On the other hand, by corollary of lemma 13, if $q_\theta(\Lambda)$ satisfies the equation (34), Λ is closed. Set $\Lambda=dh$, we get $f=P_{ad,-\theta}(h)$ if $h=f-I[f,\theta]$. Then by $(8)_{ad}'$, we have

$$df = [f,\theta] + P_{ad,-\theta}(\Lambda+I[\Theta,f]).$$

Hence we have $P_{ad,-\theta}(\Lambda)=df-[f,\theta]$ if $I[\Theta,f]=0$.

References

[1] Agrachev,V.A. and Gamkrelidze,R.V.: The exponential representation of flows and the chronological calculus, Mat. Sb.,107 (1978), 467-532, Math. USSR Sb.,35(1979), 727-785.

[2] Agrachev,V.A., Vakhrameev,S.A. and Gamkrelidze,R.V.: Differential-geometric and group-theoretical methods in optimal control theory, Problems in Geometry 14(Itogi Nauki i Tekh.,) 1983, 3-56, J. Sov. Math.,28(1985), 145-182.

[3] Andersson,S.I.: Pseudodifferential operators and characteristic classes for non-abelian cohomology, Lecture Notes in Math. 1045 Springer, Berlin-New York, 1984, 1-10.

[4] Asada,A.: Curvature forms with singularities and non-integral characteristic classes, Lecture Notes in Math. 1139, Springer, Berlin-New York, 1985, 152-168.

[5] Asada,A.: Flat connections of differential operators and related characteristic classes, To appeare in Lecture Notes in Math., Springer, Berlin-New York.

[6] Asada,A.: Non-abelian de Rham theories, To appeare in Proc. of International Colloquim of Differential Geometry, Hajduszoboslo, 1984.

[7] Gaveau,B.: Intégrales harmoniques non abéliennes, Bull. Sc. math., 2e série, 106(1982), 113-169.

[8] Gribov,V.N.: Quantization on non-abelian gauge theories, Nucl. Phys.,B139(1978),1-19.

[9] Gross,L.: A Poincaré lemma for connection forms, Preprint, Cornell University, 1985.

[10] Maeda,Y.: Pointwise convergence of the product integral for a certain integral transformation associated with a Riemannian metric, Kodai Math.J.,8(1985), 193-223.

[11] Oniščik,A.: On classification of fibre spaces, Sov. Math., Doklady 2(1961), 1561-1564.

[12] Oniščik,A.: Connections with zero curvature and de Rham theorem, Sov. Math. Doklady 5(1964), 1654-1657.

[13] Peano,G.: Integration par séries des équations differentielles, Math. Ann.,32(1888), 450-456.

[14] Rasch,G.: Zur Theorie und Anwendung des Produktintegrals, J.für reine und angew. Math.,171(1934), 65-119.

[15] Spivak,M.: Calculus on Manifolds, Benjamin, New York, 1965.

[16] Vassiliou,E.: Tatal differential equations and the structure of fibre bundles, To appeare.

[17] Witten,E.: Supersymmetry and Morse theory, J. Differential Geometry,17(1982), 661-692.

[18] Ziolkowski,R.W. and Deschamps,G.A.: The asymptotic Poincaré lemma and its applications, SIAM J. Math. Anal.,15(1984), 535-558.

LE PROBLÈME DE YAMABE CONCERNANT LA COURBURE SCALAIRE

Thierry Aubin
Université Pierre et Marie Curie
75230-Paris Cedex 05

Introduction.

Soit (V_n, g) une variété Riemannienne compacte de dimension $n \geq 3$, R sa courbure scalaire. Le Problème de Yamabe est le suivant :

Existe-t-il une métrique $\tilde{g}$ conforme à g telle que la courbure scalaire $\tilde{R}$ (correspondant à $\tilde{g}$) soit constante ?

En 1960 Yamabe [9] affirmait qu'une telle métrique $\tilde{g}$ existait toujours. Mais une erreur s'était glissée dans sa démonstration. 25 ans plus tard, il n'est pas superflu de faire le point sur ce problème.

Si on écrit le changement conforme de métrique sous la forme $\tilde{g} = \varphi^{4/(n-2)} g$ la courbure scalaire $\tilde{R}$ s'exprime en fonction de φ comme suit :

(E) $\quad 4(n-1)(n-2)^{-1} \Delta\varphi + R\varphi = \tilde{R}\varphi^{(n+2)/(n-2)}$ avec $\Delta = -g^{ij}\nabla_{ij}$.

Le Problème revient à prouver l'existence d'une fonction $\varphi > 0$, $\varphi \in C^\infty$ vérifiant l'équation (E) avec $\tilde{R} = Cte$.

Posons $L = \Delta + (n-2)R/4(n-1)$ et soit λ la première valeur propre de L. Il est bien connu que l'espace propre associé à λ est de dimension un et qu'il existe une fonction propre $\psi > 0$ ($L\psi = \lambda\psi$).

Nous allons distinguer , suivant la valeur de λ, trois cas . Nous admettons, sans nuire à la généralité que le volume de (V_n, g) est égal à 1 .

1/ Le cas négatif $\lambda < 0$.

Considérons pour $\varphi \in H_1$ et $N = 2n/(n-2)$ la fonctionnelle

$$J(\varphi) = 4(n-1)(n-2)^{-1}\int_V |\nabla\varphi|^2 dV + \int_V R\varphi^2 dV + \int_V |\varphi|^N dV .$$

Cette fonctionnelle a un sens car d'après le théorème de Sobolev $H_1 \subset L_N$.

Posons $\nu = \inf\{J(\varphi)$ pour tout $\varphi \in H_1, \varphi \geq 0\}$.

Comme $J(\varphi) = J(|\varphi|)$, $\nu = \inf\{J(\varphi)/\varphi \in H_1\}$. Montrons que $\nu < 0$. Pour

$k \in R^+$

$$J(k\psi) = 4(n-1)(n-2)^{-1} \lambda k^2 \int_V \psi^2 dV + k^N \int_V |\psi|^N dV .$$

Si k est suffisamment petit, $J(k\psi) < 0$. D'où $\nu < 0$.
D'autre part ν est fini car $|\int_V R\varphi^2 dV| \leq \sup|R| \|\varphi\|_N^2$ et $N > 2$.
Soit $\{\varphi_i\}$, $\varphi_i > 0$ une suite minimisante de H_1 ($J(\varphi_i) \to \nu$ quand $i \to \infty$) . Cette suite est bornée dans H_1 et un raisonnement standard montre qu'il existe une sous suite qui converge dans H_1 vers une fonction $\varphi_o \in H_1$, $\varphi_o \geq 0$. Comme $J(\varphi_o) = \nu$, φ_o vérifie faiblement dans H_1 l'équation d'Euler de la fonctionnelle $J(\varphi)$ qui est l'équation (E) avec $\tilde{R} = Cte < 0$. D'après Trüdinger (7) $\varphi_o \in C^\infty$ et d'après le principe du maximum , comme $\varphi_o \geq 0$, deux cas seulement peuvent se produire : $\varphi_o \equiv 0$ ou $\varphi_o > 0$. Le premier cas est exclu car $J(\varphi_o) = \nu < 0$. Ainsi le Problème de Yamabe est résolu si $\lambda < 0$.

2/ Le cas $\lambda = 0$.

Ce cas est très facile, il suffit de considérer le changement de métrique conforme $\tilde{g} = \psi^{4/(n-2)} g$. Comme $L\psi = 0$, $\tilde{R} = 0$.

Si $\lambda > 0$ en faisant ce même changement de métrique nous obtenons $\tilde{R} = 4(n-1)(n-2)^{-1} \lambda \psi^{2-N} > 0$. D'où le

THÉORÈME 1. - *Sur une variété Riemannienne compacte* (V_n, g) , *de dimension* $n \geq 3$, *on peut, par un changement conforme de métrique, rendre la courbure scalaire partout positive ou égale à une constante négative ou nulle. Les trois cas s'excluent.*

En intégrant (E) on établit ce dernier point.
Ce théorème ne résout pas le Problème de Yamabe dans le cas positif, cas que nous allons aborder dans toute la suite.

3/ Le cas positif $\lambda > 0$.

Considérons la fonctionnelle de Yamabe définie par :

$$I(\varphi) = [4(n-1)(n-2)^{-1} \int_V |\nabla\varphi|^2 dV + \int_V R\varphi^2 dV] \|\varphi\|_N^{-2}$$

et posons $\mu = \inf \{I(\varphi)$ pour tout $\varphi \in H_1$, $\varphi \geq 0$, $\varphi \neq 0\}$. Un calcul montre que μ est un invariant conforme. Nous allons établir le théorème fondamental.

THÉORÈME 2. - *μ vérifie $\mu \leq n(n-1)\,\omega_n^{2/n}$. Si $\mu < n(n-1)\,\omega_n^{2/n}$, il existe une fonction C^∞ : $\tilde{\varphi} > 0$ solution de (E) avec $\tilde{R} = Cte$, (ω_n est le volume de la sphère S_n de rayon 1) .*

Depuis la démonstration de ce théorème dans Aubin [3] , de nombreuses démonstrations ont été établies par différents auteurs. La plus simple d'entre elles, à mon avis, est celle de Vaugon [8] que je vais exposer maintenant. Nous montrerons ultérieurement (théorème 3) que $\mu \leq n(n-1)\,\omega_n^{2/n}$.

D'après le théorème 1 et comme μ est un invariant conforme, on peut, sans nuire à la généralité, supposer que $R > 0$.

Supposons $\mu < n(n-1)\,\omega_n^{2/n}$, il existe alors une fonction $\varphi_o > 0$, $\|\varphi_o\|_N = 1$, $\varphi_o \in C^\infty$, telle que $I(\varphi_o) < n(n-1)\,\omega_n^{2/n}$. Considérons la suite de fonctions C^∞ strictement positives $\{\varphi_i\}$ définie par récurrence pour $i \geq 1$, comme solution de l'équation :

$$L\varphi_i = \lambda_i\,\varphi_{i-1}^{N-1} \quad , \quad \|\varphi_i\|_N = 1 \quad , \quad \lambda_i > 0 \ .$$

Cette suite est parfaitement définie car l'opérateur L est inversible puisque $\lambda > 0$. λ_i est défini par la normalisation $\|\varphi_i\|_N = 1$.

Enfin le principe du maximum entraîne $\varphi_i > 0$.

Posons $B(\varphi,\gamma) = \int_V \varphi L\gamma\, dV = \int_V \gamma L\varphi\, dV$. Il existe un réel $a > 1$ tel que $a^{-1}\|\varphi\|_{H_1} \leq B(\varphi,\varphi) \leq a\|\varphi\|_{H_1}$.

L'inégalité de Hölder entraîne pour $i \geq 1$, $B(\varphi_i,\varphi_i) \leq \lambda_i$. Et comme $B(\varphi_{i+1},\varphi_i) = \lambda_{i+1}$,

$$0 \leq B(\varphi_{i+1}-\varphi_i\, ,\, \varphi_{i+1}-\varphi_i) \leq \lambda_{i+1} + \lambda_i - 2\lambda_{i+1} = \lambda_i - \lambda_{i+1} \ .$$

La suite $\{\lambda_i\}$ est donc décroissante, elle tend vers un réel $\tilde{\lambda}$. De plus lorsque $i \to \infty$

$$B(\varphi_{i+1}-\varphi_i\, ,\, \varphi_{i+1}-\varphi_i) \to 0 \text{ d'où } \|\varphi_{i+1}-\varphi_i\|_{H_1} \to 0 \ .$$

Comme la suite $\{\varphi_i\}$ est bornée dans H_1 ($\|\varphi_i\|_{H_1} \leq a\lambda_1$) , on peut appliquer les théorèmes de Banach et de Kondrakov. Il existe une suite croissante d'entiers positifs $\{j\}$ et une fonction $\tilde{\varphi} \in H_1$ telles que $\varphi_j \to \tilde{\varphi}$ faiblement dans H_1 , fortement dans L_2 et presque partout lorsque $j \to \infty$. Comme l'inclusion $H_1 \subset L_N$ est continue, la suite $\{\varphi_j\}$ est bornée dans L_N, d'où $\|\varphi_j^{N-1}\|_{N/(N-1)} \leq Cte$ et comme $\varphi_j^{N-1} \to \tilde{\varphi}^{N-1}$ presque partout, on en déduit que $\varphi_j^{N-1} \to \tilde{\varphi}^{N-1}$ faiblement dans $L_{N/(N-1)}$. De plus nous avons vu que $\|\varphi_{j+1} - \varphi_j\|_{H_1} \to 0$ lorsque $j \to \infty$, d'où $\varphi_{j+1} \to \tilde{\varphi}$ faiblement dans H_1 et fortement dans L_2 lorsque $j \to \infty$

En passant à la limite dans $L\varphi_{j+1} = \lambda_{j+1}\varphi_j^{N-1}$, on trouve que $\tilde{\varphi}$ vérifie faiblement dans H_1 l'équation (E) avec $\tilde{R} = 4(n-1)(n-2)^{-1}\tilde{\lambda}$. D'après Trudinger [7] $\tilde{\varphi} \in C^\infty$. De plus le principe du maximum entraîne que deux cas seulement sont possibles soit $\tilde{\varphi} > 0$ partout et le problème de Yamabe est résolu soit $\tilde{\varphi} \equiv 0$. Il s'agit donc d'éliminer ce dernier cas.

Pour ceci nous allons utiliser la meilleure constante dans l'inclusion de Sobolev $H_1 \subset L_N$ (voir Aubin [1] ou [2]). Pour tout $\varepsilon > 0$ il existe A tel que tout $\varphi \in H_1$ vérifie :

$$\|\varphi\|_N^2 \leq (K^2 + \varepsilon)\|\nabla\varphi\|_2^2 + A\|\varphi\|_2^2 \quad \text{avec} \quad K^{-2} = n(n-2)\,\omega_n^{2/n}/4 .$$

Nous pouvons écrire :

$$1 = \|\varphi_j\|_N^2 \leq (K^2 + \varepsilon)\,[B(\varphi_j, \varphi_j) - (n-2)\int_V R\varphi_j^2 dV/4(n-1)] + A\|\varphi_j\|_2^2 .$$

Et il existe une constante C telle que :

$$1 - (K^2 + \varepsilon)\lambda_j \leq C\|\varphi_j\|_2^2 .$$

De plus nous avons d'après l'hypothèse :

$$0 \leq B(\varphi_1 - \varphi_o, \varphi_1 - \varphi_o) \leq \lambda_1 - 2\lambda_1 + B(\varphi_o, \varphi_o) \leq n(n-2)\,\omega_n^{2/n}/4 - \lambda_1 .$$

D'où $K^2\lambda_j \leq K^2\lambda_1 < 1$. Par conséquent si ε est choisi assez petit nous montrons que $\|\varphi_j\|_2^2 \geq Cte > 0$. Or $\varphi_j \to \tilde{\varphi}$ fortement dans L_2. Donc $\|\tilde{\varphi}\|_2 \neq 0$ et $\tilde{\varphi} = 0$ est exclu.

Dans Aubin [1], ou [3], il est montré aussi que pour la sphère (S_n, g_c), $\mu = n(n-1)\,\omega_n^{2/n}$, g_c étant la métrique canonique. D'où

<u>Conjecture</u>. *Pour toutes les variétés riemanniennes compactes (V_n, g), $n \geq 3$, non conformes à (S_n, g_c), $\mu < n(n-1)\,\omega_n^{2/n}$.*

C'est par la démonstration de cette conjecture et donc l'utilisation du Théorème 2 qu'on résoudra le Problème de Yamabe. Pour démontrer la conjecture , on mettra en évidence une fonction Ψ telle que $I(\psi) < n(n-1)\,\omega_n^{2/n}$

4/ THÉORÈME 3 , Aubin [3]. - *Si la variété riemannienne compacte* $(V_n,\tilde{g})$, $n > 5$, *n'est pas localement conformément plate,* $\mu < n(n-1)\,\omega_n^{2/n}$.

Sous l'hypothèse du Théorème 3, il existe un point P où le tenseur de Weyl $W^j_{ik\ell}(P)$ n'est pas nul : $\|W(P)\| \neq 0$. D'autre part on peut voir qu'il existe une métrique g conforme à $\tilde{g}$ pour laquelle le tenseur de Ricci $R_{ij}(P) = 0$.
C'est dans cette métrique que nous ferons les calculs . Soit $\delta > 0$ inférieur au rayon d'injectivité en P . Considérons la suite $\{\psi_k\}$ de fonctions appartenant à H_1 définie par :

$$\begin{cases} \psi_k(Q) = 0 \quad \text{si} \quad r = d(P,Q) \geq \delta \\ \psi_k(Q) = (r^2 + 1/k)^{1-n/2} - (\delta^2 + 1/k)^{1-n/2} \quad \text{pour} \quad 0 < r < \delta . \end{cases}$$

Un calcul montre que $I(\psi_k) \to n(n-1)\,\omega_n^{2/n}$ lorsque $k \to \infty$.
Ceci prouve la première partie du Théorème 2. Un développement limité en k donne , C_1 et C_2 étant des constantes positives qui dépendent que de n :

$$I(\psi_k) = n(n-1)\,\omega_n^{2/n} - C_1 \|W(P)\|^2 k^{-2} + o(k^{-2}) \quad \text{si} \quad n > 6$$

$$I(\psi_k) = 30\,\omega_6^{1/3} - C_2 \|W(P)\|^2 k^{-2} \operatorname{Log} k + o(k^{-2}) \quad \text{si} \quad n = 6 .$$

Ceci prouve le Théorème 3 . On peut démontrer aussi voir Aubin [3] le

THÉORÈME 4. - *Si la variété riemannienne compacte* (V_n,g) , $n \geq 3$, *est à groupe de Poincaré fini et localement conformément plate, sans être conforme à* (S_n,g_c) , *alors* $\mu < n(n-1)\,\omega_n^{2/n}$.

Ainsi que le théorème suivant, en faisant $\varphi \equiv 1$ dans $I(\varphi)$:

THÉORÈME 5. - *Si la variété riemannienne compacte* (V_n,g) , $n \geq 3$, *satisfait l'inégalité* $[\int_V dV]^{2/n-1} \int_V R\,dV \leq n(n-1)\,\omega_n^{2/n}$, *alors il existe une métrique conforme à* g *pour laquelle la courbure scalaire est constante.*

5/ Les résultats récents.

D'après les résultats précédents, il reste à étudier, outre les dimensions 3 , 4, 5, les variétés localement conformément plates à groupe de Poincaré infini pour lesquelles l'inégalité du théorème 5 n'est pas vérifiée.

Olga Gil-Medrano [4] a résolu le cas des variétés localement conformément plates connues en mettant en évidence une fonction ψ très simple vérifiant $I(\psi) < n(n-1)\,\omega_n^{2/n}$ pour les variétés produit ou fibré, et en montrant que pour les sommes connexes. $V_1 \# V_2 = V$, $\mu \leq \inf(\mu_1,\mu_2)$. Je ne m'étendrais pas plus car Olga Gil-Medrano doit faire un exposé sur ses résultats.

R.Schoen [5] a eu l'idée de considérer la fonction de Green G_L de L . Comme nous sommes dans le cas $\lambda > 0$, L est inversible , $G_L > 0$. Supposons que dans une boule B_δ de centre P et de rayon δ , la métrique g soit plate. Posons $r = d(P,Q)$ et $G = (n-2)\,\omega_{n-1}\, G_L$. Dans un voisinage de P :

$$G(P,Q) = r^{2-n} + A + \alpha(Q) \quad \text{avec} \quad \alpha(P) = 0 .$$

Et considérons la fonction ψ définie par :

$$\psi = \begin{cases} u_\varepsilon(r) = (\varepsilon + r^2/\varepsilon)^{1-n/2} & \text{pour} \quad r \leq \rho \leq \delta/2 \\ \varepsilon_o[G - f(r)\,\alpha(Q)] & \text{pour} \quad \rho < r \leq 2\rho \\ \varepsilon_o\, G & \text{pour} \quad r > 2\rho \end{cases}$$

ε_o et ρ étant choisis petits comme il sera dit plus loin, pour que ψ soit lipschitzienne et ainsi appartienne à H_1 , ε doit vérifier :

$$(\varepsilon + \rho^2/\varepsilon)^{1-n/2} = \varepsilon_o(\rho^{2-n} + A)$$

ce qui est possible et f est une fonction C^∞ de r , égale à 1 pour $r < \rho$ et nulle pour $r > 2\rho$. Calculons $I(\psi)$:

$$\|\psi\|_N^2 \; I(\psi) = 4(n-1)(n-2)^{-1} \int_{B_\rho} |\nabla u_\varepsilon|^2 dE + \varepsilon_o^2 \int_{V-B_\rho} [4(n-1)(n-2)^{-1}\, |\nabla G|^2 + RG^2]dV + O(\rho\varepsilon_o^2)$$

où dE est l'élément de volume euclidien. Remarquons que $\Delta u_\varepsilon = n(n-2)u_\varepsilon^{N-1}$ et rappelons que $LG = 0$. En faisant des intégrations par partie nous trouvons :

$$\|\psi\|_N^2 \, I(\psi) = 4n(n-1) \int_{B_\rho} u_\varepsilon^N \, dE + 4(n-1)(n-2)^{-1} \int_{S_\rho} (u_\varepsilon \partial_r u_\varepsilon - \varepsilon_o^2\, G\, \partial_r G)d\sigma + O(\rho\varepsilon_o^2)$$

$d\sigma$ étant l'élément d'aire euclidien sur $S_\rho = \partial B_\rho$. Comme $\int_{R^n} u_\varepsilon^N \, dE = 2^{-n}\omega_n$, un calcul montre qu'il existe une constante $C_3 > 0$ telle que :

$$I(\psi) \leq n(n-1)\,\omega_n^{2/n} - C_3 A \;\varepsilon_o^2 + O(\rho\varepsilon_o^2) .$$

D'où si $A > 0$, on pourra choisir ρ puis ε_o assez petits pour que $I(\omega) < n(n-1)\,\omega_n^{2/n}$. Or Schoen et Yau [6] , après avoir démontré en dimension 3 le résultat qui suit, ont annoncé qu'en toute dimension :

On a toujours $A \geq 0$ *et si* $A = 0$, (V_n, g) *est conforme à* (S_n, g_c) .

Moyennant ce résultat le problème de Yamabe est résolu pour les variétés localement conformément plates, ainsi que pour celles de dimension 3 , car si $n = 3$;

$G(P,Q) = 1/r + A + \alpha(Q)$ avec $\alpha(P) = 0$ et α lipschitzienne ,

et la même démonstration est applicable. Reste les dimensions 4 et 5 .
Pour celles-ci, Schoen approxime la métrique g par une métrique plate dans B_ρ et égale à g hors de $B_{2\rho}$. Il peut alors appliquer sa démonstration mais elle est beaucoup plus compliquée.

En conclusion, lorsque Schoen et Yau auront donné une démonstration de leur caractérisation de la sphère mentionnée plus haut, le Problème de Yamabe sera entièrement résolu.

BIBLIOGRAPHIE

[1] Aubin T. Nonlinear Analysis on Manifolds. Monge-Ampère Equations. Springer-Verlag, New-York (1982).

[2] Aubin T. Espaces de Sobolev sur les variétés riemanniennes. Bull. Sc. Math. 100 (1976), 149-173.

[3] Aubin T. Equations différentielles non linéaires et Problème de Yamabe concernant la courbure scalaire. J. Math. Pures et Appl. 55 (1976), 269-296.

[4] Gil-Medrano O. On the Yamabe Problem concerning the compact locally conformally flat manifolds. A paraître dans J. Funct. Ana.

[5] Schoen R. Conformal deformation of a riemannian metric to constant curvature. J. Diff. Geo. 20 (1984), 479-495.

[6] Schoen R., Yau S.T. On the proof of the positive mass conjecture in General Relativity, Comm. Math. Phys. 65 (1979), 45-76.

[7] Trüdinger N. Remarks concerning the conformal deformation of riemannian structures on compact manifolds. Ann. Scuola Norm. Sup. Pisa 22 (1968), 265-274.

[8] Vaugon à paraître.

[9] Yamabe H. On the deformation of riemannian structures on compact manifolds. Osaka Math. J. 12 (1960), 21-37.

FINITE TYPE SPHERICAL SUBMANIFOLDS

Manuel Barros
Departmento de Geometria
Universidad de Granada
Granada, Spain

and

Bang-yen Chen
Department of Mathematics
Michigan State University
East Lansing, Michigan 48824
U.S.A.

1. Introduction.

Let M be a compact, connected, Riemannian manifold and Δ the Laplacian of M acting on smooth functions in $C^\infty(M)$. Then Δ has an infinite sequence of eigenvalues: $0 = \lambda_0 < \lambda_1 < \cdots < \lambda_k < \cdots \uparrow \infty$. For each λ_k, the associated eigenspace V_k is finite-dimensional. If one defines an inner product on $C^\infty(M)$ by $(fg,g) = \int fg dV$, then the decomposition $\Sigma_{t\geq 0} V_t$ is orthogonal and dense in $C^\infty(M)$. Therefore, for each $f \in C^\infty(M)$, one can consider its spectral decomposition : $f = \Sigma_{t\geq 0} f_t$, $\Delta f_t = \lambda_t f_t$, which is convergent in L^2-sense. Since M is compact, V_0 is one-dimensional. Thus, for each non-constant function $f \in C^\infty(M)$, there is a positive integer $p \geq 1$ such that $f_p \neq 0$ and $f = f_0 + \Sigma_{t\geq p} f_t$. If there are infinitely many f_t's which are nonzero, we put $q = \infty$. Otherwise, we let q be the largest integer such that $f_q \neq 0$. Hence, we have the following spectral decomposition:

$$(1.1) \qquad f = f_0 + \sum_{t=p}^{q} f_t.$$

This decomposition can be extended to $\mathbb{R}^{m+1}$-valued smooth functions on M.

Let M be a compact, connected, n-dimensional submanifold of $\mathbb{R}^{m+1}$. Then, with respect to the induced metric, M is a compact Riemannian manifold. Let x denote the position vector M in $\mathbb{R}^{m+1}$. Then we have the following spectral decomposition of x:

$$(1.2) \qquad x = x_0 + \sum_{t=p}^{q} x_t, \qquad \Delta x_t = \lambda_t x_t.$$

It is easy to see that x_0 is nothing but the center of mass of M in $\mathbb{R}^{m+1}$. The pair $[p,q]$ is called the *order* of M in $\mathbb{R}^{m+1}$. It is clear that p is a positive integer and q is either ∞ or an integer $\geq p$. The submanifold M is said to be of *finite type* if q is finite. And M is of *infinite type* if $q = \infty$. M is said to be *of k-type* if there are exactly k nonzero x_t's in the decomposition of x (except x_0). (The concept of order and finite type submanifolds were introduced in [7]. For general results on finite type submanifolds, see [8,10].) A submanifold M of a hypersphere S^m in $\mathbb{R}^{m+1}$ is called *mass-symmetric* in S^m if the center of mass of M is the center of the hypersphere S^m in $\mathbb{R}^{m+1}$.

In terms of finite type submanifolds, a well-known result of Takahashi [19] says that a 1-type submanifold M in $\mathbb{R}^{m+1}$ is nothing but a minimal submanifold of a hypersphere S^m of $\mathbb{R}^{m+1}$. Furthermore, M is always mass-symmetric in S^m.

In this paper, we will give some results on finite type spherical submanifolds in $\mathbb{R}^{m+1}$. Furthermore, we will study submanifolds with homothetic or conformal Weingarten map. An application to finite type submanifolds is also given in Section 8.

2. Preliminaries.

Let M be a compact, connected, n-dimensional submanifold of a Euclidean (m+1)-space $\mathbb{R}^{m+1}$. Denote by ∇, D and h the induced Riemannian connection on M, the normal connection and the second fundamental form of M in $\mathbb{R}^{m+1}$, respectively. We denote by A and H the Weingarten map and the mean curvature vector of M in $\mathbb{R}^{m+1}$, respectively.

Let $e_1,\dots,e_n,\xi_{n+1},\dots,\xi_{m+1}$ be an orthonormal frame field such that $e_1,\dots,e_n$ are tangent to M and $\xi_{n+1},\dots,\xi_{m+1}$ are normal to M. We put

$$(2.1) \qquad \operatorname{tr}(\bar{\nabla}A_H) = \Sigma\{(\nabla_{e_i}A_H)e_i + A_{D_{e_i}H}e_i\}.$$

Then we have the following.

Lemma 1 [9]. *Let* M *be an n-dimensional submanifold of* $\mathbb{R}^{m+1}$. *Then we have*

$$(2.2) \qquad \operatorname{tr}(\bar{\nabla}A_H) = \frac{n}{2}\operatorname{grad}\alpha^2 + 2\operatorname{tr}A_{DH},$$

where $\alpha^2 = \langle H,H\rangle$ *and* $\langle\, ,\rangle$ *the inner product in* $\mathbb{R}^{m+1}$.

For any normal vector field ξ, if we choose $\xi_{n+1},\ldots,\xi_{m+1}$ such that ξ_{n+1} is parallel to ξ, then the *allied vector field* is defined by [4, p. 203]

$$(2.3)\qquad \mathcal{a}(\xi) = \sum_{r=n+2}^{m+1} (\mathrm{tr}(A_\xi A_r))\xi_r.$$

The *allied mean curvature vector* is defined by $\mathcal{a}(H)$. A submanifold M is called an $\mathcal{a}$-*submanifold* if $\mathcal{a}(H) = 0$. The allied mean curvature vector and the concept of $\mathcal{a}$-submanifold can be defined in the same way for a submanifold in a Riemannian manifold (cf. [4,8]).

Let Δ and Δ^D denote the Laplacian of M and the Laplacian associated with the normal connection D. Then we have the following.

Lemma 2 [7,8]. *Let* M *be a submanifold of* $\mathbb{R}^{m+1}$. *Then we have*

$$(2.4)\qquad \Delta H = \Delta^D H + \|A_{n+1}\|^2 H + \mathcal{a}(H) + \mathrm{tr}(\bar{\nabla} A_H).$$

If M is a submanifold of the unit hypersphere $S^m(1)$ centered at the origin of $\mathbb{R}^{m+1}$, then the mean curvature vector H' of M in $S^m(1)$ is given by $H' = H + x$, where x is the position vector of M in $\mathbb{R}^{m+1}$. Let A', h', and D' denote the Weingarten map, the second fundamental form and the normal connection of M in $S^m(1)$. Then we have $A_\eta = A'_\eta$ for any vector η normal to M in $S^m(1)$. Let $\mathcal{a}'$ denote the allied operator of M in $S^m(1)$. Then we have the following fundamental formula.

Lemma 3 [8, p. 273]. *Let* M *be an n-dimensional submanifold of* $S^m(1)$ *in* $\mathbb{R}^{m+1}$. *Then we have*

$$(2.5)\qquad \Delta H = \Delta^{D'} H' + \mathcal{a}'(H') + \mathrm{tr}(\bar{\nabla} A_H) + (\|A_\xi\|^2 + n)H' - n\alpha^2 x,$$

where $H' = \alpha'\xi$ *and* $\alpha^2 = 1 + (\alpha')^2$.

We mention the following results for later use.

Theorem 1 [8, p. 274]. *If* M *is a mass-symmetric, 2-type submanifold of* $S^m(1)$ *in* $\mathbb{R}^{m+1}$, *then*

(1) *the mean curvature* α *is constant and it is given by*

$$(2.6)\qquad \alpha^2 = \frac{1}{n}(\lambda_p + \lambda_q) - \frac{1}{n^2}\lambda_p\lambda_q \qquad \textit{and}$$

(2) $\operatorname{tr}(\bar{\nabla} A_H) = 0$.

In particular, if $m = n + 1$, we also have the following.

Theorem 2 [8, p. 276]. *If M is a mass-symmetric, 2-type hypersurface of* $S^{n+1}(1)$, *then*

(1) *the scalar curvature* τ *of* M *is constant and it is given by*

$$\tau = \frac{1}{n}(\lambda_p + \lambda_q) - \frac{1}{n(n-1)}\lambda_p\lambda_q \tag{2.7}$$ and

(2) *the length of second fundamental form* h *is constant and it is given by*

$$\|h\|^2 = \lambda_p + \lambda_q. \tag{2.8}$$

Theorem 3 [8,9]. *Let* M *be a compact hypersurface of* S^{n+1} *in* $\mathbb{R}^{n+2}$ *such that* M *is not a small hypersphere of* S^{n+1}. *Then* M *is mass-symmetric and of 2-type if and only if* M *has nonzero constant mean curvature* α' *and constant scalar curvature* τ.

Theorem 4 [8, p. 279]. *Let* M *be a compact surface in* S^3. *Then* M *is the product of two plane circles of different radii if and only if* M *is mass-symmetric and of 2-type.*

By using Lemma 3, it follows that if a 2-type submanifold of S^m in $\mathbb{R}^{m+1}$ has constant mean curvature, then either M is mass-symmetric in S^m or M is mass-symmetric in a small hypersphere of S^m ([3]). Thus, we have the following.

Lemma 4 [3]. *Let* M *be a 2-type hypersurface of* S^{n+1} *in* $\mathbb{R}^{n+2}$. *Then* M *is a mass-symmetric in* S^{n+1} *if and only if* M *has constant mean curvature.*

By applying Lemmas 1,3 and 4, one may prove the following.

Theorem 5 [3]. *Let* M *be a compact hypersurface of* S^{n+1} *in* $\mathbb{R}^{n+2}$. *Then*

(1) *if* M *is of 2-type,* M *is mass-symmetric in* S^{n+1} *and it has constant mean curvature, and*

(2) *if* M *is of 3-type,* M *has non-constant mean curvature.*

Theorem 5 shows that the assumption "M is mass-symmetric" in Theorems 2,3 and 4 holds automatically if M is a 2-type hypersurface. Thus it can be omitted.

We also need the following.

Lemma 5. *If* M *is a mass-symmetric, 2-type submanifold of* $S^m(1)$ *in* $\mathbb{R}^{m+1}$, *then we have*

$$\|dH'\|^2 = \{\lambda_p + \lambda_q - n\}\{n(\lambda_p + \lambda_q) - \lambda_p\lambda_q - n^2\}/n^2, \tag{2.9}$$

$$\mathcal{A}'(H') = \alpha' \sum_{r=n+2}^{m} \{tr(\nabla\omega_{n+1}{}^r) - \langle D'\xi, D'\xi_r\rangle\}\xi_r, \tag{2.10}$$

where $\xi = \xi_{n+1},\ldots,\xi_n$ *is an orthonormal normal basis of* M *in* $S^m(1)$ *such that* $H' = \alpha'\xi$ *and* $D'\xi_t = \Sigma\, \omega_t{}^s\xi_s$.

Proof. Under the hypothesis, M has constant mean curvature (Theorem 1). Thus, we have

$$\Delta^D H = \Delta^{D'} H' = \alpha' \sum_{r=n+2}^{m} \{\langle D'\xi, D'\xi_r\rangle - tr(\nabla\omega_{n+1}{}^r)\}\xi_r \tag{2.11}$$

$$+ \langle D'\xi, D'\xi\rangle H'.$$

Since M is mass-symmetric and of 2-type, we also have $\Delta H = (\lambda_p + \lambda_q)H + (\lambda_p\lambda_q/n)x$. Thus, by applying Lemma 3, Theorem 1 and (2.11), we obtain (2.9) and (2.10). (Q.E.D.)

Theorem 6 ([7],[8, p. 307]). *Let* M *be an n-dimensional, finite-type, mass-symmetric submanifold of* $S^m(1)$ *in* $\mathbb{R}^{m+1}$. *If* M *is not of 1-type, then we have*

$$0 < \lambda_p < n < \lambda_q < \infty, \tag{2.12}$$

where [p,q] *is the order of* M.

3. A Non-existence Theorem.

Theorem 4 shows that the product of two plane circles gives examples of mass-symmetric, 2-type surfaces in S^3. In this section, we give the following *non-existence theorem*.

Theorem 7. *There exist no mass-symmetric, 2-type surfaces which lie fully in* $S^4(1)$ *in* $\mathbb{R}^5$.

Proof (Outlined). Assume that M is a mass-symmetric, 2-type surface in $S^4(1)$. Then $\langle D\xi, D\xi_4\rangle = 0$. Thus, by Lemma 5, we obtain $\mathfrak{a}'(H') = \alpha' \operatorname{tr}(\nabla\omega_3{}^4)\xi_4$. Hence, we have

$$(3.1) \qquad \operatorname{tr}(A_3 A_4) = \operatorname{tr}(\nabla\omega_3{}^4).$$

On the other hand, Theorem 1 and Lemma 1 give $\operatorname{tr} A_{D\xi} = 0$. Now, we may prove that $A_4\omega_3{}^4 = 0$ and M is an $\mathfrak{a}$-surface of $S^4(1)$. Furthermore, by applying these, we may prove that ξ_4 is a geodesic direction, i.e., $A_4 = 0$. From these, together with structure equations, we may show that M is a flat surface which lies fully in a great hypersphere $S^3(1)$ of $S^4(1)$. (Q.E.D.)

4. Examples of Mass-Symmetric, 2-type Surfaces in S^5.

Let $f : M \to M'$ be an isometric immersion of a surface M into an m-dimensional Riemannian manifold M'. We denote by α' and R' the mean curvature of f and the sectional curvature of M' with respect to the tangent space of M. We define $\tau(f)$ by

$$(4.1) \qquad \tau(f) = \int_M ((\alpha')^2 + R')dV.$$

It was proved in [6] that $\tau(f)$ is invariant under conformal changes of metric of M' (cf. also [8, p. 207]). $\tau(f)$ is called the *conformal total mean curvature*. Surfaces which are critical points of $\tau(f)$ are called *stationary*. Related to Chen-Willmore's problem, Weiner asked in [20] whether minimal surfaces of S^m are the only stationary, mass-symmetric surfaces of S^m? Ejiri gave in [12] a counter example to Weiner's question. Ejiri's example is given by the isometric immersion f of the flat torus $S^1(1) \times S^1(\sqrt{1/3})$ into S^5 defined by

$$(4.2) \qquad f((x,y),(z,w)) = (\sqrt{2/3}\, x,\ xz, xw,\ \sqrt{2/3}\, y,\ yz, yw).$$

It is easy to see that Ejiri's example is a 2-type mass-symmetric surface in S^5.

5. Stationary 2-type Surfaces in S^m.

Theorem 5 shows that a 2-type surface in S^3 is always mass-symmetric. So, Theorem 4 classifies all 2-type surfaces in S^3. On the other hand, Theorem 7 shows that there exist no mass-symmetric, 2-type surfaces which lie fully in S^4. In this section, we give the following.

Theorem 8. *Let* M *be a stationary, mass-symmetric, 2-type surface in* $S^m(1)$. *Then* M *is a flat surface which lies fully in a totally geodesic* $S^5(1)$ *or in a totally geodesic* $S^7(1)$ *of* $S^m(1)$.

Proof (Outlined). Let M be a stationary surface in $S^m(1)$. Then we have [20]

$$(5.1)\qquad \Delta^{D'} H' = -2(\alpha')^2 H' + \|A_\xi\|^2 H' + \mathcal{a}'(H').$$

Assume that M is mass-symmetric and of 2-type. Then we also have

$$(5.2)\qquad \Delta H = (\lambda_p + \lambda_q)H' + (\lambda_p\lambda_q/2 - (\lambda_p + \lambda_q))x.$$

From Lemma 3 and (5.1) we find

$$(5.3)\qquad \Delta H = 2\,\mathcal{a}'(H') + 2(\|A_\xi\|^2 - (\alpha')^2 + 1)H' - 2(\alpha')^2 x.$$

Thus, by (5.2) and (5.3) we find

$$(5.4)\qquad \mathcal{a}'(H') = 0, \quad \text{i.e.,} \quad M \text{ is an } \mathcal{a}\text{-surface in } S^4(1),$$

$$(5.5)\qquad \|A_\xi\|^2 = \lambda_p + \lambda_q - 2 - \frac{\lambda_p\lambda_q}{4}.$$

From Lemma 5 and (5.5) we obtain

$$(5.6)\qquad \|D'\xi\|^2 = \lambda_p\lambda_q/4.$$

Moreover, (5.4) and Lemma 5 imply

$$(5.7)\qquad \mathrm{tr}(\nabla\omega_3{}^r) = \langle D'\xi, D'\xi_r\rangle, \qquad r = 4,\ldots,m.$$

From Theorem 1, Theorem 2, and Theorem 6, we have

$$(5.8)\qquad (\alpha')^2 = (2 - \lambda_p)(\lambda_q - 2)/4 \neq 0,$$

$$(5.9)\qquad \mathrm{tr}(\bar{\nabla}A_H) = 0.$$

Combining (5.5) and (5.7) we obtain

(5.10) M is not pseudo-umbilical in $S^m(1)$.

We choose e_1, e_2 which diagonalize A_3. Then from (5.4), (5.6), and (5.10), we may choose $\xi_3, \ldots, \xi_m$ in such a way that we have

$$A_3 = \begin{bmatrix} \beta & 0 \\ 0 & \gamma \end{bmatrix}, \quad A_4 = \begin{bmatrix} 0 & b \\ b & 0 \end{bmatrix}, \quad A_6 = \begin{bmatrix} 0 & c \\ c & 0 \end{bmatrix},$$

$$(5.11)\qquad A_5 = A_7 = \ldots = A_m = 0,$$

$$D'\xi_3 = \omega_3{}^4\xi_4 + \omega_3{}^5\xi_5,$$

where β and γ are unequal constants. From (5.8), (5.9), and (5.11) we may obtain $b\omega_3{}^4 = 0$. If $b = 0$, $A_4 = 0$. Then by taking exterior differentiation of $\omega_1{}^3 = \beta\omega^1$ and $\omega_2{}^3 = \gamma\omega^2$, we obtain $\omega_1{}^2 = 0$. If $b \neq 0$, $\omega_3{}^4 = 0$. Thus (5.11) and structure equations imply $\omega_1{}^2 = 0$. In both cases, we have

(5.12) $\omega_1{}^2 = 0$ and M is flat.

Consequently, there is an adapted orthonormal frame field $\{e_1, e_2, \xi_3, \ldots, \xi_m\}$ such that one of the following two cases occur.

<u>Case</u> (1). With respect to the adapted frame field, we have

$$(5.13)\qquad A_3 = \begin{bmatrix} \beta & 0 \\ 0 & \gamma \end{bmatrix}, \quad A_4 = \begin{bmatrix} 0 & b \\ b & 0 \end{bmatrix}, \quad A_5 = \ldots = A_m = 0,$$

$$D'\xi_3 = \omega_3{}^5\xi_5, \quad \omega_1{}^2 = 0, \quad b \neq 0, \quad \text{or}$$

<u>Case</u> (2). With respect to the adapted frame field, we have

$$A_3 = \begin{bmatrix} \beta & 0 \\ 0 & \gamma \end{bmatrix}, \quad A_4 = A_5 = 0, \quad A_6 = \begin{bmatrix} 0 & b \\ b & 0 \end{bmatrix},$$

$$(5.14)\qquad A_7 = \ldots = A_m = 0,$$

$$D'\xi_3 = \omega_3{}^4\xi_4 + \omega_3{}^5\xi_5, \quad \omega_1{}^2 = 0.$$

In both cases, β, γ, and b are constant with $b^2 = 1 + \beta\gamma$.

If case (1) occurs, by applying structure equations and (5.7), we may prove that $\mathrm{Span}\{\xi_3, \xi_4, \xi_5\}$ is a parallel normal subbundle with respect to D'. Since the first normal subbundle $\mathrm{Im}\, h$ is contained in $\mathrm{Span}\{\xi_3, \xi_4, \xi_5\}$, we may conclude that M is contained in a totally geodesic $S^5(1)$ in $S^m(1)$. If case (2) occurs, (5.14) and structure equations yield $b \neq 0$ and $\omega_6{}^7 = \ldots = \omega_6{}^m = 0$.

If we put

$$(5.15) \qquad \omega_3{}^4 = \alpha_1\omega^1 + \alpha_2\omega^2, \qquad \omega_3{}^5 = \delta_1\omega^1 + \delta_2\omega^2.$$

Then by structure equations and (5.14) we may obtain

$$(5.16) \qquad \omega_6{}^4 = (\frac{\beta\alpha_2}{b})\omega^1 + (\frac{\gamma\alpha_1}{b})\omega^2, \qquad \omega_6{}^5 = (\frac{\beta\delta_2}{b})\omega^1 + (\frac{\beta\delta_1}{b})\omega^2.$$

Since $\omega_3{}^6 = 0$, (5.7) gives $\langle D'\xi_3, D'\xi_6\rangle = 0$. Thus, (5.15) and (5.16) give

$$(5.17) \qquad \alpha_1\alpha_2 + \delta_1\delta_2 = 0.$$

In the following, we may choose ξ_4 in such a way that

$$(5.18) \qquad D'_{e_1}\xi_3 = \omega_3{}^4(e_1)\xi_4, \quad \delta_1 = 0.$$

Thus, we obtain $\omega_3{}^5 = \delta_2\omega^2$, $\omega_6{}^5 = (\beta\delta_2/b)\omega^1$. Since $\delta_1 = 0$, (5.17) gives $\alpha_1 = 0$ or $\alpha_2 = 0$. If $\alpha_1 = 0$, we have

$$(5.19) \qquad D'_{e_1}\xi_3 = 0.$$

In this case, we may choose ξ_4 in such a way that $D'\xi_3 = \omega_3{}^4\xi_4$ and $\omega_3{}^4 = \alpha_2\omega^2$. Thus we have $\omega_3{}^5 = 0$. Hence, by interchanging ξ_4 and ξ_6 we obtain case (1). If $\delta_2 = 0$, the same argument holds. Consequently, we see that if M is not a flat surface in a $S^5(1)$, then, with respect to a suitable orthonormal frame $\{e_1, e_2, \xi_3, \ldots, \xi_m\}$, we have

$$A_3 = \begin{bmatrix} \beta & 0 \\ 0 & \gamma \end{bmatrix}, \quad A_4 = A_5 = 0, \quad A_6 = \begin{bmatrix} 0 & b \\ b & 0 \end{bmatrix},$$

$$(5.20) \qquad A_7 = \ldots = A_m = 0, \quad \omega_1{}^2 = 0, \quad \omega_3{}^4 = \alpha_1\omega^1, \quad \omega_3{}^5 = \delta\omega^2,$$

$$\omega_3{}^6 = \dots = \omega_3{}^m = 0, \quad \omega_4{}^6 = -(\gamma\alpha_1/b)\omega^2, \quad \omega_5{}^6 = -(\beta\delta/b)\omega^1,$$

$$\omega_6{}^7 = \dots = \omega_6{}^m = 0, \quad \beta\gamma + 1 = b^2 \neq 0, \quad \alpha_1\delta \neq 0.$$

By using (5.20) and structure equations, we may prove that α_1 and $\delta = \delta_2$ are constant. And

(5.21) $$\omega_4{}^7(e_1) = \omega_5{}^7(e_2) = 0,$$

(5.22) $$\omega_4{}^5 = 0.$$

Thus, we may choose ξ_7 such that

(5.23) $$D_{e_2}\xi_4 = \omega_4{}^6(e_2)\xi_6 + \omega_4{}^7(e_2)\xi_7.$$

In this way, we obtain $\omega_4{}^r = 0$, $r = 8,\dots,m$. With respect to this adapted frame field, we may prove that $\mathrm{Span}\{\xi_3,\dots,\xi_7\}$ is parallel with respect to D'. Since this subbundle contains the first normal subbundle, we see that M is contained in a totally geodesic $S^7(1)$ of $S^m(1)$. From the connection form $(\omega_A{}^B)$, we see that M lies fully in $S^5(1)$ for case (1) and it lies fully in $S^7(1)$ for case (2). (Q.E.D.)

6. Examples and Remarks.

Let $\mathbb{R}^2$ be the Euclidean plane. Let u,v and w be real numbers with $u,v > 0$. We define the following lattice

(6.1) $$\Lambda = \{(2n\pi u,\ 2m\pi v + 2n\pi w \mid n,m \in Z\}.$$

The dual lattice of Λ is given by

(6.2) $$\Lambda^* = \{(\frac{h}{2\pi u} - \frac{kw}{2\pi uv},\ \frac{k}{2\pi v}) \mid h,k \in Z\}.$$

Let T_{uvw} be the flat torus given by $\mathbb{R}^2/\Lambda$. Then the spectrum of T_{uvw} is given by

(6.3) $$\{(\frac{h}{u} - \frac{kw}{uv})^2 + \frac{k^2}{v^2} \mid h,k \in Z\}.$$

For any two natural numbers h and $\bar{h}$ satisfying

$$h \neq w, \quad w \leqq (2h^2 + \bar{h}^2)/2, \tag{6.4}$$

we define an isometric immersion y from $\mathbb{R}^2$ into $S^5(1)$ in $\mathbb{R}^6$ by

$$\begin{aligned} y(s,t) = (&v \cos \frac{\epsilon s}{u} \cos \frac{t}{v}, \; v \cos \frac{\epsilon s}{u} \sin \frac{t}{v}, \; e \cos \frac{\bar{\epsilon} s}{u}, \\ &v \sin \frac{\epsilon s}{u} \cos \frac{t}{v}, \; v \sin \frac{\epsilon s}{u} \sin \frac{t}{v}, \; e \sin \frac{\bar{\epsilon} s}{u}), \end{aligned} \tag{6.5}$$

where ϵ, $\bar{\epsilon}$, u, v and e are given by

$$\epsilon = h - \frac{w}{v}, \quad \bar{\epsilon} = \bar{h}. \tag{6.6}$$

$$\begin{aligned} &u = \sqrt{3}\ \epsilon\bar{\epsilon}/(2\epsilon^2 + \bar{\epsilon}^2)^{\frac{1}{2}}, \qquad v = \bar{\epsilon}/(2\epsilon^2 + \bar{\epsilon}^2)^{\frac{1}{2}}, \\ &e = \sqrt{2}\ \epsilon/(2\epsilon + \bar{\epsilon}^2)^{\frac{1}{2}}. \end{aligned} \tag{6.7}$$

Then the immersion y induces an isometric immersion from T_{uvw} into $S^5(1)$. We denote it by x:

$$x : T_{uvw} \longrightarrow S^5(1) \subset \mathbb{R}^6. \tag{6.8}$$

It is easy to verify that x is a stationary, mass-symmetric, isometric immersion. Furthermore, x is of 1-type if and only if $\bar{\epsilon}^2 = 4\epsilon^2$. Otherwise, x is of 2-type with $\{\lambda_p, \lambda_q\} = \{(\bar{\epsilon}/u)^2, (\epsilon/u)^2 + (1/v)^2\}$.

<u>Remark</u> 1. By determining the connection form of a stationary, mass-symmetric, 2-type surface in $S^5(1)$ and by applying the Fundamental Theorem of Submanifolds, we may prove that stationary mass-symmetric, 2-type surfaces in $S^5(1)$ "are obtained in this way", (see, [2] for details).

<u>Remark</u> 2. By applying Theorem 6, we may prove that if M is a stationary, mass-symmetric, 2-type surface in $S^5(1)$, then $2/3 < \lambda_p < 2$. Moreover, for any $2/3 < c \leqq 4/3$, there is "one" such surface with $\lambda_p = c$ and for any $4/3 < c < 2$, there are "two" such surfaces with $\lambda_p = c$. (For the details, see [2]).

<u>Remark</u> 3. For any real number $d \in (2,\infty)$, there is a real number $c \in (2/3,2)$ and a stationary, mass-symmetric, 2-type flat surface in $S^7(1)$ with $(\lambda_p, \lambda_q) = (c,d)$, (see, [2] for details.)

7. Submanifolds with Conformal Weingarten Map.

Let M be an n-dimensional submanifold of a Riemannian manifold $\bar{M}$ of dimension m. Then the Weingarten map A of M in $\bar{M}$ is a well-defined linear map from the normal bundle $T^{\perp}$ into S_nT;

$$A : T^{\perp} \longrightarrow S_nT, \tag{7.1}$$

where $S_n(T)$ is the space of self-adjoint endormorphisms on the tangent bundle $T = T(M)$. On S_nT, there is a canonical metric $\langle\langle\ ,\ \rangle\rangle$ defined by

$$\langle\langle B,C\rangle\rangle = \frac{1}{n}\,\mathrm{tr}(BC). \tag{7.2}$$

It is clear that $T^{\perp}$ admits an induced inner product $\langle\ ,\ \rangle$ from $\bar{M}$. The Weingarten map A *is said to be conformal if* $\langle\langle A_\xi, A_\eta\rangle\rangle = \rho^2\langle\xi,\eta\rangle$ for some positive function ρ on M. If ρ is a positive constant, A is called *homothetic*.

It is clear that a non-totally geodesic submanifold M in $\bar{M}$ has conformal Weingarten map if and only if M is a *strong $\mathcal{A}$-submanifold*, i.e., the allied operator $\mathcal{A}$ vanishes identically; and the Weingarten map A is homothetic if and only if M is a strong $\mathcal{A}$-submanifold with nonzero constant length of second fundamental form.

If we define

$$M_n = \{B \in S_nT \mid \mathrm{tr}\, B = 0\}, \tag{7.3}$$

then M_n is of codimension one in S_nT. Moreover, we have

$$S_nT = M_n \oplus \mathbb{R}I_n\,, \quad M_n \perp \mathbb{R}I_n\,, \tag{7.4}$$

where I_n is the identity endormorphism on TM. Since S_nT is a $n(n+1)/2$-vector bundle over M, (7.4) gives the following.

Lemma 6. *Let M be an n-dimensional submanifold of an m-dimensional Riemannian manifold $\bar{M}^m$. If the Weingarten map $A : T^{\perp} \longrightarrow S_nT$ is conformal, then $m \leq n(n+3)/2$. In particular, if $m = n(n+3)/2$, M is pseudo-umbilical.*

If $\bar{M}$ is a real-space-form, we have the following.

Proposition 1. *Let* M *be an n-dimensional* $(n > 2)$ *submanifold of a real-space-form* $\bar{M}^m(c)$ *of constant sectional curvature* c. *If the Weingarten map* $A : T^\perp \longrightarrow S_nT$ *is conformal and surjective, then* (1) $m = n(n+3)/2$, (2) A *is homothetic, and* (3) M *is a real-space-form of curvature* $< c$.

Proof. Assume that the Weingarten map A is conformal and surjective. Then A is bijective. Thus, $m = n(n+3)/2$. In this case, we may choose an orthonormal frame field $e_1, \ldots, e_n, \xi_{n+1}, \ldots, \xi_m$ such that with respect to this frame field, the Weingarten map satisfies

$$A_{n+1} = \rho \begin{bmatrix} \sqrt{n} & 0 & \cdots & 0 \\ 0 & & & \\ \vdots & & \bigcirc & \\ 0 & & & \end{bmatrix}, \quad \ldots, \quad A_{2n} = \rho \begin{bmatrix} & & & 0 \\ & \bigcirc & & \vdots \\ & & & 0 \\ 0 & \cdots & 0 & \sqrt{n} \end{bmatrix},$$

(7.5)

$$A_{n+[i,j]} = \rho \begin{bmatrix} & \vdots & & \vdots & \\ \cdots & 0 & \cdots & \sqrt{n/2} & \cdots \\ & \vdots & & \vdots & \\ \cdots & \sqrt{n/2} & \cdots & 0 & \cdots \\ & \vdots & & \vdots & \end{bmatrix}, \qquad 1 \leqq i < j \leqq n.$$

Thus, by applying the equation of Gauss and Schur's lemma, we obtain (2) and (3). (Q.E.D.)

The following result can be obtained in a similar way.

Proposition 2. *Let* M *be an n-dimensional* $(n > 2)$ *minimal submanifold of* $\bar{M}^m(c)$. *If the Weingarten map* $A : T^\perp \longrightarrow M_n$ *is conformal and surjective, then* (1) $m = n(n+3)/2 - 1$, (2) A *is homothetic, and* (3) M *is a real-space-form.*

8. Surfaces with Homothetic Weingarten Map and Application.

In this section, we consider surfaces with homothetic Weingarten map and apply our results to some 2-type surfaces.

Theorem 9. *Let* M *be a surface in a 5-dimensional real-space-form* $\bar{M}^5(c)$. *If the Weingarten map* $A : T^\perp \longrightarrow S_2T$ *is homothetic, then* $c > 0$ *and* M *is locally a Veronese surface in a totally* *umbilical hypersurface* $N^4(\bar{c})$ *of* $\bar{M}^5(c)$ *with* $\bar{c} = 3c/2$.

Proof. We put

$$(8.1) \qquad B_1 = \begin{bmatrix} \sqrt{2} & 0 \\ 0 & 0 \end{bmatrix}, \quad B_2 = \begin{bmatrix} 0 & 0 \\ 0 & \sqrt{2} \end{bmatrix}, \quad B_3 = \begin{bmatrix} 0 & 1 \\ 1 & 0 \end{bmatrix}.$$

Then B_1, B_2, B_2 form an orthonormal basis of the space of symmetric 2×2-matrices with respect to the inner product defined by (7.2). Thus, if M is a surface in $\bar{M}^5(c)$ with homothetic Weingarten map $A : T^{\perp} \to S_2T$, there exists an orthonormal frame field $e_1, e_2, \bar{\xi}_3, \bar{\xi}_4, \xi_5$ such that with to this field, the Weingarten map A satisfies

$$(8.2) \qquad A_3 = \rho \begin{bmatrix} \sqrt{2} & 0 \\ 0 & 0 \end{bmatrix}, \quad A_4 = \rho \begin{bmatrix} 0 & 0 \\ 0 & \sqrt{2} \end{bmatrix}, \quad A_5 = \rho \begin{bmatrix} 0 & 1 \\ 1 & 0 \end{bmatrix},$$

where ρ is a positive constant. We put

$$(8.3) \qquad \xi_3 = \frac{1}{\sqrt{2}}(\bar{\xi}_3 + \bar{\xi}_4), \qquad \xi_4 = \frac{1}{\sqrt{2}}(\bar{\xi}_3 - \bar{\xi}_4).$$

Thus, we obtain

$$(8.4) \qquad \omega_1{}^3 = \rho\omega^1, \qquad \omega_2{}^3 = \rho\omega^2,$$

$$(8.5) \qquad \omega_1{}^4 = \rho\omega^1, \qquad \omega_2{}^4 = -\rho\omega^2,$$

$$(8.6) \qquad \omega_1{}^5 = \rho\omega^2, \qquad \omega_2{}^5 = \rho\omega^1,$$

where ρ is a positive constant. By taking exterior differentiation of (8.4) and (8.5), we get

$$(8.7) \qquad \omega_4{}^3 \wedge \omega^1 = -\omega_5{}^3 \wedge \omega^2, \qquad \omega_4{}^3 \wedge \omega^2 = \omega_5{}^3 \wedge \omega^1,$$

$$(8.8) \qquad \omega_4{}^3 \wedge \omega^1 = \omega_5{}^4 \wedge \omega^2 - 2\,\omega_2{}^1 \wedge \omega^2,$$

$$(8.9) \qquad \omega_4{}^3 \wedge \omega^2 = \omega_5{}^4 \wedge \omega^1 - 2\,\omega_2{}^1 \wedge \omega^1.$$

Similarly, from (8.6), we find

(8.10) $\omega_5^{\ 3} \wedge \omega^1 = -\omega_5^{\ 4} \wedge \omega^1 + 2\, \omega_2^{\ 1} \wedge \omega^1$

(8.11) $\omega_5^{\ 3} \wedge \omega^2 = \omega_5^{\ 4} \wedge \omega^2 + 2\, \omega_2^{\ 1} \wedge \omega^2.$

By using (8.7) - (8.11), we may obtain

(8.12) $\omega_3^{\ 4} = \omega_3^{\ 5} = 0, \quad \omega_4^{\ 5} = 2\, \omega_1^{\ 2}.$

By taking exterior differentiation of $\omega_4^{\ 5} = 2\, \omega_1^{\ 2}$ and by applying structure equations and (8.12) we may obtain

(8.13) $c = 2\,\rho^2 > 0.$

On the other hand, (8.2), (8.3) and (8.12) imply that M is pseudo-umbilical with parallel mean curvature vector. Thus, by applying a result of Yano-Chen (cf. [8, p. 133]), we see that M lies in a totally umbilical hypersurface $N^4(\bar{c})$ of constant curvature $\bar{c}$ as a minimal surface. Since the Weingarten map $\bar{A}$ of M in $N^4(\bar{c})$ is given by

(8.14) $$\bar{A}_3 = \begin{bmatrix} \rho & 0 \\ 0 & -\rho \end{bmatrix}, \qquad \bar{A}_4 = \begin{bmatrix} 0 & \rho \\ \rho & 0 \end{bmatrix}$$

with respect to $e_1, e_2, \bar{\xi}_3, \bar{\xi}_4$ with $\bar{\xi}_3 = \xi_4$, $\bar{\xi}_4 = \xi_5$, equation of Gauss implies that the Gauss curvature G of M is given by $G = \bar{c} - 2\rho^2$. On the other hand, (8.2) gives $G = c - \rho^2$. Thus, $\bar{c} - c = \rho^2$. Combining this with (8.13), we get $G = c/2$ and $\bar{c} = 3c/2$. Thus, by applying a result of [11], we conclude that M is a locally Veronese surface in $N^4(3c/2)$ in $\bar{M}^5(c)$. (Q.E.D.)

Theorem 10. *Let* M *be a surface with constant mean curvature in a 4-dimensional real-space-form* $\bar{M}^4(c)$. *If the Weingarten map* $A : T^\perp \longrightarrow S_2T$ *is homothetic, then one of the following three cases occurs:*

(a) $c = 0$, M *is pseudo-umbilic in* $\bar{M}^4(0)$, *and it is locally a Clifford torus in a totally umbilical hypersurface* $N^3(\bar{c})$ *of* $\bar{M}^4(0)$;

(b) $c > 0$ *and* M *is of constant Gauss curvature* $c/2$ *which lies in* $\bar{M}^4(c)$ *as a non-pseudo-umbilical,* $\mathcal{A}$*-surface;*

(c) $c > 0$, M *is minimal in* $\bar{M}^4(c)$, *and* M *is locally a Veronese surface in* $\bar{M}^4(c)$.

Proof. Let M be a surface in $\bar{M}^4(c)$ with constant mean curvature. If the Weingarten map A is homothetic, then M is an $\mathcal{A}$-surface and, moreover, the principal curvatures of A_H are constant. Thus, one of the following two cases holds.

Case (1). M is pseudo-umbilical and with respect to a suitable frame field e_1, e_2, ξ_3, ξ_4, the Weingarten map satisfies

$$(8.15) \qquad A_3 = \begin{bmatrix} \alpha & 0 \\ 0 & \alpha \end{bmatrix}, \qquad A_4 = \begin{bmatrix} \alpha & 0 \\ 0 & -\alpha \end{bmatrix}$$

for some positive constant α, or

Case (2). M is not pseudo-umbilical and, with respect to a suitable frame field e_1, e_2, ξ_3, ξ_4, we have

$$(8.16) \qquad A_3 = \begin{bmatrix} \beta & 0 \\ 0 & \gamma \end{bmatrix}, \quad A_4 = \begin{bmatrix} 0 & \delta \\ \delta & 0 \end{bmatrix}, \quad \beta \neq \gamma$$

for some constants β, γ and δ with $\beta_2 + \gamma^2 = 2\delta^2$.

If case (1) holds, we have

$$(8.17) \qquad \omega_1{}^3 = \alpha\omega^1, \quad \omega_2{}^3 = \alpha\omega^2, \quad \omega_1{}^4 = \alpha\omega^1, \quad \omega_2{}^4 = \alpha\omega^2.$$

From (8.17) and structure equations, we may prove that $\omega_3{}^4 = 0$, i.e., the mean curvature vector of M in $M^4(c)$ is parallel. Thus, M lies in a totally umbilical hypersurface $N^3(\bar{c})$ as a minimal surface. Since, M is of constant Gauss curvature G,. Corollary 1 of [5] implies $\bar{c} > 0$ and M is locally a Clifford torus. From (8.15) we get $0 = G = c$. Thus, case (a) occurs.

If case (2) holds, then we have

$$(8.18) \qquad \omega_1{}^3 = \beta\omega^1, \quad \omega_2{}^3 = \gamma\omega^2, \quad \omega_1{}^4 = \delta\omega^2, \quad \omega_2{}^4 = \delta\omega^1.$$

By taking exterior differentiation of (8.18) and applying structure equations, we may obtain

$$(8.19) \qquad (\beta-\gamma)\omega_2{}^1 = \delta\omega_4{}^3,$$

$$(8.20) \qquad 2\delta\omega_1{}^2 \wedge \omega^1 = \beta\omega_3{}^4 \wedge \omega^1, \qquad 2\delta\omega_2{}^1 \wedge \omega^2 = \gamma\omega_3{}^4 \wedge \omega^2.$$

Combining (8.19) and (8.20) we get

$$(8.21)\qquad \beta(\beta+\gamma)\omega_1{}^2 \wedge \omega^2 = 0, \qquad \gamma(\beta+\gamma)\omega_1{}^2 \wedge \omega^1 = 0.$$

If $\beta+\gamma = 0$, M is minimal in $\overline{M}^4(c)$. In this case, we have

$$(8.22)\qquad A_3 = \begin{bmatrix} \delta & 0 \\ 0 & -\delta \end{bmatrix}, \qquad A_4 = \begin{bmatrix} 0 & \delta \\ \delta & 0 \end{bmatrix}$$

Therefore, the structure equations imply $\omega_3{}^4 = 2\omega_1{}^2$ and $c = 3\delta^2$. Thus, by a result of [11], M is locally a Veronese surface. So, case (c) occurs. Now, assume that $\omega_1{}^2 = 0$. Then (8.19) gives $\omega_3{}^4 = 0$. This implies $0 = d\omega_3{}^4 = \delta(\gamma-\beta)\omega^1 \wedge \omega^2$. Since M is not pseudo-umbilical, $\delta = 0$. This contradicts to the assumption that the Weingarten map being homothetic. So, we have $\omega_1{}^2 \neq 0$. Consequently, if $\beta+\gamma \neq 0$, then $\beta\gamma = 0$. Without loss of generality, we may assume that $\beta = 0$ and $\gamma \neq 0$. In this case, we may put

$$(8.23)\qquad \omega_1{}^2 = \mu\omega^1, \qquad \delta\omega_3{}^4 = -\gamma\mu\omega^1.$$

By taking exterior differentiation of $\omega_1{}^2 = \mu\omega^1$ we may obtain

$$(8.24)\qquad e_2\mu = c + \mu^2 - \delta^2.$$

From the second equation of (8.23) we have

$$(8.25)\qquad e_2\mu = \mu^2 + \delta^2.$$

(8.24) and (8.25) imply $c = 2\delta^2 > 0$. Thus, $G = c/2$. Since the Weingarten map is homothetic, M is an $\mathcal{O}$- surface. (Q.E.D.)

From Theorem 10, we obtain the following.

Corollary 1. *Let* M *be a compact surface with constant mean curvature in* $\overline{M}^4(c)$. *Then the Weingarten map is homothetic if and only if one of the following two cases occurs:*

(a) $c = 0$, M *is pseudo-umbilical in* $\overline{M}^4(0)$, *and it is locally a Clifford torus in a totally umbilical hypersurface* $N^3(\bar{c})$ *of* $\overline{M}^4(0)$;

(b) $c > 0$, M *is minimal in* $\overline{M}^4(c)$, *and* M *is locally a Veronese surface in* $\overline{M}^4(c)$.

Proof. It is easy to verify that surfaces given in (a) and (b) have constant mean curvature and homothetic Weingarten map. So, it suffices to show that case (b) of Theorem 10 cannot occur if M is compact. If M is a compact surface of case (b) in Theorem 10, then (8.25) holds. Let u be a point on M which gives a global maximum for μ^2. Then, $e_2\mu = 0$ at u. This implies $\mu^2 + \delta^2 = 0$ at u by virtue of (8.25) which is impossible. The converse of this is easy to verify. (Q.E.D.)

Let $\psi : M \longrightarrow S^m(1)$ be a minimal immersion of an n-dimensional, compact, Riemannian manifold M into the hypersphere $S^m(1)$. Denote by $f : S^m(1) \longrightarrow E^{m+m(m+1)/2}$ the second standard immersion of $S^m(1)$. In [18], Ros proved the following.

Theorem 11 [18]. *Let $\psi : M \longrightarrow S^m(1)$ be an isometric immersion of a compact Riemannian manifold M into $S^m(1)$ such that ψ is full and minimal. Then the immersion $f \circ \psi$ is of 2-type if and only if M is Einsteinian and $tr(A_\xi A_\eta) = k\langle \xi,\eta\rangle$ for all normal vectors ξ, η of M in $S^m(1)$, where k is constant.*

It is easy to see that the Weingarten map of M in $S^m(1)$ is homothetic if and only if $tr(A_\xi, A_\eta) = k\langle \xi,\eta\rangle$ for a positive constant k. Thus, by applying Corollary 1 and Ros' result, we obtain the following.

Corollary 2. *The Clifford torus in $S^3(1)$ and the Veronese surface in $S^4(1)$ are the only compact minimal surfaces in $S^m(1)$ such that the composite immersion $f \circ \psi$ are of 2-type.*

Similarly, Proposition 2 and Theorem 11 imply the following.

Corollary 3. *Let $\psi : M \longrightarrow S^m(1)$ be an isometric immersion of an n-dimensional Riemannian manifold M into $S^m(1)$ such that ψ is full and minimal. If the immersion $f \circ \psi$ is of 2-type, then $m \leqq n(n+3)/2 - 1$. In particular, if $m = n(n+3)/2 - 1$, then M is a real-space-form.*

Remark 4. Let $M = S^1(a) \times S^1(b)$ be a product surface of two plane circles of different radii. If $a^2 + b^2 = 1$, M lies in $S^3(1)$. It can be proved that $f \circ i : M \longrightarrow S^3(1) \longrightarrow E^9$ is of 4- or 5-type in E^9 where i denotes the inclusion map. Moreover, M is non-mass-symmetric in the hypersphere of E^9 which $S^3(1)$ lies in.

9. Local Immersions of $S^2(\sqrt{2})$ into $S^4(1)$ with A Homothetic.

In this section, we construct infinitely many local isometric immersions of $S^2(\sqrt{2})$ into $S^4(1)$ with homothetic Weingarten map.

We begin with an open disc $D(\epsilon)$ of radius ϵ about $(0,0)$ in the (s,θ)-plane, where ϵ is a sufficiently small positive number. Let $\phi = \phi(s,\theta)$ be a differentiable function of s and θ. Consider the following quasi-linear partial differential equation:

$$(9.1) \qquad \cos\phi \frac{\partial\phi}{\partial s} - \frac{\sin\phi}{\cos s}\frac{\partial\phi}{\partial\theta} = \tan s \sin\phi .$$

Let I be the interval in $D(\epsilon)$ given by $I = \{(s,\theta) \in D(\epsilon) | s = 0\}$ and let f be a differentiable function from $(-\epsilon,\epsilon)$ into $(-\pi/2,\pi/2)$. Then, by the Existence Theorem of first order quasi-linear partial differential equations (cf. [14, p. 11]), there exists one and only one solution $\phi = \phi(s,\theta)$ defined in a simply-connected neighborhood U of I which satisfies (9.1) and the initial condition $\phi(0,\theta) = f(\theta)$. (For example, if $f = 0$ on $(-\epsilon,\epsilon)$, we obtain $\phi = 0$ and the function μ of (9.3) becomes $\mu = \tan s/\sqrt{2}$.)

On U we define a Riemannian metric g by $g = 2ds^2 + 2\cos^2 s\, d\theta^2$. Then (U,g) is of constant Gauss curvature $1/2$. Using $\phi = \phi(s,\theta)$ we define an orthonormal frame field e_1, e_2 by

$$(9.2) \qquad \begin{aligned} e_1 &= \frac{\sin\phi}{\sqrt{2}}\frac{\partial}{\partial s} + \frac{\cos\phi}{\sqrt{2}\cos s}\frac{\partial}{\partial\theta}, \\ e_2 &= \frac{\cos\phi}{\sqrt{2}}\frac{\partial}{\partial s} - \frac{\sin\phi}{\sqrt{2}\cos s}\frac{\partial}{\partial\theta}. \end{aligned}$$

Then, by a straight-forward computation, we may prove that $\omega_1{}^2 = \mu\omega^1$, where μ is given by

$$(9.3) \qquad \mu = \frac{1}{\sqrt{2}}\left\{\frac{\cos\phi}{\cos s}\left(\frac{\partial\phi}{\partial\theta} + \sin s\right) + \sin\phi\frac{\partial\phi}{\partial s}\right\}.$$

We put $E = U \times \mathbb{R}^2$. Then E can be considered as the total space of a Riemannian vector bundle over U. Let ξ_3, ξ_4 be the natural orthonormal frame in $\mathbb{R}^2$. We define a metric connection D on $\mathbb{R}^2$ by $D\xi_3 = -\sqrt{2}\mu\omega^1\xi_4$, where ω^1, ω^2 denote the dual frame of e_1, e_2. We also define a bilinear map h by

(9.4) $\quad h(e_1,e_1) = 0, \quad h(e_1,e_2) = \frac{1}{\sqrt{2}}\,\xi_4, \quad h(e_2e_2) = \xi_3.$

Then, E together with D and h satisfies equations of Gauss, Codazzi and Ricci. Thus, by applying the Fundamental Theorem of Submanifolds, there is an isometric immersion from U into $S^4(1)$ with h as its second fundamental form. It is clear that such a surface has constant mean curvature, constant Gauss curvature 1/2, and homothetic Weingarten map.

References.

1. M. Barros and B.Y. Chen, Classification of stationary, 2-type surfaces of hyperspheres, C.R. Math. Rep. Acad. Sci. Canada, **7** (1985).

2. M. Barros and B.Y. Chen, Stationary 2-type surfaces in a hypersphere, to appear.

3. M. Barros, B.Y. Chen and O.J. Garay, Spherical finite type hypersurfaces, to appear.

4. B.Y. Chen, *Geometry of Submanifolds*, Marcel Dekker, New York, 1973.

5. ________, Minimal surfaces with constant Gauss curvature, Proc. Amer. Math. Soc., **34** (1972), 504-508.

6. ________, Some conformal invariants of submanifolds and their applications, Boll. Un. Mat. Ital., **10** (1974), 380-385.

7. ________, On the total curvature of immersed manifolds; IV, Bull. Inst. Math. Acad. Sinica, **7** (1979), 301-311; _______; VI, ibid, **11** (1983), 309-328.

8. ________, *Total Mean Curvature and Submanifolds of Finite Type*, World Scientific, 1984.

9. ________, 2-type submanifolds and their applications, to appear.

10. ________, *Finite Type Submanifolds and Generalizations*, Quad. Sem. Top. Alg. Diff., Instituto "Castelnuovo", Rome, 1985.

11. S.S. Chern, M. doCarmo and S. Kobayashi, Minimal submanifolds of a sphere with second fundamental form of constant length, Functional Analysis and Related Fields, 1970, 59-75, Springer.

12. N. Ejiri, A counter example for Weiner's open question, Indiana Math. J., **31** (1982), 209-211.

13. J. Erbarcher, Reduction of the codimension of an isometric immersion, J. Diff. Geom., 5 (1971), 333-340.

14. F. John, *Partial Differential Equations*, Springer, 1971.

15. K. Kenmotsu, Minimal surfaces with constant curvature in 4-dimensional space forms, Proc. Amer. Math. Soc., **89** (1983), 133-138.

16. H.B. Lawson, Jr., Complete minimal surfaces in S^3, Ann. of Math., **92** (1970), 335-374.

17. A. Ros, On spectral geometry of Kaehler submanifolds, J. Math. Soc. Japan, **36** (1984), 433-448.

18. ______, Eigenvalue inequalities for minimal submanifolds and P-manifolds, Math. Z. **187** (1984), 393-404.

19. T. Takahashi, Minimal immersionss of Riemannian manifolds, J. Math. Soc. Japan, **18** (1966), 380-385.

20. J.L. Weiner, On a problem of Chen, Willmore, et al., Indiana Math. J., **27** (1978), 19-35.

MAPPINGS BETWEEN MANIFOLDS WITH

CARTAN CONNECTIONS

R.A. Blumenthal
Department of Mathematics
Saint Louis University
St. Louis, MO 63103

Apart from its intrinsic interest, the question of when a submersion is a fibration is an important one in the study of foliations. Frequently one encounters a complicated foliation $\mathfrak{F}$ of a manifold M which has the property that the lift $\widetilde{\mathfrak{F}}$ of $\mathfrak{F}$ to the universal cover $\widetilde{M}$ of M is defined by a submersion $f : \widetilde{M} \to N$ where N is some smooth manifold (which we may assume is simply connected). If one knows that f is a fibration, then there is an action of $\pi_1(M)$ on N such that the leaf space $M/\mathfrak{F}$ is identified with the orbit space of this action. From this one can obtain significant global information concerning the influence of the topology of M upon the structure of $\mathfrak{F}$. In this paper we address the question of when a submersion is a fibration.

W. Ambrose [1] showed that a local isometry defined on a complete Riemannian manifold is a covering and N. Hicks [7] proved a similar result for local affine isomorphisms. The present author has obtained analogous results for local projective and conformal isomorphisms [3]. R. Hermann [6] showed that a Riemannian submersion defined on a complete Riemannian manifold is a locally trivial fiber bundle (thus generalizing the classical result of C. Ehresmann [5] that a submersion defined on a compact manifold is a locally trivial fiber bundle) and the present author has shown that an affine submersion defined on a complete affinely connected manifold is a fibration [4]. All of the geometric structures occurring in the results quoted above (Riemannian, affine, conformal and projective geometries) can be treated in a uniform fashion under the rubric of Cartan connections. We consider submersions between manifolds with Cartan connections and we give sufficient ·conditions for such maps to be fibrations. In a subsequent paper we shall apply these results to the study of foliations whose transverse structure is modeled on a Cartan geometry.

We briefly recall some generalities concerning Cartan connections. Let M be a manifold. Let G be a Lie group and $H \subset G$ a closed subgroups such that $\dim G/H = \dim M$, and let $\pi : P \to M$ be a principal H-bundle. Let $\mathcal{G}$ and $\mathcal{H}$ be the Lie algebras of G and H, respectively and for each $A \in \mathcal{H}$ let A^* be the corresponding fundamental vertical vector field on P. A Cartan connection in P is a $\mathcal{G}$-valued one-form ω on P satisfying

    i) $\omega(A^*) = A$ for all $A \in \mathcal{H}$,

    ii) $(R_a^*)\omega = \mathrm{ad}(a^{-1})\omega$ for all $a \in H$ where R_a denotes the right translation

           by a acting on P and $\mathrm{ad}(a^{-1})$ is the adjoint action of a^{-1} on

$\mathfrak{g}$, and

iii) $\omega_u : T_u(P) \to \mathfrak{g}$ is an isomorphism for all $u \in P$.

One says ω is complete if each vector field X on P such that $\omega(X)$ is constant is complete. Recall that G/H is weakly reductive if there is a subspace $\mathfrak{m} \subset \mathfrak{g}$ such that $\mathfrak{g} = \mathfrak{m} \oplus \mathfrak{h}$ and $\mathrm{ad}(H)\mathfrak{m} \subset \mathfrak{m}$. In this case a curve σ in M is said to be a geodesic of ω if $\sigma = \pi \circ \gamma$ where γ is an integral curve of a vector field X on P such that $\omega(X) \in \mathfrak{m}$ is constant and ω is complete if and only if each geodesic is infinitely extendable [8]. Let M and M' be manifolds. Let G and G' be Lie groups and let H and H' be closed subgroups of G and G' , respectively with $\dim G/H = \dim M$ and $\dim G'/H' = \dim M'$. Let $\pi : P \to M$ (respectively, $\pi' : P' \to M'$) be a principal H (respectively, H')-bundle and let ω and ω' be Cartan connections in P and P' , respectively.

<u>Definition</u>. A Cartan map is a bundle homomorphism $f : M \to M'$, $F : P \to P'$, $\varphi : H \to H'$ and a homomorphism $\Phi : G \to G'$ satisfying $F^*\omega' = \Phi_* \circ \omega$ where $\Phi_* : \mathfrak{g} \to \mathfrak{g}'$ is the induced homomorphism between Lie algebras. If $G = G'$, $H = H'$, and $\Phi = \mathrm{Id}$, we say f is a local Cartan isomorphism.

<u>Theorem 1</u>. Let (f, F, Φ) be a Cartan map with $\Phi_* : \mathfrak{g} \to \mathfrak{g}'$ onto. If ω is complete then f is a Serre fibration, F is a locally trivial fiber bundle, and ω' is complete.

<u>Proof</u>. Let M and M' be manifolds. Let G and G' be Lie groups with closed subgroups H and H' respectively such that $\dim G/H = \dim M$ and $\dim G'/H' = \dim M'$. Let $\pi : P \to M$ be a principal H-bundle, let $\pi' : P' \to M'$ be a principal H'-bundle, and let ω and ω' be Cartan connections in P and P' respectively with ω complete. Let $f : M \to M'$, $F : P \to P'$, $\Phi : G \to G'$ be a Cartan map with $\Phi_* : \mathfrak{g} \to \mathfrak{g}'$ onto. Let $u \in P$ and let $Z' \in T_{F(u)}(P')$. Let $Y = \omega'(Z') \in \mathfrak{g}'$ and choose $X \in \mathfrak{g}$ such that $\Phi_*(X) = Y$. Letting Z be the unique vector in $T_u(P)$ satisfying $\omega(Z) = X$ we have that $\omega'(F_*(Z)) = (F^*\omega')(Z) = \Phi_*\omega(Z) = Y$ and so $F_*(Z) = Z'$ thus showing that F (and hence also f) is a submersion. Let $\mathfrak{F}$ be the foliation of P defined by F and let $E = \mathrm{kernel}(F_*) \subset T(P)$. Let $A_1, \ldots, A_r$, $B_1, \ldots, B_s$ be a basis of $\mathfrak{g}$ such $A_1, \ldots, A_r$ is a basis of $\mathrm{kernel}(\Phi_*)$. Let $X_1, \ldots, X_r$, $Y_1, \ldots, Y_s$ be smooth vector fields on P satisfying $\omega(X_i) = A_i$, $\omega(Y_j) = B_j$. Then $X_1, \ldots, X_r$ span E . Let $Q \subset T(P)$ be the subbundle spanned by $Y_1, \ldots, Y_s$. We may regard Q as the normal bundle of $\mathfrak{F}$. Let Z_j be the unique vector field on P' satisfying $\omega'(Z_j) = \Phi_*(B_j)$. Then $\omega'(F_*(Y_j)) = \Phi_*\omega(Y_j) = \omega'(Z_j)$ and so Y_j is F-related to Z_j thus showing that $Y_1, \ldots, Y_s$ are parallel along the leaves of $\mathfrak{F}$. Since ω is complete we have that $Y_1, \ldots, Y_s$ are complete and so $\mathfrak{F}$ is a transversely

complete foliation of P. Since the leaves of $\mathfrak{F}$ are closed, the space of leaves $P/\mathfrak{F}$ is a smooth Hausdorff manifold and the natural projection $q : P \to P/\mathfrak{F}$ is a locally trivial fiber bundle [9]. Now F induces a local diffeomorphism $h : P/\mathfrak{F} \to P'$ such that $F = h \circ q$ and $Y_1, \ldots, Y_s$ project to complete vector fields on $P/\mathfrak{F}$ which are h-related to $Z_1, \ldots, Z_s$. Hence by Lemma A below, h is a covering map and so F is a locally trivial fiber bundle. Since $f \circ \pi$ is a Serre fibration and π is a fiber bundle, it follows that f is a Serre fibration [4]. Clearly ω' is complete and so Theorem 1 is proved.

Corollary 1. Let M and M' be connected manifolds of the same dimension each with a Cartan connection and let $f : M \to M'$ be a local Cartan isomorphism. If M is complete, then f is a covering map.

Proof. Note that Φ_* is the identity and so f is a Serre fibration by Theorem 1. Hence given any path σ in M' and $p_0 \in f^{-1}\{\sigma(0)\}$, there is a unique lift of σ to M starting at p_0 whence f is a covering projection.

We remark that the results quoted above concerning local isometries and local affine, projective, and conformal isomorphisms are special cases of Corollary 1.

One can show that any affine submersion is part of a Cartan map and so we obtain

Corollary 2. (c.f. [4]). Let M and M' be affinely connected manifolds and let $f : M \to M'$ be an affine submersion. If M is complete, then f is a Serre fibration.

Assume now that G/H and G'/H' are weakly reductive.

Theorem 2. Let $f : M \to M'$ be a submersion such that for each geodesic σ of ω, the curve $f \circ \sigma$ is a geodesic of ω'. If ω is complete, then f is a Serre fibration.

Proof. Let M and M' be manifolds and let G/H and G'/H' be weakly reductive homogeneous spaces with $\dim G/H = \dim M$ and $\dim G'/H' = \dim M'$. Let $\pi : P \to M$ (respectively, $\pi' : P' \to M'$) be a principal H (respectively, H')-bundle and let ω and ω' be Cartan connections in P and P' respectively with ω complete. Let $f : M \to M'$ be a submersion which sends geodesics of ω to geodesics of ω'. The geodesics of ω and ω' determine sprays X and X' on $T(M)$ and $T(M')$ [8] relative to which f is spray-preserving (that is, X and X' are f_*-related). There exist linear connections ∇ and ∇' on M and M' giving rise to the sprays X and X' [2] and with respect to these linear connections f is affine. Since ω is complete, X is complete (in fact, ω is complete if and only if X is complete [8]) and so ∇ is complete. Hence by Corollary 2, f is a Serre fibration which proves Theorem 2.

Definition. A subbundle $Q \subset T(M)$ is totally geodesic if whenever a geodesic of ω is tangent to Q at one point, it is tangent to Q at all its points. A geodesic tangent to Q will be called a horizontal geodesic and we say that ω is horizontally complete if each horizontal geodesic is infinitely extendable.

Theorem 3. Let $f : M \to M'$ be a submersion. Let $E \subset T(M)$ be the kernel of f_* and let $Q \subset T(M)$ be a complementary totally geodesic subbundle such that for each horizontal geodesic σ of ω, the curve $f \circ \sigma$ is a geodesic of ω'. If ω is horizontally complete (e.g., if ω is complete), then f is a locally trivial fiber bundle.

Proof. We first recall some generalities concerning sprays. Let M be a smooth manifold and let $\pi : T(M) \to M$ be the tangent bundle of M. Let X be a vector field on $T(M)$. Then X is a spray if $\pi_* \circ X = \mathrm{Id}_{T(M)}$ and $X_{cv} = c\mu_{c_*}(X_v)$ for $v \in T(M)$, $c \in \mathbb{R}$ where μ_c is multiplication by c. For $v \in T(M)$ let σ_v be the integral curve of X through v and let $\alpha_v = \pi \circ \sigma_v$. Then X is a spray if and only if $\alpha_v' = \sigma_v$ and $\alpha_{cv}(t) = \alpha_v(ct)$ in which case the curves α_v are the geodesics of X and the exponential map at a point $p \in M$ given by $\exp(v) = \alpha_v(1)$ maps a neighborhood of 0 in $T_p(M)$ diffeomorphically onto a neighborhood of p in M. We say that a subbundle Q of $T(M)$ is totally geodesic if Q is a union of integral curves of X. Theorem 3 will follow from the following lemma.

Lemma A. Let M and M' be connected manifolds with sprays X and X' respectively and let $f : M \to M'$ be a submersion. Let $E \subset T(M)$ be the kernel of f_* and let $Q \subset T(M)$ be a complementary totally geodesic subbundle such that $X|Q$ is a f_*-related to X'. If $X|Q$ is complete then f is onto, $f : M \to M'$ is a locally trivial fiber bundle, and X' is complete.

Proof of Lemma A. Since f is a submersion, $f(M)$ is open in M'. To show f is onto it suffices to show that $f(M)$ is also closed. Let $q \in \overline{f(M)}$. Let V be a neighborhood of q in M' and U a neighborhood of 0 in $T_q(M')$ such that $\exp : U \to V$ is a diffeomorphism. Let $z \in V \cap f(M)$. Let $u \in U$ be such that $\exp(u) = z$. Then $\alpha_u(0) = q$, $\dot{\alpha}_u(0) = u$, and $\alpha_u(1) = z$. Let $p \in f^{-1}\{z\}$ and let $v \in Q_p$ be the unique vector satisfying $f_*(v) = -\dot{\alpha}_u(1)$. Let $\rho(t) = \alpha_u(1-t)$. Then ρ is a geodesic in M' satisfying $\rho(0) = z$, $\dot{\rho}(0) = -\dot{\alpha}_u(1)$ and so $\rho = \alpha_{f_*(v)} = f \circ \alpha_v$ since $X|Q$ and X' are f_*-related. Now $\alpha_v(1)$ is defined (since $X|Q$ is complete) and $f(\alpha_v(1)) = q$. Then $q \in f(M)$ and so $f(M)$ is closed. Clearly X' is complete.

Let $q \in M'$. Let V be a neighborhood of q in M' and U a neighborhood of 0 in $T_q(M')$ such that $\exp : U \to V$ is a diffeomorphism. Let $L = f^{-1}\{q\}$. Define $\Phi : V \times L \to M$ as follows. Let $(z,p) \in V \times L$. Let $u \in U$ be the unique vector satisfying $\exp(u) = z$, let $v \in Q_p$ be the unique vector satisfying $f_*(v) = u$, and set $\Phi(z,p) = \exp(v)$. Clearly $\Phi : V \times L \to f^{-1}(V)$ and $f \circ \Phi$ is projection onto the first factor. Define $\Psi : f^{-1}(V) \to V \times L$ as follows. Let $x \in f^{-1}(V)$. Let $y = f(x) \in V$. Then $y = \exp(u)$ where $u \in U$. Let τ be the unique geodesic in M' satisfying $\tau(0) = q$, $\dot{\tau}(0) = u$. Then $\tau(1) = y$. Let $v \in Q_x$ be the unique vector satisfying $f_*(v) = -\dot{\tau}(1)$. Let σ be the unique geodesic in M satisfying $\sigma(0) = x$, $\dot{\sigma}(0) = v$. Then $\sigma(1) \in L$. Indeed, if we let $\rho(t) = \tau(1-t)$, then $f \circ \sigma$ and ρ are geodesics in M' satisfying the same initial condition whence $\rho = f \circ \sigma$ and so $f(\sigma(1)) = \rho(1) = q$. Set $\Psi(x) = (f(x), \sigma(1))$. Then Ψ is inverse to Φ (this uses that Q is totally geodesic) and so Φ is a diffeomorphism which completes the proof of Lemma A.

Corollary 3. Let M and M' be affinely connected manifolds such that the holonomy group of M is completely reducible and let $f : M \to M'$ be an affine submersion. If M is complete, then f is a locally trivial fiber bundle.

Proof. Let $E \subset T(M)$ be the kernel of f_*. Then E is a holonomy-invariant distribution on M. Since the holonomy group of M is completely reducible, there exists a complementary holonomy-invariant distribution $Q \subset T(M)$. Since Q is totally geodesic, Corollary 3 now follows from Theorem 3.

Corollary 4. (R. Hermann [6]). Let M and M' be connected Riemannian manifolds and let $f : M \to M'$ be a Riemannian submersion. If M is complete, then f is a locally trivial fiber bundle.

Proof. Let ω and ω' be the Cartan connections is the orthnormal frame bundles of M and M' respectively arising from the Riemannian connections in these bundles. Let $Q = E^{\perp}$. Then Q is totally geodesic with respect to ω and for each geodesic σ of ω which is tangent to Q, the curve $f \circ \sigma$ is a geodesic of ω' [10]. If M is complete, then ω is horizontally complete and so by Theorem 3 f is a locally trivial fiber bundle.

References

[1] W. Ambrose, "Parallel translation of Riemannian curvature", Ann. of Math. 64 (1956), 337-363.

[2] W. Ambrose, R.S. Palais, and I.M. Singer, "Sprays", An. Acad. Bras. Ciênc. 32 (1960), 163-178.

[3] R.A. Blumenthal, "Local isomorphisms of projective and conformal structures", Geom. Ded. 16 (1984), 73-78.

[4] R.A. Blumenthal, "Affine submersions", Ann. Global Analysis and Geom. 3 (1985), 275-287.

[5] C. Ehresmann, "Sur les espaces fibrés différentiables", C.R. Acad. Sci., Paris 224 (1947), 1611-1612.

[6] R. Hermann, "A sufficient condition that a mapping of Riemannian manifolds be a fiber bundle", Proc. A.M.S. 11 (1960), 236-242.

[7] N. Hicks, "A theorem on affine connections", Ill. J. Math. 3 (1959), 242-254.

[8] S. Kobayashi, "Theory of connections", Annali di Mat. Pura Appl. 43 (1957), 119-194.

[9] P. Molino, "Etude des feuilletages transversalement complets et applications", Ann. Scient. Éc. Norm. Sup 10 (1977), 289-307.

[10] B. Reinhart, "Foliated manifolds with bundle-like metrics", Annals of Math. 69 (1959), 119-132.

INVARIANTS INTEGRAUX FONCTIONNELS POUR DES EQUATIONS AUX DERIVEES PARTIELLES D'ORIGINE GEOMETRIQUE

Jean Pierre BOURGUIGNON

Centre de Mathématiques, Unité Associée au CNRS n° 169
Ecole Polytechnique, F-91128 PALAISEAU Cedex
(France)

Le maintenant classique problème de Nirenberg s'énonce : "étant donnée une fonction f sur la sphère à deux dimensions, existe-t-il une métrique g conforme à la métrique standard c dont la courbure de Gauss s_g soit f ?" Pour que ce problème ait une solution, on voit facilement qu'il est nécessaire que la fonction f soit positive quelque part sur la sphère à cause du théorème de Gauss-Bonnet, selon lequel l'intégrale $\int_{S^2} s_g v_g$ (où v_g désigne l'élément de volume de la métrique g) ne dépend pas de la métrique et vaut 4π. Dans [5], J.L. Kazdan et F. Warner donnent d'autres conditions nécessaires sur la fonction f pour que le problème ait une solution, conditions qui peuvent s'exprimer ainsi : "Si ξ est une première harmonique sphérique, alors

$$\int_{S^2} (\nabla^c \xi, \nabla^c s_g)\, v_g = 0.$$

où ∇^c désigne le gradient dans la métrique c".

La preuve initiale qu'ils en ont donnée se fait par un calcul explicite utilisant le fait que les premières harmoniques sphériques sont les restrictions à la sphère S^2 des fonctions linéaires sur $\mathbb{R}^3$. Ces relations fournissent des exemples non triviaux de fonctions qui ne peuvent être la courbure scalaire d'une métrique conforme à c (on dit qu'elles sont "interdites") tels les fonctions dépendant de façon monotone de la distance à un point.

Si on remplace la courbure de Gauss par la courbure scalaire, le problème de Nirenberg garde un sens en dimension supérieure, et pour le problème ainsi étendu les obstructions précitées demeurent. (Pour des développements récents sur ce sujet, on peut consulter [1] et [3].)

Dans cet exposé nous nous proposons de montrer que ces identités ne sont en rien attachées à la sphère, mais sont en fait universelles. Dans un cadre approprié que nous décrivons dans la Section 1, leur démonstration devient alors élémentaire. De plus on peut obtenir de nouvelles identités dont nous donnons des exemples dans la Section 2. Il devient alors possible de relier précisément ce problème avec d'autres comme la recherche de métriques d'Einstein sur les variétés kählériennes à première classe de Chern positive pour laquelle A. Futaki a donné dans [4] des obstructions qui ressemblent étrangement à celles de Kazdan-Warner. Nous abordons ce point dans la Section 3. Nous concluons par quelques remarques sur le lien de notre approche avec la théorie des anomalies qui intéressent beaucoup les physiciens théoriciens en ce moment.

1. Le cadre géométrique

Nous allons travailler sur un espace $\mathcal{O}$ d'objets géométriques, sur lequel un groupe de Lie G agit. Nous allons concentrer notre attention sur certaines formes différentielles fermées α sur $\mathcal{O}$ invariantes par l'action du groupe G. Une application élémentaire de la théorie des invariants intégraux d'Elie Cartan nous amène à faire les constatations suivantes : à cause de l'invariance de la forme différentielle α, pour tout élément X de l'algèbre de Lie $\mathfrak{g}$, $\mathcal{L}_X\alpha = 0$; par ailleurs, d'après la relation fondamentale $\mathcal{L}_X = i_X \circ d + d \circ i_X$ appliquée à la forme fermée α et à un élément X de l'algèbre de Lie, on voit que $di_X\alpha = 0$, autrement dit que la fonction $i_X\alpha$ est constante.

Nous affirmons que les relations d'intégrabilité de Kazdan-Warner ainsi que celles d'A. Futaki proviennent toutes de l'application de ce calcul élémentaire après qu'on ait choisi convenablement l'espace des objets géométriques, le groupe qui agit sur lui et les formes différentielles fermées à considérer.

Nous expliquons ces choix dans le cas du problème de Nirenberg généralisé. Nous partons d'une variété différentielle compacte M de dimension n. Nous fixons une classe conforme Γ de métriques riemanniennes sur M, autrement dit nous considérons toutes les métriques g qui se déduisent de l'une d'entre elles g_0 par multiplication par une fonction positive e^{2f}. (Lorsque $n \geq 3$, il est classique, et pratique, d'écrire e^{2f} sous la forme $u^{\frac{4n}{n-2}}$ en supposant $u > 0$). Ce sera notre espace $\mathcal{O}$ d'objets géométriques.

Le groupe de Lie G sera le groupe des transformations conformes de la variété conforme (M,Γ) que nous faisons opérer naturellement sur Γ comme suit : pour φ dans G et g dans Γ, nous posons $\varphi(g) = \varphi^*g$ où dans le membre de droite les objets prennent le sens qu'ils ont habituellement en dimension finie. Si X est un élément de l'algèbre de Lie $\mathfrak{g}$ de G (autrement dit un champ de vecteur conforme), le champ de vecteurs $\bar{X}$ qu'il induit sur $\mathcal{O}$ est donc défini par $\bar{X}(v) = \mathcal{L}_X v$.

Il sera commode d' identifier Γ à l'ensemble des éléments de volume définis sur M, une métrique g étant représentée par son élément de volume v_g (à cause de cette identification, si $g = e^{2f}g_0$, alors $v_g = e^{nf}v_{g_0}$). Grâce à cette identification, on peut voir l'espace tangent à Γ en un point g identifié à v_g comme l'ensemble des n-formes différentielles sur M (pour simplifier la présentation, nous supposons M orientable), et par suite l'espace cotangent $T^*\Gamma$ apparaît comme l'espace des fonctions sur Γ. Une forme différentielle sur Γ associe donc à une métrique g de la classe conforme Γ une fonction sur M. Dans le cadre de notre problème, la forme différentielle à considérer est la forme σ qui à g associe sa courbure scalaire, donc définie par $\sigma(g) = s_g$.

LEMME 1.- <u>Sur</u> Γ, <u>la forme différentielle</u> σ <u>est fermée</u>.

<u>Preuve</u> : Pour vérifier que σ est fermée, il suffit d'évaluer $d\sigma$ sur deux vecteurs tangents V_1 et V_2 en un point g de Γ. Comme Γ a été identifié à un ouvert de l'espace vectoriel des n-formes différentielles sur M, on peut supposer que V_1 et V_2 ont été étendus au voisinage de g comme des champs de vecteurs constants. Par application de la formule donnant le cobord d'une forme différentielle, nous avons donc

$$\begin{aligned} d\sigma(V_1,V_2) &= V_1.\sigma(V_2) - V_2.\sigma(V_1) \\ &= \frac{d}{dt}\left(\int_M s_{(1+2ft_1)g}V_2\right)_{|t=0} - \frac{d}{dt}\left(\int_M s_{(1+2ft_2)g}V_1\right)_{|t=0} \end{aligned}$$

où nous avons utilisé le fait que la courbe de métriques $(1+2tf_i)g$ a pour vecteur-vitesse $V_i = n\,f_i\,v_g$ $(i=1,2)$ dans la représentation que nous avons choisie. Comme on peut permuter $\int_M$ et $\frac{d}{dt}$ et que la linéarisation de la courbure scalaire est donnée par la formule

$$\frac{d}{dt}\left(s_{(1+2ft_i)g}\right)_{|t=0} = 2(n-1)\Delta_g f_i - 2s_g f_i$$

(où Δ_g désigne le laplacien de la métrique g), nous obtenons donc

$$\begin{aligned} d\sigma(V_1,V_2) &= \int_M [2(n-1)\Delta_g f_1 - 2s_g f_1]\,n\,f_2\,v_g \\ &\quad - \int_M [2(n-1)\Delta_g f_2 - 2s_g f_2]\,n\,f_1\,v_g \\ &= \int_M 2n(n-1)[(\Delta_g f_1)f_2 - (\Delta_g f_2)f_1]\,v_g \\ &= 0, \end{aligned}$$

car Δ_g est auto-adjoint.∎

Il nous faut maintenant démontrer la naturalité.

LEMME 2.- <u>Sur</u> Γ, <u>la forme différentielle</u> σ <u>est invariante par le groupe</u> G <u>des transformations conformes de</u> Γ.

<u>Preuve</u> : Pour cela, pour un élément φ du groupe G, nous évaluons la forme $\varphi^*(\sigma)$ au point $v = v_g$ sur le vecteur tangent V. Nous avons

$$\begin{aligned} \varphi^*(\sigma)_v(V) &= \sigma_{\varphi^* v}(T\varphi(V)) \\ &= \int_M s_{\varphi^* g}\,\varphi^* V \\ &= \int_M (s_g \circ \varphi)\,\varphi^* V \\ &= \int_M s_g V \\ &= s(V). \end{aligned}$$

Dans ce calcul, il faut prendre garde que nous avons utilisé la même notation pour l'image réciproque des formes sur la variété de dimension finie M et sur la variété $\mathcal{O}=\Gamma$ de dimension infinie. Les seules propriétés utilisées sont la naturalité de la courbure ($s_{\varphi^* g} = s_g \circ \varphi$) et le théorème de changement de variable dans les intégrales sur une variété sans bord.∎

De la discussion que nous avons présentée au début de ce paragraphe, il ressort que, pour tout champ de vecteurs conforme X, la fonction $i_X\sigma$ est constante, ce qui nous conduit directement à la relation de Kazdan-Warner comme l'établit la proposition suivante.

PROPOSITION 3.- Soit (M,Γ) une variété conforme et G son groupe de transformations conformes. Pour tout champ de vecteurs conforme X et pour toute métrique g dans la classe conforme Γ,

$$\int_M X.s_g v_g = 0 \ .$$

Preuve : Si nous traduisons la constance de la fonction $i_X\sigma$, nous obtenons pour toutes métriques g et g' de Γ,

$$\int_M s_g \mathcal{L}_X v_g = \int_M s_{g'} \mathcal{L}_X v_{g'}$$

ce qui par intégration par partie donne

$$\int_M X.s_g v_g = \int_M X.s_{g'} v_{g'} \ .$$

Il ne nous reste donc plus qu'à démontrer que cette intégrale est nulle. Ce fait est spécial au groupe des transformations conformes d'une variété compacte. En effet, si G est un groupe compact, pour évaluer l'intégrale on peut prendre une métrique dans la classe conforme obtenue par moyenne sur le groupe compact G, donc telle que G soit pour elle un groupe d'isométries. Par suite la courbure scalaire de cette métrique est invariante par l'action de G, donc en particulier annihilée par les champs de vecteurs X de $\mathfrak{g}$. Si le groupe G est non compact, alors par un théorème dû à M. Obata (cf [7]) et à J. Lelong-Ferrand (cf [6]), la classe conforme Γ est nécessairement la classe standard sur la sphère. Il suffit alors de prendre une métrique à courbure constante pour évaluer l'intégrale.∎

Une autre preuve de cette relation est donnée dans [2]. Pour $n \geqslant 3$, elle est obtenue directement par intégration sur la variété M de l'identité que satisfait un champ de vecteurs conforme X pour toute métrique g de la classe conforme Γ, à savoir

$$\Delta_g(\mathrm{div}_g X) = \frac{1}{n-1} s_g \mathrm{div}_g X + \frac{n}{2(n-1)} X.s_g \ .$$

Le cas $n=2$ nécessite une analyse plus fine, et met en jeu la définition de deux actions du groupe conforme sur l'espace des fonctions (cf [2]).

2. De nouvelles relations intégrales

Dans ce paragraphe, nous donnons de nouveaux exemples de relations intégrales obtenues comme invariants intégraux fonctionnels par le schéma que nous avons décrit dans la Section 1.

D'abord il va de soi que pour toute fonction F définie sur l'espace $\mathcal{O}$ d'objets géométriques (i.e. la classe conforme Γ dans le cas qui nous occupe) invariante par le groupe G, sa différentielle dF donne naissance à un invariant intégral fonctionnel. Cela donne une infinité d'exemples qui sont a priori tautologiques mais qui se révèlent être cependant intéressants. Donnons-en quelques exemples dans le cas particulier considéré pour illustrer notre propos.

La 1-forme différentielle sur Γ considérée dans la Section précédente est en fait une différentielle exacte pour $n \geqslant 3$, car on a alors $\sigma = d\Sigma$ avec $\Sigma = \frac{n-2}{n}\int_M s_g v_g$. Cela "explique" pourquoi la preuve donnée dans [2] se simplifie directement dans ce cas-là.

Parmi les fonctions F géométriques, il est naturel de considérer la famille Σ_k pour $k \in \mathbb{N}$ définie par

$$\Sigma_k(v_g) = \int_M s_g^k v_g ,$$

de telle sorte que $\Sigma_1 = \Sigma$. Par un calcul direct, on trouve que

$$d\Sigma_k = (n-2)k\,\Delta(s^{k-1}) + (n-2k)s^k ,$$

où on rappelle qu'il faut interpréter le membre de droite comme une 1-forme différentielle sur Γ. Quelques cas particuliers méritent d'être mentionnés : pour $n = 4$ par exemple, on voit que la 1-forme différentielle $v_g \longmapsto \Delta_g s_g$ est exacte. Il lui est donc associée une loi de conservation qui s'énonce comme suit :

PROPOSITION 4.- *Pour tout champ de vecteurs conforme* X *et pour toute métrique* g *d'une classe conforme* Γ *définie sur la variété* M *de dimension* 4,

$$\int_M X.\Delta_g s_g v_g = 0 .$$

Un énoncé analogue peut bien sûr être donné pour une variété de dimension 2k en remplaçant s par s^{k-1}.

Un autre exemple d'invariant intégral fonctionnel qui se déduit de la considération de la différentielle d'une fonction invariante est le suivant. Soit λ_1 la première valeur propre de l'opérateur de Yamabe $4\frac{n-1}{n-2}\Delta_g + s_g$ (le fait de considérer la première valeur propre n'intervient que pour s'assurer qu'elle est simple, donc facilement dérivable en tant que fonction de la métrique). La principale différence avec les fonctionnelles considérées plus haut réside dans le fait que λ_1 n'est pas une fonctionnelle locale. On démontre que $d\lambda_1 = \psi_g^2$ où ψ_g désigne la fonction propre de l'opérateur de Yamabe pour la métrique g associée à cette valeur propre. La relation de conservation qui s'en déduit s'énonce alors comme suit :

PROPOSITION 5.- <u>Pour tout champ de vecteurs conforme</u> X <u>et pour toute métrique</u> g <u>de la classe conforme</u> Γ,

$$\int_M X.\psi_g^2 \, v_g = 0 \ .$$

Il est intéressant de noter qu'à la différence de la Proposition 4, dont une démonstration peut être obtenue par une intégration astucieuse à partir de l'identité que vérifie un champ de vecteurs conforme, il ne semble pas facile d'obtenir la relation de la Proposition 5 par un calcul direct. Ce phénomène est probablement dû au caractère global de la fonction λ_1.

Il y a encore une autre famille de 1-formes différentielles fermées sur Γ qu'il est particulièrement intéressant de considérer. Cette famille est reliée aux invariants caractéristiques de la variété M. Il est bien connu que, par la théorie de Chern-Weil, on peut exprimer ces invariants comme intégrales de polynômes en la courbure de n'importe quelle métrique riemannienne sur M.

Ainsi si la dimension de M est 2k, il existe un polynôme en la courbure $\chi_{2k}(R^g)$ tel que la caractéristique d'Euler de M, soit $\chi(M)$, s'exprime comme $\chi(M) = \int_M \chi_{2k}(R^g)\, v_g$. Indépendamment de la dimension n de M, il est donc naturel de considérer sur Γ la forme différentielle que nous notons encore χ_{2k} définie par $v_g \longmapsto \chi_{2k}(R^g)$. On peut noter que si $n < 2k$, alors pour toute métrique riemannienne g sur M, $\chi_{2k}(R^g) \equiv 0$.

A cause de sa nature géométrique, la forme χ_{2k} est évidemment invariante par le groupe G, mais nous avons aussi

LEMME 6.- <u>La 1-forme différentielle</u> χ_{2k} <u>est fermée sur</u> Γ. <u>Si</u> $2k < n$, χ_{2k} <u>est, à un coefficient près, la différentielle de la fonction</u> $v_g \longmapsto \int_M \chi_{2k}(R^g)\, v_g$.

Il est donc possible d'appliquer aux 1-formes différentielles fermées et invariantes χ_{2k} le raisonnement de la Section 1 et d'obtenir ainsi de nouveaux invariants intégraux fonctionnels.

THEOREME 7.- <u>Pour tout champ de vecteurs conforme</u> X <u>et pour toute métrique</u> g <u>de la classe conforme</u> Γ,

$$\int_M X.\chi_{2k}(R^g)\, v_g = 0 \ .$$

On peut noter que la forme σ n'est rien d'autre que χ_2 de telle sorte que la relation d'intégrabilité de Kazdan-Warner est un cas particulier du Théorème 7.

Comme nous concentrons notre attention sur une classe conforme et que les nombres de Pontryaguine de M s'expriment avec le tenseur de courbure conforme de Weyl qui est invariant par changement conforme de métrique (quand on lui donne la bonne variance), il ne semble pas possible de tirer d'information intéressante de cette famille.

3. Une nouvelle approche de l'obstruction de Futaki

Dans [4], A. Futaki trouve un nouvel invariant d'une classe de Kähler d'une variété complexe M de dimension complexe m, qui présente l'intérêt d'être une obstruction à l'existence de métriques de Kähler-Einstein dans cette classe. Si, pour une métrique kählérienne ω, on désigne par s_ω la courbure et si on introduit la fonction f_ω d'intégrale nulle par rapport à l'élément de volume ω^m telle que $\Delta_\omega f_\omega = s_\omega - s_\omega^0$ (où s_ω^0 désigne la moyenne de s_ω), A. Futaki montre que "pour tout champ de vecteurs holomorphe X sur M, $\int_M X.f_\omega \omega^m$ est indépendant de la forme de Kähler prise dans la classe que l'on considère". Cette intégrale est bien sûr nulle s'il existe une métrique à courbure scalaire constante dans la classe de Kähler, a fortiori s'il existe une métrique de Kähler-Einstein. Il remarque aussi que cette relation est une généralisation directe de la relation d'intégrabilité de Kazdan-Warner.

Cette relation peut être décrite dans le formalisme que nous avons utilisé pour l'étude d'une classe conforme de métriques de la façon suivante. Nous prenons comme espace $\mathcal{O}$ d'objets géométriques une classe de Kähler K que nous identifions à l'ouvert de l'espace des fonctions ψ définies sur M telles que pour une métrique de Kähler ω_0 prise comme origine $\omega_0 + \partial\bar\partial\psi$ est une forme de Kähler. Le groupe G est le groupe des transformations holomorphes de la variété complexe M.

Il s'agit maintenant de trouver des formes différentielles fermées et invariantes sur K. On peut bien sûr considérer des 1-formes qui se trouvent être des différentielles exactes de fonctions invariantes. Dans notre description, un vecteur tangent à K est une fonction définie sur M, et un vecteur cotangent n'est rien d'autre qu'une 2m-forme différentielle définie sur M. Si à une forme de Kähler ω, on associe la 2m-forme différentielle $\rho_\omega \wedge \omega^{m-1}$ (où ρ_ω désigne la forme de Ricci de la métrique kählérienne ω qui comme il est bien connu représente à un coefficient universel près la première classe de Chern de M), alors on retrouve l'invariant intégral fonctionnel de Futaki. Il faut bien sûr vérifier que cette forme est fermée, car son invariance est automatique à cause de sa nature géométrique et du fait que les transformations holomorphes commutent avec l'opérateur $\partial\bar\partial$. L'apparence différente de la relation de Futaki tient au fait que les éléments de l'espace K que nous considérons doivent être dérivés deux fois pour définir les véritables objets géométriques qui importent, à savoir les classes de Kähler.

La fin du calcul se présente donc comme suit. Il s'agit de trouver la forme que prend le champ de vecteurs X tangent à l'espace K. Pour cela, il suffit de remarquer que la 2-forme de type (1,1) $\mathcal{L}_X\omega$ appartient en chaque point ψ de K à l'image de l'opérateur $\partial\bar\partial$ et que l'on peut poser $\mathcal{L}_X(\omega_0 + \partial\bar\partial\psi) = \partial\bar\partial\psi_X$ pour une fonction ψ_X que l'on peut prendre d'intégrale nulle. On a alors $X(\psi) = \psi_X$. Le développement de la Section 1 affirme alors que pour toutes formes de Kähler ω et ω' appartenant à la même classe et pour tout champ de vecteurs holomorphe X,

$$\int_M \psi_X \rho_\omega \wedge \omega^{m-1} = \int_M \psi'_X \rho_{\omega'} \wedge \omega'^{m-1} .$$

En décomposant la 2-forme fermée ρ_ω en sa partie harmonique et sa partie exacte qui se trouve être $\partial\bar\partial f_\omega$ (où f_ω est la fonction de Futaki), on obtient

$$\begin{aligned}
\int_M \psi_X \rho_\omega \wedge \omega^{m-1} &= \int_M \psi_X \, s_\omega \\
&= s^0_\omega \int_M \psi + \int_M \psi_X \Delta_\omega f_\omega \, \omega^m \\
&= \int_M \Delta_\omega \psi_X \, f_\omega \, \omega^m \\
&= -\int_M \mathrm{div}_\omega X \, f_\omega \, \omega^m \\
&= \int_M X.f_\omega \, \omega^m ,
\end{aligned}$$

ce qui donne la relation trouvée par A. Futaki.

On peut noter que la 1-forme différentielle considérée sur M est, à un coefficient universel près, l'intégrand du nombre caractéristique kählérien obtenu en intégrant sur M le cup-produit de la première classe de Chern par la puissance qu'il faut de la forme de Kähler. La méthode développée permet de trouver des généralisations de la relation de Futaki. Nous ne les exposerons pas ici.

4. Un lien avec les anomalies de jauge ?

Nous ne ferons que de brefs commentaires sur l'analogie qui semble exister entre les développements que nous venons de faire et les anomalies de jauge considérées par les physiciens.

Rappelons que les physiciens désignent sous le terme d'anomalies certains termes de nature cohomologique apparaissant dans le remplacement d'une fonctionnelle classique par son analogue quantique et qui rendent cette opération non consistante par brisure de symétrie. (La non-nullité de ces termes les amène à rejeter le modèle en question, d'où le terme d'anomalie.) Les espaces considérés sont soit l'espace de toutes les connexions adaptées à un groupe de symétrie (anomalies chirales), soit l'espace des métriques riemanniennes sur une variété (anomalies gravitationnelles). Dans ces situations, il y a un groupe agissant sur ces espaces dit groupe des symétries locales (le groupe de jauge dans le cas des connexions, le groupe des difféomorphismes dans le cas gravitationnel), ce qui nous rapproche du cadre que nous avons développé. De la même façon que pour les deux exemples que nous avons présentés, les espaces considérés sont contractiles et le groupe qui agit est non compact.

La cohomologie considérée par les physiciens n'est pas la cohomologie ordinaire, car ils ne s'intéressent qu'à des objets définis localement sur l'espace de configuration, d'où le nom souvent avancé de cohomologie locale sans qu'une définition mathématique vraiment rigoureuse n'ait été donnée. Les invariants caractéristiques jouent aussi un grand rôle dans les fonctionnelles qu'ils considèrent.

Malgré des ressemblances frappantes, il est cependant encore difficile d'énoncer des résultats plus précis, décrivant la relation que les invariants intégraux fonctionnels présentés dans cet exposé entretiennent avec les anomalies.

REFERENCES

[1] A. BAHRI, J.M. CORON, Une théorie des points critiques à l'infini pour l'équation de Yamabe et le problème de Kazdan-Warner, C.R. Acad. Sci. Paris 300 (1985), 513-516.

[2] J.P. BOURGUIGNON, J.P. EZIN, Scalar curvature functions in a conformal class of metrics and conformal transformations, Preprint Ecole Polytechnique.

[3] J.F. ESCOBAR, R. SCHŒN, Conformal metrics with prescribed scalar curvature, Preprint, U.C. San Diego, 1985.

[4] A. FUTAKI, An obstruction to the existence of Einstein Kähler metrics, Inventiones Math. 73 (1983), 437-443.

[5] J.L. KAZDAN, F. WARNER, Curvature functions on compact 2-manifolds, Ann. of Math. 99 (1974), 14-47.

[6] J. LELONG-FERRAND, Transformations conformes et quasi-conformes des variétés riemanniennes compactes (démonstration de la conjecture de Lichnerowicz), Acad. Roy. Belg., Cl. Sci. Mémoire XXXIX, 5 (1971).

[7] M. OBATA, The conjectures on conformal transformations of Riemannian manifolds, J. Differential Geometry 6 (1971), 247-258.

Centre de Mathématiques
Unité Associée au CNRS n° 169
Ecole Polytechnique
F-91128 PALAISEAU Cedex
(France)

HERMITIAN NATURAL DIFFERENTIAL OPERATORS

F. J. Carreras, A. Ferrández and V. Miquel
Departamento de Geometría y Topología
Universidad de Valencia
Burjasot (Valencia) SPAIN

§1.Introduction

In [Gi 1] P. Gilkey studies the invariants of Riemannian manifolds with values in forms and the same is done in [A-B-P] using a more elegant approach. Epstein [E] introduces and elucidates the concept of natural tensor as a generalization of invariant with values in forms.

In later papers, [Gi 2], [Do 1], [G-H], the concept of hermitian invariant is introduced and, following [E], the hermitian natural tensors are studied in [F-M].

In a similar way as in [E], natural differential operators are defined in [S]. In this paper we give the general notion of hermitian natural differential operator (briefly, HNDO) on almost hermitian manifolds.

The main tool (Theorem 2.10) to classify HNDO's is the same as in the Riemannian case, with the only modifications introduced by the fact that there are many hermitian natural connections (see [F-M]) and then the expression of a HNDO is not unique. The essential contribution of this paper is providing a list of examples of HNDO's and showing that there are some relations between the almost hermitian geometry and the spectrum of some of them.

In §2 we recall the necessary background and state the classification theorem for HNDO's.

In §3 and §4 we give some examples of HNDO's of type $D\colon \Gamma(\Lambda^p M) \longrightarrow \Gamma(\Lambda^{p+1} M)$ and $D\colon \Gamma(\Lambda^p M) \longrightarrow \Gamma(\Lambda^{p-1} M)$ of order one and obtain all those which are homogeneous of maximal weight when p = 0, 1.

In §5 some examples of HNDO's $D\colon \Gamma(\Lambda^p M) \longrightarrow \Gamma(\Lambda^p M)$ are given and we get all the homogeneous of maximal weight when p = 0. There is a HNDO, for each p, that will play a

----

Work partially supported by C.A.I.C.Y.T. 1985-87, Nº 120.

prominent role in this paper; namely, given a homogeneous hermitian regular connection D on the tangent bundle, we can define the associated D-laplacian $\Delta^D = d^D d^{D*} + d^{D*} d^D$ (d^{D*} is the adjoint of d^D (see remark 5.6)). This operator is used in §6, where we apply the techniques of [Gi 3] and [Gi 4] in order to determine the first two terms in the asymptotic expansion of Δ^D acting on 1-forms. This shows that the spectrum of Δ^D on functions and 1-forms allows us to know when an almost Kaehler or a nearly Kaehler manifold is Kaehler. In [Do 2] and [Gi 3] the Kaehler condition is found out from the spectrum of different operators acting on (p,q)-forms on a hermitian manifold. As far as we know, our results can be considered as a starting point for the study of the spectrum on almost hermitian manifolds which are not complex; and, on the other hand, as an attempt of getting at the Kaehler condition from the spectrum of real operators. For the geometry of nearly Kaehler manifolds see [GR 1,2] and interesting examples of almost Kaehler manifolds are in [C-F-G].

In a forthcoming paper we shall deal with the complex laplacian as the restriction to Hermitian manifolds of a HNDO on almost hermitian manifolds, working on the complexified tangent space.

After the completion of this paper we became aware of the recent work of Donnelly ([Do 3]), where he obtains the formula of Theorem 6.4 by using different methods.

§2. Hermitian natural tensors and hermitian natural differential operators.

2.1. Let E be a functor from the category of hermitian vector spaces (V,g,J) into itself (see [E-K] or [S]) satisfying

(i) $E(V) \subset \otimes^r V$, for any (V,g,J); and

(ii) E(V) is invariant under the action of J induced on $\otimes^r V$.

We suppose also given

(iii) an ordered basis $E(v_i)$ of E(V), for each ordered basis (v_i) of a vector space V; and

(iv) $E(fv_i) = (Ef)(E(v_i))$, for vector spaces V, W and any isomorphism $f \in Hom(V,W)$.

We denote the dual vector space $(EV)^* \subset \otimes^r V^*$ by E^*V and we consider on E^*V the restriction E^*g of the metric induced on $\otimes^r V^*$ and the restriction E^*J of the endomorphism induced by J on $\otimes^r V$.

If (v_i) is an ordered basis of V and $E(v_i) = w_i$, we define the ordered basis $E(v^i)$ of E^*V to be the ordered basis (w^k) so that $w^k(w_j) = \delta^k{}_j$

A functor E satisfying (i) will be called a functor of rank r.

2.2. Given an almost hermitian manifold (M,g,J), a functor E as in 2.1 induces riemannian bundles (EM,Eg) and (E^*M, E^*g) over M, which are riemannian subbundles of ($\otimes^r TM$,g') and ($\otimes^r T^*M$, g'), respectively, where g' is the riemannian structure induced by g. On these bundles we have the endomorphisms of fibre bundles EJ: EM $\longrightarrow$ EM and E^*J: $E^*M \longrightarrow E^*M$ which are the restrictions to EM and E^*M of the endomorphisms J': $\otimes^r TM \longrightarrow \otimes^r TM$ and J': $\otimes^r T^*M \longrightarrow \otimes^r T^*M$, respectively, induced by J. They verify $(EJ)^2 = (E^*J)^2 = (-1)^r id$, and $Eg(EJ\,\bullet, EJ\,\bullet) = Eg(\bullet,\bullet)$ and $E^*g(E^*J\,\bullet, E^*J\,\bullet) = E^*g(\bullet,\bullet)$. Furthermore, it follows from 2.1(iii) that a local coordinate system x determines unique local bases of sections $E(\partial/\partial x^i)$, $E(dx^i)$ for EM, E^*M, respectively.

2.3.DEFINITION: Let E, F be functors as in 2.1. A hermitian natural tensor field t of type (E,F) assigns to each almost hermitian manifold (M,g,J) a tensor field $t_{(M,g,J)} \in \Gamma(EM \otimes F^*M)$ such that if f: (M,g,J) $\longrightarrow$ (M',g',J') is a holomorphic ($J' \circ f_* = f_* \circ J$) isometry of M onto an open subset of M', then $f_* t_{(M,g,J)} = t_{(M',g',J')}|_{f(M)}$. t is said to be homogeneous of weight w if $t_{(M,c^2 g,J)} = c^w\, t_{(M,g,J)}$, c being a non-zero real number.

As it is pointed out by Epstein [E], the problem of classifying all natural tensor fields becomes very complicated; however, there is a natural concept of regularity for such tensor fields, which was introduced in [Gi 1] and [A-B-P]. In order to settle the same concept for almost hermitian manifolds, we first need the following:

2.4.DEFINITION: Let (M,g,J) be an almost hermitian manifold of real dimension 2n and let p be a point of M. A coordinate system x centered at p will be called a J-coordinate system if $(\partial/\partial x^{n+i})(p) = J(\partial/\partial x^i)(p)$, $i = 1,\dots,n$.

Then, we have

2.5.DEFINITION: A hermitian natural tensor field t of type (E,F) is said to be regular if for each almost hermitian manifold (M,g,J) and each J-coordinate system x on an open subset U of M, the coefficients of $t_{(M,g,J)}$ with respect to the local basis $E \otimes F\,(\partial/\partial x^i \otimes dx^j)$ are given by universal polynomials in

$$g_{ij},\quad \Omega_{rs},\quad g^{kl},\quad \partial^{|\alpha|}(g_{ij})/\partial x^{\alpha},\quad \partial^{|\beta|}(\Omega_{rs})/\partial x^{\beta}.$$

where α, β are multiindices and Ω is the Kaehler form defined as usual by $\Omega(X,Y) = g(JX,Y)$.

Next theorem summarizes some results of [F-M] in order to apply them for computing hermitian natural differential operators.

<u>2.6.THEOREM:</u> Let t be a regular hermitian natural tensor of type (E,F). Then $t_{(M,g,J)} \in \Gamma(EM \otimes F^*M)$ is the restriction of an element of $\Gamma(\otimes^r TM \otimes \otimes^s T^*M)$, (r = rank E, s = rank F), which is a linear combination of the elementary monomials

$$m(\Omega,R) = \Sigma\, g_{i_1 i_2} \cdots g_{i_{2k-2} i_{2k}}\, g^{j_1 j_2} \cdots g^{j_{2l-1} j_{2l}}\, \Omega_{\alpha_1} \cdots \Omega_{\alpha_p}\, R_{\beta_1} \cdots_{\beta_q}$$

where each α_i (resp. β_j) is a multi-index $\alpha_i = (u_1,\dots,u_{n_{\alpha_i}})$, $(\beta_j = (v_1,\dots,v_{n_{\beta_j}}))$, $\Omega_{\alpha_i} = \nabla^{n_{\alpha_i}}_{u_3 \dots u_{n_{\alpha_i}}}(\Omega)_{u_1 u_2}$, $R_{\beta_j} = \nabla^{n_{\beta_j}}_{v_5 \dots v_{n_{\beta_j}}}(R)_{v_1 \dots v_4}$, and, if $N = 2k + 2p + 4q + \Sigma_{i=1}^{p} \varepsilon_i + \Sigma_{j=1}^{q} \eta_j$ ($\varepsilon_i = n_{\alpha_i} - 2$ = number of covariant derivatives in Ω_{α_i}, $\eta_j = n_{\beta_j} - 4$ = number of covariant derivatives in R_{β_j}), we have N-s contractions of upper and lower indices (and possible alternations or symmetrizations in the upper or lower indices non-contracted). Notice that $r = 2l - N + s$. Furthermore, the weight of such a monomial is $w(m(\Omega,R)) = s - r - \Sigma\,\varepsilon_i - \Sigma\,\eta_j - 2q$.

Similarly to hermitian natural tensors we can define hermitian natural differential operators as follows:

<u>2.7.DEFINITION:</u> Let E, F, G, H be functors as in 2.1. A hermitian natural differential operator $\mathcal{D}$ of type (E,F,G,H) assigns to each almost hermitian manifold (M,g,J) a differential operator $\mathcal{D}_{(M,g,J)} : \Gamma(EM \otimes F^*M) \longrightarrow \Gamma(GM \otimes H^*M)$ such that if $f: (M,g,J) \longrightarrow (M',g',J')$ is a holomorphic isometry of M onto an open set of M', then $\mathcal{D}_{(M,g,J)} = f^* \mathcal{D}_{(M',g',J')}$.

Now, we are going to express the regularity condition for HNDO's. Let (M,g,J) be an almost hermitian manifold and x a local J-coordinate system on an open subset U of M. Then x determines local bases of sections $(e_\alpha)_{\alpha \in A}$, $(f^\beta)_{\beta \in B}$, $(g_\gamma)_{\gamma \in C}$ and $(h^\delta)_{\delta \in D}$ for EM, F*M, GM and H*M, respectively. Let $\mathcal{D}: \Gamma(EM \otimes F^*M) \longrightarrow \Gamma(GM \otimes H^*M)$ be a differential operator of order k. Then, locally, we can write

$$\mathcal{D}(s^\alpha{}_\beta\, e_\alpha \otimes f^\beta) = \Sigma_{r=0}{}^{k}\, a^{\beta\gamma i_1 \dots i_r}{}_{\alpha\delta}\, (\partial^r s^\alpha{}_\beta / \partial x^{i_1} \dots \partial x^{i_r})\, g_\gamma \otimes h^\delta\ ,$$

where the functions $a^{\beta\gamma i_1 \dots i_r}_{\alpha\delta}$ are symmetric in $i_1,\dots,i_r$.

<u>2.8.DEFINITION:</u> A HNDO $\mathcal{D}$ is said to be regular if the coefficients $a^{\beta\gamma i_1 \dots i_r}_{\alpha\delta}$ of $\mathcal{D}_{(M,g,J)}$, in any local J-coordinate system, are given by universal polynomials in g_{ij}, g^{kl}, Ω_{rs}, $\partial^{|\alpha|}(g_{ij})/\partial x^{\alpha}$, $\partial^{|\beta|}(\Omega_{rs})/\partial x^{\beta}$.

The weight of a HNDO is defined as in the case of hermitian natural tensors.

In order to get a general expression of a HNDO, we need the concept of hermitian natural connection, which we take from [F-M].

<u>2.9.DEFINITION:</u> A hermitian natural connection is a map which assigns to each almost hermitian manifold (M,g,J) a linear connection $D^{(M,g,J)}$ on TM such that if $f:(M,g,J) \longrightarrow (M',g',J')$ is a holomorphic isometry of M onto an open set of M', then

$D^{(M,g,J)}{}_X Y = D^{(M',g',J')}{}_{f_*X} f_* Y$ for every vector fields X,Y on M.

We shall say that a hermitian natural connection D is regular if, for every local J-coordinate system, the Christoffel symbols of D are universal polynomials in the components of the metric tensor, the Kaehler form, their derivatives and the components of the metric induced on T*M.

In [F-M] a list of all the homogeneous hermitian natural connections is given.

Let E, F be functors as in 2.1. Let (M,g,J) be an almost hermitian manifold and let D be a homogeneous (of weight zero) regular hermitian natural connection on TM. Then, D induces another connection D on $EM \otimes F^*M$ in a natural way. We write $D^k = D \circ \overset{k}{\cdots} \circ D$, and define differential operators D_k making commutative the diagrams

$$\begin{array}{ccc} \Gamma(EM \otimes F^*M) & \xrightarrow{\quad D_k \quad} & \Gamma(S^k(T^*M) \otimes EM \otimes F^*M) \\ & {\scriptstyle D^k} \searrow \qquad \nearrow {\scriptstyle S^k \otimes 1} & \\ & \Gamma(\otimes^k T^*M \otimes EM \otimes F^*M) & \end{array}$$

where $S^k : \otimes^k T^*M \longrightarrow S^k(T^*M)$ is defined by $S^k(v_1 \otimes \dots \otimes v_k) = (1/k!) \sum_{\sigma \in S} v_{\sigma(1)} \otimes \dots \otimes v_{\sigma(k)}$.

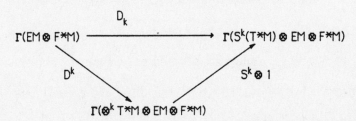

Then, it is easy to see that the symbol $\gamma_k(D_k) \in \mathrm{Hom}(S^k(T^*M) \otimes EM \otimes F^*M,\ S^k(T^*M) \otimes EM \otimes F^*M)$ of D_k is the identity map . (For the definition of the symbol $\gamma_r(\mathcal{D})$ of an operator $\mathcal{D}$ of order r see [S]). Furthermore, D_k is a homogeneous regular HNDO of order k and weight zero.

Therefore, the proof of theorem 3.7 in [S] works also here to show the following

2.10.THEOREM: Let $\mathcal{D}$ be a HNDO of type (E,F,G,H) and order k. Then, for each homogeneous regular hermitian connection D, there exist k+1 unique hermitian natural bundle maps $t_r: \Gamma(S^r(T^*M) \otimes EM \otimes F^*M) \longrightarrow \Gamma(GM \otimes H^*M)$, $0 \leq r \leq k$, such that

$$\mathcal{D} = \Sigma_{r=0}^{k} t_r \circ D_r,$$

and the t_r are regular if and only if $\mathcal{D}$ is. Furthermore,

$$t_r = \gamma_r(\mathcal{D} - \Sigma_{l=r+1}^{k} t_l \circ D_l).$$

2.11.REMARK: A bundle map t_r as that given in 2.10 can be identified , in a natural way, with a tensor field $t_r \in \Gamma(S^r(TM) \otimes E^*M \otimes FM \otimes GM \otimes H^*M)$ and so, saying that t_r is a hermitian natural bundle map means that it is hermitian natural when considered as a tensor field.

Given $\mathcal{D} = \Sigma_{r=0}^{k} t_r \circ D_r$ as in 2.10, $\mathcal{D}$ is homogeneous of weight w if and only if each t_r is homogeneous of weight w. From 2.6, the maximal weight of t_k is a+d-b-c-k (a=rankE, b=rank F, c=rank G, d=rank H). Then we shall say that $\mathcal{D}$ has maximal weight if it is homogeneous of weight a+d-b-c-k.

§3. The set of HNDO's of type ($\mathbf{R}$, $\Lambda^p TM$, $\mathbf{R}$, $\Lambda^{p+1} TM$) and order one.

3.1. In this section we shall deal with homogeneous regular HNDO's $\mathcal{D}: \Gamma(\Lambda^p T^*M) \longrightarrow \Gamma(\Lambda^{p+1} T^*M)$ of order one and maximal weight. By 2.10 such operators have the form

$$\mathcal{D} = t_0 + t_1 \circ D, \tag{3.1}$$

where t_0 and t_1 are homogeneous regular natural tensors of weight $w = p+1-p-1 = 0$, $t_0 \in \Gamma(\Lambda^p TM \otimes \Lambda^{p+1} T^*M)$ and $t_1 \in \Gamma(TM \otimes \Lambda^p TM \otimes \Lambda^{p+1} T^*M)$. Then, the classification of the operators $\mathcal{D}$ reduces to that of the tensors t_0 and t_1.

First of all, we study the space of tensors t_1.

3.2.PROPOSITION: The space of tensors t_1 appearing in (3.1) is spanned by the tensors t_1 whose action on $\omega \in \Gamma(T^*M \otimes \Lambda^p T^*M)$, with components $\omega_{k_1 \ldots k_{p+1}}$ in an orthonormal local J-frame $\{e_i\}$, is given by

$$(t_1\omega)_{j_1\ldots j_{p+1}} = \sum_{\sigma\in S_{p+1}} \mathrm{sgn}(\sigma)\Big(\sum_{i_0\ldots i_r} \omega_{i_0 i_0 i_1 i_1^* \ldots i_r i_r^* j^*_{\sigma(1)} \ldots j^*_{\sigma(s)} j_{\sigma(s+1)} \ldots j_{\sigma(s+u)}}\Big)$$

$$\times\, \Omega_{j_{\sigma(s+u+1)} j_{\sigma(s+u+2)}} \ldots \Omega_{j_{\sigma(p)} j_{\sigma(p+1)}},$$

where $0 \le r \le (p-1)/2$, $0 \le s,u \le p-1$, or by

$$(t_1\omega)_{j_1\ldots j_{p+1}} = \sum_{\sigma\in S_{p+1}} \mathrm{sgn}(\sigma)\Big(\sum_{i_1\ldots i_r} \omega_{j^*_{\sigma(1)} \ldots j^*_{\sigma(s)} j_{\sigma(s+1)} \ldots j_{\sigma(s+u)} i_1 i_1^* \ldots i_r i_r^*}\Big)\times$$

$$\times\, \Omega_{j_{\sigma(s+u+1)} j_{\sigma(s+u+2)}} \ldots \Omega_{j_{\sigma(p)} j_{\sigma(p+1)}},$$

where $0 \le r \le p/2$, $1 \le s \le p+1$, $0 \le u \le p$, or by

$$(t_1\omega)_{j_1\ldots j_{p+1}} = \sum_{\sigma\in S_{p+1}} \mathrm{sgn}(\sigma)\Big(\sum_{i_1\ldots i_r} \omega_{j_{\sigma(1)} \ldots j_{\sigma(s)} j^*_{\sigma(s+1)} \ldots j^*_{\sigma(s+u)} i_1 i_1^* \ldots i_r i_r^*}\Big)$$

$$\times\, \Omega_{j_{\sigma(s+u+1)} j_{\sigma(s+u+2)}} \ldots \Omega_{j_{\sigma(p)} j_{\sigma(p+1)}},$$

where $0 \le r \le p/2$, $1 \le s \le p+1$, $0 \le u \le p$.

<u>Proof:</u> As we know, $w(t_1) = 0$ and then, from 2.6, $\varepsilon_i = \eta_j = q = 0$ and the space of tensors t_1 will be spanned by elementary monomials of the form

(3.2) $$\Sigma^{*p}_{p+1}\; g_{i_1 i_2} \ldots g_{i_{2k-1} i_{2k}}\, g^{j_1 j_2} \ldots g^{j_{2l-1} j_{2l}}\, \Omega_{u_1 u_2} \ldots \Omega_{u_{2b-1} u_{2b}},$$

where $2k+2b = 2l$ and $\Sigma^*{}_{p+1}{}^p$ means that $2l-(p+1)$ upper indices are contracted with $2l-(p+1)$ lower ones and p upper and $p+1$ lower indices are skewsymmetrized.

The possible contractions using $g^{\cdot\cdot}$, $g_{\cdot\cdot}$ and $\Omega_{\cdot\cdot}$ give us elements of the form $g^{\cdot\cdot}$, $g_{\cdot\cdot}$, $\Omega_{\cdot\cdot}$, $\Omega^{\cdot\cdot}$, $J^{\cdot}_{\cdot}$, $\delta^{\cdot}_{\cdot}$, and, thus, the monomials (3.2) can be written as

$$\mathrm{Alt}^{p}_{p+1}\{g^{i_1 i_2} \ldots g^{i_{2r-1} i_{2r}}\; \Omega^{i_{2r+1} i_{2r+2}} \ldots \Omega^{i_{2s-1} i_{2s}}\; J^{i_{2s+1}}_{j_1} \ldots J^{i_{2s+u}}_{j_u} \times$$

$$\times\; \delta^{i_{2s+u+1}}_{j_{u+1}} \ldots \delta^{i_{p+1}}_{j_{u+v}}\, g_{j_{u+v+1} j_{u+v+2}} \ldots g_{j_{u+v+2w-1} j_{u+v+2w}} \times$$

$$\times\; \Omega_{j_{u+v+2w+1} j_{u+v+2w+2}} \ldots \Omega_{j_p j_{p+1}}\},$$

where $\mathrm{Alt}_{p+1}{}^{p}$ means that p upper indices and p+1 lower indices are skewsymmetrized. Therefore, the elementary monomials with some $g_{..}$ or more than one $g^{..}$ vanish and so the monomials (3.2) can be finally written as

$$\mathrm{Alt}^{p}_{p+1}\{g^{i_0 i_1}\Omega^{i_2 i_3}\ldots\Omega^{i_{2r}i_{2r+1}}J^{i_{2r+2}}_{j_1}\ldots J^{i_{2r+s+1}}_{j_s}\delta^{i_{2r+s+2}}_{j_{s+1}}\ldots\delta^{i_p}_{j_{s+u}}\Omega_{j_{s+u+1}j_{s+u+2}}\ldots\Omega_{j_p j_{p+1}}\}$$

where $0 \leq r \leq (p-1)/2$, $0 \leq s,u \leq p-1$, and it is not necessary that $g^{i_0 i_1}$ appears in (3.3). Notice also that only one index of $g^{..}$ can be skew-symmetrized if the monomial is not zero. Also, the indices of all the $\Omega^{..}$ must be skewsymmetrized, because, if not, (3.3) represents a zero map or a map defined only on $\Lambda^{p+1}T^*M$ and not on all $T^*M \otimes \Lambda^p T^*M$.

Let t_1 be the tensor given by (3.3) (with $g^{..}$ appearing in its expression) and let $\omega \in \Gamma(T^*M \otimes \Lambda^p T^*M)$. Then

$$(t_1\omega)_{j_1\ldots j_{p+1}} = \sum_{i_0\ldots i_p}\sum_{\sigma\in S_{p+1}}\mathrm{sgn}(\sigma)\sum_{\tau\in S_p}\mathrm{sgn}(\tau)\, g^{i_0 i_{\tau(1)}}\Omega^{i_{\tau(2)}i_{\tau(3)}}\ldots\Omega^{i_{\tau(2r)}i_{\tau(2r+1)}}\times$$
$$\times J^{i_{\tau(2r+2)}}_{j_{\sigma(1)}}\ldots J^{i_{\tau(2r+s+1)}}_{j_{\sigma(s)}}\delta^{i_{\tau(2r+s+2)}}_{j_{\sigma(s+1)}}\ldots\delta^{i_{\tau(p)}}_{j_{\sigma(s+u)}}\times$$
$$\times\,\Omega_{j_{\sigma(s+u+1)}j_{\sigma(s+u+2)}}\ldots\Omega_{j_{\sigma(p)}j_{\sigma(p+1)}}\,\omega_{i_0 i_1\ldots i_p},$$

where $0 \leq r \leq (p-1)/2$, $0 \leq s,u \leq p-1$.

If in (3.3) the $g^{..}$ does not appear, we have the following two possible expressions for $t_1\omega$:

$$(t_1\omega)_{j_1\ldots j_{p+1}} = \sum_{i_0\ldots i_p}\sum_{\sigma\in S_{p+1}}\mathrm{sgn}(\sigma)\sum_{\tau\in S_p}\mathrm{sgn}(\tau)\,\Omega^{i_{\tau(1)}i_{\tau(2)}}\ldots\Omega^{i_{\tau(2r-1)}i_{\tau(2r)}}\times$$
$$\times J^{i_0}_{j_{\sigma(1)}}J^{i_{\tau(2r+1)}}_{j_{\sigma(2)}}\ldots J^{i_{\tau(2r+s-1)}}_{j_{\sigma(s)}}\delta^{i_{\tau(2r+s)}}_{j_{\sigma(s+1)}}\ldots\delta^{i_{\tau(p)}}_{j_{\sigma(s+u)}}\times$$
$$\times\,\Omega_{j_{\sigma(s+u+1)}j_{\sigma(s+u+2)}}\ldots\Omega_{j_{\sigma(p)}j_{\sigma(p+1)}}\,\omega_{i_0 i_1\ldots i_{p+1}},$$

or

$$(t_1\omega)_{j_1\dots j_{p+1}} = \sum_{i_0\dots i_p} \sum_{\sigma\in S_{p+1}} \mathrm{sgn}(\sigma) \sum_{\tau\in S_p} \mathrm{sgn}(\tau)\, \Omega^{i_{\tau(1)} i_{\tau(2)}} \dots \Omega^{i_{\tau(2r-1)} i_{\tau(2r)}} \times$$

$$\times\ J^{i_{\tau(2r+1)}}_{j_{\sigma(1)}} \dots J^{i_{\tau(2r+s)}}_{j_{\sigma(s)}}\, \delta^{i_0}_{j_{\sigma(s+1)}}\, \delta^{i_{\tau(2r+s+1)}}_{j_{\sigma(s+2)}} \dots \delta^{i_{\tau(p)}}_{j_{\sigma(s+u)}} \times$$

$$\times\ \Omega_{j_{\sigma(s+u+1)} j_{\sigma(s+u+2)}} \dots \Omega_{j_{\sigma(p)} j_{\sigma(p+1)}}\, \omega_{i_0 i_1 \dots i_p}$$

where $0 \le r \le p/2$, $1 \le s \le p+1$, $0 \le u \le p$, which, when we take a J-orthonormal frame, are the required formulas.

<u>3.3.</u> The tensors t_1 given in Proposition 3.2 are listed in Table I, where we have used the following notations:

$(J\omega)(X_1, \dots, X_{p+1}) = (-1)^{p+1}\,\omega(JX_1, \dots, JX_{p+1})$,

$(c_{11}\omega)(X_1, \dots, X_{p-1}) = \Sigma_{i=1}{}^{2n}\,\omega(e_i, e_i, X_1, \dots, X_{p-1})$,

$(c_J{}^p\,\omega)(X,Y) = \Sigma_{i_1\dots i_{(p-1)/2}}\,\omega(X, Y, e_{i_1}, Je_{i_1}, \dots, e_{i_{(p-1)/2}}, Je_{i_{(p-1)/2}})$ when $p+1$ is even,

$(J^s\omega)(X_1, \dots, X_{p+1}) = (-1)^s\,\omega(X_1, JX_2, \dots, JX_{s+1}, X_{s+2}, \dots, X_{p+1})$,

$(J_1{}^s\omega)(X_1, \dots, X_{p+1}) = (-1)^s\,\omega(JX_1, \dots, JX_s, X_{s+1}, \dots, X_{p+1})$,

$(c_J{}^r\omega)(X_1, \dots, X_{p+1-r}) = \Sigma_{i_1\dots i_{r/2}}\,\omega(X_1, \dots, X_{p+1-r}, e_{i_1}, Je_{i_1}, \dots, e_{i_{r/2}}, Je_{i_{r/2}})$.

<u>3.4.</u> In Table II we have listed the expressions of some of the operators $t_1{\circ}D$, when they have a simple form. We have used the following notations: if D is a linear connection in TM, the operators $d^D : \Gamma(\Lambda^p T^*M) \longrightarrow \Gamma(\Lambda^{p+1}T^*M)$ and $\delta^D : \Gamma(\Lambda^p T^*M) \longrightarrow \Gamma(\Lambda^{p-1}T^*M)$ are defined by

$$d^D\omega = \mathrm{Alt}(D\omega) \qquad\qquad \delta^D\omega = -\,c_{11}(D\omega).$$

They satisfy $(d^D)^2 = 0 = (\delta^D)^2$ if and only if D is symmetric (and then $d^D = d = d^\nabla$, and $\delta^D = \delta = \delta^\nabla$, where ∇ is the Levi-Civita connection). Moreover, d^D is a skew-derivation.

Now we study the space of tensors t_0.

<u>3.5.PROPOSITION:</u> The space of tensors t_0 in (3.1) is spanned by those tensors t_0 whose action on an $\omega \in \Gamma(\Lambda^p T^*M)$, with components $\omega_{k_1\dots k_p}$ with respect to an orthonormal J-frame $\{e_i\}$ is given by one of the following expresions:

$$(t_0\omega)_{j_1\dots j_{p+1}} = \sum_{\sigma\in S_{p+1}} \operatorname{sgn}(\sigma) \sum_{i_1\dots i_r} \omega_{i_1 i_1^* \dots i_r i_r^* j_{\sigma(1)}^* \dots j_{\sigma(s)}^* j_{\sigma(s+1)} \dots j_{\sigma(s+u)}} \times$$

$$\times\, \Omega_{j_{\sigma(s+u+1)} j_{\sigma(s+u+2)}} \dots \Omega_{j_{\sigma(p-3)} j_{\sigma(p-2)}} \nabla_{j_{\sigma(p-1)}} \Omega_{j_{\sigma(p)} j_{\sigma(p+1)}},$$

where $2r+s+u = p$;

$$(t_0\omega)_{j_1\dots j_{p+1}} = \sum_{\sigma\in S_{p+1}} \operatorname{sgn}(\sigma) \sum_{i_1\dots i_r i_p} \omega_{i_1 i_1^* \dots i_r i_r^* j_{\sigma(1)}^* \dots j_{\sigma(s)}^* j_{\sigma(s+1)} \dots j_{\sigma(s+u)} i_p} \times$$

$$\times\, \Omega_{j_{\sigma(s+u+1)} j_{\sigma(s+u+2)}} \dots \Omega_{j_{\sigma(p-2)} j_{\sigma(p-1)}} \nabla_{i_p} \Omega_{j_{\sigma(p)} j_{\sigma(p+1)}},$$

where $2r+s+u = p-1$;

$$(t_0\omega)_{j_1\dots j_{p+1}} = \sum_{\sigma\in S_{p+1}} \operatorname{sgn}(\sigma) \sum_{i_1\dots i_r i_p} \omega_{i_1 i_1^* \dots i_r i_r^* j_{\sigma(1)}^* \dots j_{\sigma(s)}^* j_{\sigma(s+1)} \dots j_{\sigma(s+u)} i_p} \times$$

$$\times\, \Omega_{j_{\sigma(s+u+1)} j_{\sigma(s+u+2)}} \dots \Omega_{j_{\sigma(p-2)} j_{\sigma(p-1)}} \nabla_{j_{\sigma(p)}} \Omega_{i_p j_{\sigma(p+1)}},$$

$$(t_0\omega)_{j_1\dots j_{p+1}} = \sum_{\sigma\in S_{p+1}} \operatorname{sgn}(\sigma) \sum_{i_1\dots i_r i_{p-1} i_p} \omega_{i_1 i_1^* \dots i_r i_r^* j_{\sigma(1)}^* \dots j_{\sigma(s)}^* j_{\sigma(s+1)} \dots j_{\sigma(s+u)} i_{p-1} i_p} \times$$

$$\times\, \Omega_{j_{\sigma(s+u+1)} j_{\sigma(s+u+2)}} \dots \Omega_{j_{\sigma(p-1)} j_{\sigma(p)}} \nabla_{i_{p-1}} \Omega_{i_p j_{\sigma(p+1)}},$$

$$(t_0\omega)_{j_1\dots j_{p+1}} = \sum_{\sigma\in S_{p+1}} \operatorname{sgn}(\sigma) \sum_{i_1\dots i_r i_{p-1} i_p} \omega_{i_1 i_1^* \dots i_r i_r^* j_{\sigma(1)}^* \dots j_{\sigma(s)}^* j_{\sigma(s+1)} \dots j_{\sigma(s+u)} i_{p-1} i_p} \times$$

$$\times\, \Omega_{j_{\sigma(s+u+1)} j_{\sigma(s+u+2)}} \dots \Omega_{j_{\sigma(p-1)} j_{\sigma(p)}} \nabla_{j_{\sigma(p+1)}} \Omega_{i_{p-1} i_p},$$

$$(t_0\omega)_{j_1\dots j_{p+1}} =$$

$$\sum_{\sigma\in S_{p+1}} \operatorname{sgn}(\sigma) \sum_{i_1\dots i_r i_{p-2} i_{p-1} i_p} \omega_{i_1 i_1^* \dots i_r i_r^* j_{\sigma(1)}^* \dots j_{\sigma(s)}^* j_{\sigma(s+1)} \dots j_{\sigma(s+u)} i_{p-2} i_{p-1} i_p} \times$$

$$\times\, \Omega_{j_{\sigma(s+u+1)} j_{\sigma(s+u+2)}} \dots \Omega_{j_{\sigma(p)} j_{\sigma(p+1)}} \nabla_{i_{p-2}} \Omega_{i_{p-1} i_p},$$

$$(t_0\omega)_{j_1\dots j_{p+1}} = \sum_{\sigma\in S_{p+1}} \mathrm{sgn}(\sigma) \sum_{i_1\dots i_r} \omega_{i_1 i_1^* \dots i_r i_r^* j^*_{\sigma(1)}\dots j^*_{\sigma(s)} j_{\sigma(s+1)}\dots j_{\sigma(s+u)}} \times$$

$$\times\, \Omega_{j_{\sigma(s+u+1)} j_{\sigma(s+u+2)}} \dots \Omega_{j_{\sigma(p-1)} j_{\sigma(p)}} \delta\Omega_{j_{\sigma(p+1)}},$$

where $2r+s+u = p$;

$$(t_0\omega)_{j_1\dots j_{p+1}} = \sum_{\sigma\in S_{p+1}} \mathrm{sgn}(\sigma) \sum_{i_1\dots i_r i_p} \omega_{i_1 i_1^* \dots i_r i_r^* j^*_{\sigma(1)}\dots j^*_{\sigma(s)} j_{\sigma(s+1)}\dots j_{\sigma(s+u)} i_p} \times$$

$$\times\, \Omega_{j_{\sigma(s+u+1)} j_{\sigma(s+u+2)}} \dots \Omega_{j_{\sigma(p)} j_{\sigma(p+1)}} \delta\Omega_{i_p}$$

and the expressions obtained from the above list by adding a $*$ in one, two or three of the indices in $\nabla\Omega$ or $\delta\Omega$.

<u>Proof:</u> From 3.1 we have that $w(t_0) = 0$, and, then, from 2.6, $q = 0 = \eta_j$, and there exists one index i such that $\varepsilon_i = 1$ and $\varepsilon_k = 0$ for any $k \neq i$. Therefore, the space of tensors t_0 is spanned by elementary monomials of the form

$$m(\Omega,\nabla\Omega) = \Sigma^{*p}_{p+1}\, g_{i_1 i_2} \dots g_{i_{2k-1} i_{2k}} g^{j_1 j_2} \dots g^{j_{2l-1} j_{2l}} \Omega_{u_1 u_2} \dots \Omega_{u_{2b-1} u_{2b}} \nabla_v \Omega_{rs},$$

where $2k + 2b + 3 = 2l + 1$ and $\Sigma^*{}_{p+1}{}^p$ means that p upper indices and p+1 lower indices are skew-symmetrized, and $2l-p$ upper indices are contracted with $2l-p$ lower ones. The possible contractions using $g^{\cdot\cdot}$, $g_{\cdot\cdot}$, $\Omega_{\cdot\cdot}$ and $\nabla_\cdot\Omega_{\cdot\cdot}$ yield (up to sign) elements of the form

$g^{\cdot\cdot}$, $g_{\cdot\cdot}$, $\Omega^{\cdot\cdot}$, $\Omega_{\cdot\cdot}$, $J^{\cdot}_{\cdot}$, $\delta^{\cdot}_{\cdot}$, $\nabla_\cdot\Omega_{\cdot\cdot}$, $\nabla^\cdot\Omega_{\cdot\cdot}$, $\nabla_\cdot J^{\cdot}_{\cdot}$, $\nabla^\cdot J^{\cdot}_{\cdot}$, $\nabla_\cdot\Omega^{\cdot\cdot}$, $\nabla^\cdot\Omega^{\cdot\cdot}$,

$\nabla_{\cdot *}\Omega_{\cdot\cdot}$, $\nabla_\cdot\Omega_{\cdot *\cdot}$, $\nabla_{\cdot *}\Omega_{\cdot *\cdot}$, $\nabla^{\cdot *}\Omega_{\cdot\cdot}$, $\nabla^\cdot\Omega_{\cdot *\cdot}$, $\nabla^{\cdot *}\Omega_{\cdot *\cdot}$,

$\nabla_{\cdot *}J^{\cdot}_{\cdot}$, $\nabla_\cdot J^{\cdot}_{\cdot *}$, $\nabla_{\cdot *}J^{\cdot}_{\cdot *}$, $\nabla^{\cdot *}J^{\cdot}_{\cdot}$, $\nabla^\cdot J^{\cdot}_{\cdot *}$, $\nabla^{\cdot *}J^{\cdot}_{\cdot *}$,

$\nabla_{\cdot *}\Omega^{\cdot\cdot}$, $\nabla_\cdot\Omega^{\cdot *}$, $\nabla_{\cdot *}\Omega^{\cdot *}$, $\nabla^{\cdot *}\Omega^{\cdot\cdot}$, $\nabla^\cdot\Omega^{\cdot *}$, $\nabla^{\cdot *}\Omega^{\cdot *}$, $\delta\Omega_\cdot$, $\delta J^\cdot$, $\delta\Omega_{\cdot *}$, $\delta J^{\cdot *}$,

where $\bullet * = J\bullet$. The contractions not listed above can be reduced to these by the symmetries of $\nabla\Omega$; that is, $\nabla_i\Omega_{jk} = -\nabla_i\Omega_{kj}$ and $\nabla_i\Omega_{j*k*} = -\nabla_i\Omega_{jk}$ (see [G-H]). Then, since $g^{\cdot\cdot}$ and $g_{\cdot\cdot}$ cannot appear in a nonzero monomial (because their skew-symmetrizations are zero), the only non-vanishing elementary monomials are (up to sign):

$$\mathrm{Alt}^{p}_{p+1}\{V^{i_1\cdots i_p}_{j_1\cdots j_{p-2}}\nabla_{j_{p-1}}\Omega_{j_pj_{p+1}}\},\qquad \mathrm{Alt}^{p}_{p+1}\{V^{i_1\cdots i_{p-1}}_{j_1\cdots j_{p-1}}\nabla^{i_p}\Omega_{j_pj_{p+1}}\},$$

$$\mathrm{Alt}^{p}_{p+1}\{V^{i_1\cdots i_{p-1}}_{j_1\cdots j_{p-1}}\nabla_{j_p}J^{i_p}_{j_{p+1}}\},\qquad \mathrm{Alt}^{p}_{p+1}\{V^{i_1\cdots i_{p-2}}_{j_1\cdots j_p}\nabla^{i_{p-1}}J^{i_p}_{j_{p+1}}\},$$

$$\mathrm{Alt}^{p}_{p+1}\{V^{i_1\cdots i_{p-2}}_{j_1\cdots j_p}\nabla_{j_{p+1}}\Omega^{i_{p-1}i_p}\},\qquad \mathrm{Alt}^{p}_{p+1}\{V^{i_1\cdots i_{p-3}}_{j_1\cdots j_{p+1}}\nabla^{i_{p-2}}\Omega^{i_{p-1}i_p}\},$$

$$\mathrm{Alt}^{p}_{p+1}\{V^{i_1\cdots i_p}_{j_1\cdots j_p}\,\delta\Omega_{j_{p+1}}\},\qquad \mathrm{Alt}^{p}_{p+1}\{V^{i_1\cdots i_{p-1}}_{j_1\cdots j_{p+1}}\delta J^{i_p}\},$$

and the monomials obtained from this list when we add * to one, two or three of the indices in $\nabla\Omega$, ∇J, $\delta\Omega$ or δJ, where the tensor V is given by

$$V^{i_1\cdots i_k}_{j_1\cdots j_l} = \Omega^{i_1i_2}\cdots\Omega^{i_{2r-1}i_{2r}}J^{i_{2r+1}}_{j_1}\cdots J^{i_{2r+s}}_{j_s}\delta^{i_{2r+s+1}}_{j_{s+1}}\delta^{i_k}_{j_{s+u}}\Omega_{j_{s+u+1}j_{s+u+2}}\cdots\Omega_{j_{l-1}j_l}.$$

The proposition follows by taking an orthonormal J-frame.

<u>3.6.</u> We list in Table III the tensors t_0 given in Proposition 3.5, when p = 0, 1. If $\omega\in\Gamma(T^*M)$, $\omega^\#$ will denote the image of ω by the canonical isomorphism between T*M and TM given by the Riemannian metric g.

§4. The set of HNDO's of type ($\mathbf{R}$, Λ^pTM, $\mathbf{R}$, $\Lambda^{p-1}TM$) and order one.

<u>4.1</u> In this section we shall deal with homogeneous regular HNDO's $\mathcal{D}:\Gamma(\Lambda^pT^*M)\longrightarrow\Gamma(\Lambda^{p-1}T^*M)$ of order one and maximal weight. As we know, such a $\mathcal{D}$ has a general form

$$(4.1)\qquad \mathcal{D} = t_0 + t_1\cdot D,$$

where t_0 and t_1 are homogeneous regular hermitian natural tensors of weight $w=p-1-(p+1)=-2$, $t_0\in\Gamma(\Lambda^pTM\otimes\Lambda^{p-1}T^*M)$ and $t_1\in\Gamma(TM\otimes\Lambda^pTM\otimes\Lambda^{p-1}T^*M)$.

Similarly to §3, we have

<u>4.2.PROPOSITION:</u> The space of tensors t_1 in (4.1) is spanned by those tensors whose action on an $\omega\in\Gamma(T^*M\otimes\Lambda^pT^*M)$ (with components $\omega_{k_1\ldots k_{p+1}}$ with respect to an orthonormal local J-frame (e_i)) is given by

$$(t_1\omega)_{j_1\cdots j_{p-1}} = \sum_{\sigma\in S_{p-1}}\mathrm{sgn}(\sigma)\sum_{i_0\ldots i_r}\omega_{i_0i_0i_1i_1^*\cdots i_ri_r^*j^*_{\sigma(1)}\cdots j^*_{\sigma(s)}j_{\sigma(s+1)}\cdots j_{\sigma(s+u)}}\times$$

$$\times\ \Omega_{j_{\sigma(s+u+1)}j_{\sigma(s+u+2)}}\cdots\Omega_{j_{\sigma(p-2)}j_{\sigma(p-1)}},$$

where $0 \leq r \leq (p-1)/2$; $0 \leq s,u \leq p-1$; or by the expressions like the other two given in 3.2 but changing S_{p+1} by S_{p-1} and with $r \geq 1$.

<u>4.3.PROPOSITION:</u> The space of tensors t_0 in (4.1) is spanned by those tensors whose action on an $\omega \in \Gamma(\Lambda^p T^*M)$ is given by the same formulas as in 3.5, but changing p+1 by p-1.

Observe that the change of p+1 by p-1 can also be applied to 3.2 to get 4.2. Therefore, Table I can be considered as a list of generators of tensors t_1 in 4.2, changing p+1 by p-1 (for which, in addition, formulas with no Ω are to be deleted, and in formulas with some Ω, one of these should be deleted). The generators of the space of tensors t_0, in 4.3, when $p \leq 2$, are given in Table IV.

<u>§5. The set of HNDO's of type ($\mathbf{R}$, $\Lambda^p TM$, $\mathbf{R}$, $\Lambda^p TM$) and order two.</u>

<u>5.1.</u> In this section we will consider homogeneous regular HNDO's $D : \Gamma(\Lambda^p T^*M) \longrightarrow \Gamma(\Lambda^p T^*M)$ of order two and maximal weight. Again, by 2.10, these operators can be written as

$$(5.1) \qquad D = t_0 + t_1 \cdot D + t_2 \cdot D_2 ,$$

where t_0, t_1 and t_2 are homogeneous regular hermitian natural tensors of weight $w = p-p-2 = -2$.

As in §§3 and 4, we have the following results:

<u>5.2.PROPOSITION:</u> The space of tensors $t_2 \in \Gamma(S^2T^*M \otimes \Lambda^p TM \otimes \Lambda^p T^*M)$, appearing in (5.1), is spanned by those tensors whose action on an $\omega \in \Gamma(S^2T^*M \otimes \Lambda^p T^*M)$ (with components $\omega_{k_1 \dots k_{p+2}}$ with respect to an orthonormal local J-frame $\{e_i\}$) is given by one of the following expressions:

$$(t_2\omega)_{j_1 \dots j_p} = \sum_{\sigma \in S_p} \mathrm{sgn}(\sigma) \sum_{i_o \dots i_r} \omega_{i_o i_o i_1 i_1^* \dots i_r i_r^* j_{\sigma(1)}^* \dots j_{\sigma(s)}^* j_{\sigma(s+1)} \dots j_{\sigma(s+u)}} \times$$
$$\times \; \Omega_{j_{\sigma(s+u+1)} j_{\sigma(s+u+2)}} \dots \Omega_{j_{\sigma(p-1)} j_{\sigma(p)}} ,$$

$$(t_2\omega)_{j_1 \dots j_p} = \sum_{\sigma \in S_p} \mathrm{sgn}(\sigma) \sum_{i_o \dots i_r} \omega_{i_o i_1 i_o i_1^* i_2 i_2^* \dots i_r i_r^* j_{\sigma(1)}^* \dots j_{\sigma(s)}^* j_{\sigma(s+1)} \dots j_{\sigma(s+u)}} \times$$
$$\times \; \Omega_{j_{\sigma(s+u+1)} j_{\sigma(s+u+2)}} \dots \Omega_{j_{\sigma(p-1)} j_{\sigma(p)}} ,$$

$$(t_2\omega)_{j_1\cdots j_p} = \sum_{\sigma\in S_p} \mathrm{sgn}(\sigma) \sum_{i_0\cdots i_r} \omega_{j^*_{\sigma(1)} i_0 i_1 i^*_1 \cdots i_r i^*_r i_0 j^*_{\sigma(2)} \cdots j^*_{\sigma(s)} j_{\sigma(s+1)} \cdots j_{\sigma(s+u)}} \times$$

$$\times \; \Omega_{j_{\sigma(s+u+1)} j_{\sigma(s+u+2)}} \cdots \Omega_{j_{\sigma(p-1)} j_{\sigma(p)}} ,$$

$$(t_2\omega)_{j_1\cdots j_p} = \sum_{\sigma\in S_p} \mathrm{sgn}(\sigma) \sum_{i_0\cdots i_r} \omega_{j_{\sigma(s+1)} i_0 i_1 i^*_1 \cdots i_r i^*_r j^*_{\sigma(1)} \cdots j^*_{\sigma(s)} i_0 j_{\sigma(s+2)} \cdots j_{\sigma(s+u)}} \times$$

$$\times \; \Omega_{j_{\sigma(s+u+1)} j_{\sigma(s+u+2)}} \cdots \Omega_{j_{\sigma(p-1)} j_{\sigma(p)}} .$$

<u>5.3.PROPOSITION:</u> The space of tensors $t_1 \in \Gamma(TM \otimes \Lambda^p TM \otimes \Lambda^p T^*M)$, appearing in (5.1) is spanned by those tensors whose action on an $\omega \in \Lambda(T^*M \otimes \Lambda^p T^*M)$ is given by one of the expressions in Proposition 3.5, with the following slight modifications:

(a) $\sigma \in S_p$, and

(b) we can also take the contraction in the first two indices of ω in all the expressions.

In general, the expressions of the generators of the space of tensors $t_0 \in \Gamma(\Lambda^p TM \otimes \Lambda^p T^*M)$ are very complicated, and we shall not write them here. We shall consider only the case $p = 0$.

<u>5.4.PROPOSITION:</u> The space of tensors $t_0 \in \Gamma(\Lambda^0 TM \otimes \Lambda^0 T^*M)$ in (5.1) is spanned by the hermitian natural functions

$$\tau,\ \tau^*,\ \|\nabla\Omega\|^2,\ \|d\Omega\|^2,\ \|\delta\Omega\|^2,\ \|N\|^2,$$

where N is the Nijenhuis tensor, τ is the scalar curvature and τ^* is, as usually, defined by $\tau^* = (1/2)\sum_{i,j=1}^{2n} R_{ii^*jj^*}$.

<u>Proof:</u> Since $w(t_0) = -2$, we have , from 2.6, $q = 0 = \eta_j$ and then, either there are indices i,k such that $\varepsilon_i = \varepsilon_k = 1$ and $\varepsilon_l = 0$ for $l \neq i,k$, or there is an index i such that $\varepsilon_i = 2$ and $\varepsilon_l = 0$ for every $l \neq i$; then the result follows from [G-H, Theorem 7.1].

From 5.1 and 5.4 we have

<u>5.5.COROLLARY:</u> Every homogeneous regular HNDO $\mathcal{D}$ of maximal weight acting on functions is of the form

$$\mathcal{D}f = a\,\Delta f + b\,\delta J(f) + c\,(J\delta J)(f) + (d_1\,\tau + d_2\,\tau^* + d_3\,\|\nabla\Omega\|^2 + d_4\,\|d\Omega\|^2 + d_5\,\|\delta\Omega\|^2 + d_6\,\|N\|^2)\,f,$$

where a, b, c, d_i are real numbers and Δ is the ordinary laplacian.

Proof: When p = 0 tensors t_2 in 5.2 reduce to $t_2(\omega) = c_{11}\omega$, and the tensors t_1 in 5.3 reduce to $t_1(\omega) = \omega(\delta J)$ or $t'_1(\omega) = \omega(J\,\delta J)$. Then, taking $D = \nabla$, we get $(t_2 \cdot D_2)(f) = c_{11}(\nabla^2 f) = g^{ij}(\nabla^2_{ij} f) = \Delta f$, and $t_1(\nabla f) = df(\delta J)$ and $t'_1(\nabla f) = df(J\,\delta J)$, and the result follows from here and proposition 5.4.

5.6. For compact manifolds we consider the scalar product on $\Gamma(\Lambda T^*M)$ given by

$$(\alpha,\beta) = \int_M \alpha \wedge *\beta \,.$$

For this scalar product we have that if D is a HNDO as in 5.5 then D is selfadjoint if and only if b = c = 0 (see [McK-S], p.46).

An interesting homogeneous regular HNDO acting on p-forms is the D-laplacian, defined by

$$\Delta^D = d^D d^{D*} + d^{D*} d^D \,,$$

where D is a metric homogeneous regular hermitian natural connection on TM, and d^{D*} is the adjoint of d^D with respect to the above scalar product. It is clearly elliptic and selfadjoint, and we shall study its spectral asymptotic expansion in §6.

§6. The asymptotic expansion of Δ^D acting on 1-forms.

Within this section M will be a compact almost-hermitian manifold of real dimension m = 2n. First, we recall some well known facts.

Let E be a vector bundle over M, and $L : \Gamma(E) \longrightarrow \Gamma(E)$ a second order differential operator with symbol given by the metric tensor. Let E_x be the fibre of E over a point $x \in M$. Let us choose a smoth fibre metric $\langle\ ,\ \rangle$ on E, and let $L^2(E)$ be the completion of $\Gamma(E)$ with respect to the global integrated inner product (,). For t > 0, $\exp(-tL): L^2(E) \longrightarrow L^2(E)$ is an infinitely smoothing operator of trace class. Let $K(t,x,y,L) : E_y \longrightarrow E_x$ be the kernel of $\exp(-tL)$. If x=y, K has an asymptotic expansion as $t \to 0^+$, of the form

$$K(t,x,x,L) \sim (4\pi t)^{-m/2} \Sigma_{k=0}^{\infty} t_k H_k(x,L),$$

where the $H_k(x,L)$ are endomorphisms of E_x.

If L is selfadjoint, let $\{\lambda_i, \theta_i\}_{i \in \mathbb{Z}^+}$ be a spectral resolution of L into a complete orthonormal basis of eigenvalues λ_i and eigensections θ_i. Then,

$$\mathrm{tr}\, K(t,x,x,\mathcal{L}) = \Sigma_i \exp(-t\,\lambda_i) \langle \theta_i, \theta_i \rangle_x \sim (4\pi\, t)^{-m/2} \Sigma_{k=0}\, a_k(x,\mathcal{L}) t^k,$$

where $a_k(x,\mathcal{L}) = \mathrm{tr}\, H_k(x,\mathcal{L})$. Now, if we integrate on M, we have

$$\Sigma_i \exp(-t\,\lambda_i) \sim (4\pi\, t)^{-m/2} \Sigma_{k=0}^{\infty} a_k(\mathcal{L})\, t^k, \qquad \text{with } a_k(\mathcal{L}) = \int_M a_k(x,\mathcal{L}).$$

6.1.THEOREM ([Gi 3.4]): Let $\underline{D}$ be a connection on E, and ∇ the Levi-Civita connection, and denote also by $\underline{D}$ the connection induced by $\underline{D}$ and ∇ on $T^*M \otimes E$. Let $\mathcal{L}_{\underline{D}}$ be the reduced Laplacian defined by $\mathcal{L}_{\underline{D}} s = - g^{ij} \underline{D}^2_{ij} s$ for every $s \in \Gamma(E)$. If $\underline{D}$ is the unique connection on E such that $\mathcal{E} = \mathcal{L}_{\underline{D}} - \mathcal{L} : \Gamma(E) \longrightarrow \Gamma(E)$ is a 0th order operator, then we have

(a) $H_0 = I$

(b) $H_1 = (1/6)\,(-\tau\, I + 6\, \mathcal{E})$.

Now, let E, F be functors as in 2.1, and let $\mathcal{L}$ be a homogeneous regular HNDO of type (E,F,E,F) of order two with symbol given by the metric tensor, then it has maximal weight -2, and, according with theorem 2.10, it can be written in the form

$$\mathcal{L} = - g^{ij}\, \nabla^2_{ij} + t_1 \cdot \nabla + t_0\,,$$

where ∇ is the connection induced on $EM \otimes F^*M$ by the Levi-Civita connection. Next, we compute $\mathcal{E}$ in terms of t_0 and the tensor $\underline{B} = \underline{D} - \nabla : TM \otimes EM \otimes F^*M \longrightarrow EM \otimes F^*M$.

6.2.PROPOSITION: For every $s \in \Gamma(EM \otimes F^*M)$,

$$\mathcal{E}\, s = - g^{ij}\, (\nabla_i(\underline{B})_j s + \underline{B}_i \underline{B}_j s) - t_0 s \tag{6.2}$$

and the connection $\underline{D}$ on $EM \otimes F^*M$ such that $\mathcal{E}$ is a 0th order operator is given by the linear fibre bundle map $\underline{B} = \underline{D} - \nabla$, defined by

$$\underline{B}_X s = -(1/2)\, t_1\, (X^b \otimes s) \tag{6.3}$$

for every $X \in \Gamma(TM)$ and $s \in \Gamma(EM \otimes F^*M)$, where ^{b}: $TM \longrightarrow T^*M$ is the canonical isomorphism induced by the metric.

Proof: From the definition of $\mathcal{L}_{\underline{D}}$ we have

$$\mathcal{L}_{\underline{D}}\, s = - g^{ij}\, (\underline{D}_i \underline{D}_j s - \underline{D}_{\nabla_i j} s) = - g^{ij}\, (\nabla^2_{ij} s) - 2\, g^{ij}\, \underline{B}_i(\nabla_j s) - g^{ij}\, (\nabla_i(\underline{B})_j s + \underline{B}_i \underline{B}_j s).$$

Since, in its standard form, $\mathcal{L}_{\underline{D}}$ can be written as

$$\mathcal{L}_{\underline{D}} = - g^{ij}\, \nabla^2_{ij} + t^{\underline{D}}_1 \cdot \nabla + t^{\underline{D}}_0,$$

we have that

$$t^{\underline{D}}_1(\alpha \otimes s) = -2g^{ij}\,\underline{B}_i(\alpha_j\, s) = -2\,\underline{B}_{\alpha^*}s$$

for $s \in \Gamma(EM \otimes F^*M)$ and $\alpha \in \Gamma(T^*M)$. On the other hand, $\gamma_2(\mathcal{L}) = \gamma_2(\mathcal{L}_{\underline{D}})$, so that the condition that E be a 0th order operator is equivalent (by Theorem 2.10) to $t^{\underline{D}}_1 = t_1$; whence (6.3) follows. Then, we have $E = t^{\underline{D}}_0 - t_0 = -g^{ij}(\nabla_i(\underline{B})_j s + \underline{B}_i\underline{B}_j s) - t_0(s)$.

Next we study the spectral asymptotic expansion of the operator Δ^D, defined in 5.6. First we determine the operator d^{D*}.

<u>6.3.PROPOSITION:</u> The adjoint operator of d^D, with respect to the inner product given in 5.6, is $d^{D*} = \delta^D - \iota_{\bar{B}}$, where $\bar{B} = \Sigma_{i=1}{}^{2n} B_i i$.

<u>Proof:</u> From the fact that d^D is a skew-derivation and that $\delta^D = (-1)^{np+n+1} * d^D *$ on p-forms, it follows that, if $\alpha \in \Gamma(\Lambda^p(T^*M))$, then

$$\int_M d^D\alpha \wedge *\beta = \int_M d^D(\alpha \wedge *\beta) + \int_M \alpha \wedge *\delta^D\beta.$$

On the other hand, we have

$$d^D(\alpha \wedge *\beta) = d(\alpha \wedge *\beta) + \bar{B}^b \wedge (\alpha \wedge *\beta),$$

and, for $X \in \chi(M)$ and $\mu \in \Gamma(\Lambda^r(T^*M))$,

$$*\iota_X\mu = (-1)^{2n-1}(*\mu) \wedge X^b = (-1)^{r+1} X^b \wedge *\mu,$$

whence,

$$\bar{B}^b \wedge \alpha \wedge *\beta = (-1)^p \alpha \wedge \bar{B}^b \wedge *\beta = \alpha \wedge (\iota_{\bar{B}}\beta).$$

Then,

$$d^D(\alpha \wedge *\beta) = d(\alpha \wedge *\beta) - \alpha \wedge *\iota_{\bar{B}}\beta,$$

and, if we integrate,

$$\int_M d^D\alpha \wedge *\beta = -\int_M \alpha \wedge \iota_{\bar{B}}\beta + \int_M \alpha\, *\delta^D\beta = \int_M \alpha \wedge *(\delta^D - \iota_{\bar{B}})\beta.$$

Then, $d^{D*} = \delta^D - \iota_{\bar{B}}$.

Let $\Delta^D{}_p$ the D-laplacian Δ^D acting on p-forms (in particular, $\Delta^D{}_0 = \Delta$, the ordinary real laplacian). Then,

<u>6.4.THEOREM:</u> Let M be a compact almost hermitian manifold of real dimension 2n. Then,

$$a_1(\Delta^D{}_1) = \int_M (-((n+3)/3)\,\tau + A),$$

where $A = -(1/2)\ \Sigma\, T_{ijk}\, T_{ikj}$, T being the torsion tensor of D.

Proof: It follows in the same way as for the ordinary laplacian (see, for instance, [P]), that the action of Δ^D on a 1-form μ is given by

$$(6.4)\qquad (\Delta^D\mu)(v) = -(g^{ij} D^2{}_{ij}\,\mu)(v) - g^{ij}(R^D{}_{iv}\,\mu)_j - g^{ij}(D_{T(i,v)}\,\mu)_j - $$
$$- (D_{\bar{B}}\,\mu)(v) + (D_v\,\mu)(\bar{B}) - (\nabla_v\,\mu)(\bar{B}) + \mu(\nabla_v\bar{B}),$$

where v is a vector field on M. On the other hand, if we write $D_X\,\mu = \nabla_X\,\mu + \tilde{B}_X\,\mu$, then, $\tilde{B}$ is related with B by

$$(6.5)\qquad \tilde{B}_X\,\mu = -\,\mu\,(B(X,\bullet)).$$

(Notice that $\tilde{B}$ is analogous to the tensor $\underline{B}$ defined before 6.2, however we change the notation because we use $\tilde{B}$ for the connection D on T*M induced by the connectio n D on TM, and it is different of the connection $\underline{D}$ given in 6.1).

Now, for each point $x \in M$ we can consider a local frame $\{e_i\}$ around x which is orthonormal at x and radially parallel from x, so that $(\nabla_i(e_j))_x = 0$. Then it follows from (6.4) that, at x,

$$(6.6)\qquad \Delta^D\,\mu = -\,\nabla^2{}_{ii}\,\mu - (\nabla_i\tilde{B})_i\,\mu - 2\,\tilde{B}_i(\nabla_i\,\mu) - \tilde{B}_i(\tilde{B}_i\,\mu) + \mu(R^D{}_{i\bullet}i)$$
$$-\,(\nabla_{T(i,\bullet)}\mu)i - (\tilde{B}_{T(i,\bullet)}\,\mu)i - \mu(\nabla_\bullet\bar{B}) + (\tilde{B}_\bullet\,\mu)(\bar{B}).$$

If we write Δ^D in its canonical form (given by Theorem 2.10),

$$\Delta^D = -\,\nabla^2{}_{ii} + t_1 \circ \nabla + t_0,$$

we have

$$(6.7)\qquad t_1(\omega \otimes \mu) = -\,2\,\tilde{B}_i(\omega_i\mu) - \omega_{T(i,\bullet)}\,\mu_i$$

for every 1-form ω, and

$$(6.8)\qquad t_0\,\mu = -\,(\nabla_i\tilde{B})_i\,\mu - \tilde{B}_i(\tilde{B}_i\,\mu) + \mu(R^D{}_{i\bullet}i) - (\tilde{B}_{T(i,\bullet)}\,\mu)_i - \mu(\nabla_\bullet\bar{B}) + (\tilde{B}_\bullet\mu)\bar{B}.$$

Then, from (6.3),

$$(6.9)\qquad \underline{B}_X\,\mu = -\,(1/2)\,t_1\,(X^b \otimes \mu) = -\,\mu(B_X\bullet) + (1/2)\,T_{\mu^\#\bullet X}$$

Since our frame is parallel,

$$(6.10)\qquad -(\nabla_i \underline{B}_i)\,\mu = \mu((\nabla_i B)_i \bullet) - (1/2)\,\nabla_i (T)_{\mu^* \bullet i},$$

and

$$(6.11)\qquad -\underline{B}_i(\underline{B}_i\,\mu) = -\mu(B_i(B_i \bullet)) + (1/2)\,T_{(\mu(B(i,\bullet))^* \bullet i} + (1/2)\,T_{\mu^* B(i,\bullet)\, i} -$$

$$-(1/4)\,T_{(\langle T(\mu^*,\bullet),\, i\rangle)^* \bullet i} =$$

$$= -B_{ij\mu^*} B_{i\bullet j} + (1/2) B_{ij\mu^*} T_{j\bullet i} + (1/2) T_{\mu^* ji} B_{i\bullet j} - (1/4) T_{j\bullet i} T_{\mu^* ji}.$$

Now, from (6.10), (6.11) and (6.2) we get the trace of E

$$-\mathrm{tr}\, E = -\tau^D + (1/4)\,\|T\|^2 - \|\bar{B}\|^2 - (\nabla_k \bar{B})_k,$$

but from (4.3) in [M],

$$\tau^D = \tau^\nabla + 2\,(\nabla_k \bar{B})_k + B_{ijk} B_{jik} - \|\bar{B}\|^2,$$

and then, denoting τ^∇ by τ,

$$\mathrm{tr}\, E = -\tau + \delta \bar{B}^b - B_{ijk} B_{jik} + (1/4)\,\|T\|^2.$$

Let $A = -B_{ijk} B_{jik} + (1/4)\,\|T\|^2$. Then, since $B_{ijk} = (1/2)(T_{ijk} + T_{kij} - T_{jki})$, we have $A = -(1/2)\,T_{ijk} T_{ikj}$.

Now, from 6.1(b) and the above expressions, we have

$$a_1(\Delta^D{}_1) = \int_M (1/6)(-2n\,\tau - 6\,\tau + 6\,\delta\bar{B}^b + 6\,A) = \int_M (-((n+3)/3)\,\tau + A),$$

since $\int_M \delta \bar{B}^b = 0$.

<u>6.5.COROLLARY:</u> Let D be the characteristic connection, and let $Spc^p(M)$ the spectrum of Δ^D acting on p-forms. Let M_1 be a Kaehler manifold, and let M_2 be a nearly Kaehler or an almost Kaehler manifold. If $Spc^p(M_1) = Spc^p(M_2)$ for $p = 0,1$ then M_2 is a Kaehler manifold.

<u>Proof:</u> First, observe that, when D is the characteristic connection, we have (see [M])

$$A = (1/2)\,\|T_1\|^2 - (1/4)\,\|T_2\|^2,$$

where T_1 and T_2 are tensor fields on M satisfying that T_1 vanishes on almost Kaehler manifolds and T_2 vanishes on nearly Kaehler manifolds. On the other hand, since $\Delta^D{}_0 = \Delta_0$,

$$a_1(\Delta^D{}_0) = \int_M (1/6)(-2n\,\tau) = -(n/3)\int_M \tau.$$

Then,

$$\int_M A \;=\; a_1(\Delta^D{}_1) - ((n+3)/3)\, a_1(\Delta^D{}_0)\,.$$

Now, if $Spc^p(M_1) = Spc^p(M_2)$ for $p = 0,1$, then, we have

$$a_1(\Delta^D{}_1)(M_1) = a_1(\Delta^D{}_1)(M_2), \qquad\qquad a_1(\Delta^D{}_0)(M_1) = a_1(\Delta^D{}_0)(M_2),$$

but, if M_1 is Kaehler, $\int_{M_1} A = 0$, so that

$$a_1(\Delta^D{}_1)(M_1) - ((n+3)/3)\, a_1(\Delta^D{}_0)(M_1) = 0,$$

and then,

$$\int_{M_2} A = a_1(\Delta^D{}_1)(M_2) - ((n+3)/3)\, a_1(\Delta^D{}_0)(M_2) =$$

$$a_1(\Delta^D{}_1)(M_1) - ((n+3)/3)\, a_1(\Delta^D{}_0)(M_1) = 0.$$

If M_2 is almost Kaehler, then $T_1 = 0$, $A = -(1/4)\,\|T_2\|^2$ and $\int_{M_2} A = 0$, whence $\|T_2\|^2 = 0$, which implies that M_2 is Kaehler. Similarly, if M_2 is nearly Kaehler, then $T_2 = 0$, $A = -(1/2)\,\|T_1\|^2$ and $\int_{M_2} A = 0$, whence $\|T_1\| = 0$ and M_2 is Kaehler.

TABLE I

	p	t_1
1	$p \geq 0$	$t_1\omega = \mathrm{Alt}(\omega)$
2	$p \geq 0$	$t_1\omega = \mathrm{Alt}(J\omega)$
3	$p \geq 1$	$t_1\omega = (c_J^p c_{11}\omega)\ \Omega\wedge\overset{(p+1)/2}{\dots\ \dots}\wedge\Omega$
4	$p \geq 2$	$t_1\omega = \mathrm{Alt}(J^m\omega)$
5	$p \geq 2$	$t_1\omega = \mathrm{Alt}(J_1^m\omega)$
6	$p \geq 2$	$t_1\omega = \mathrm{Alt}(c_J^m\omega)\wedge\Omega\wedge\overset{(p+1-m)/2}{\dots\ \dots\ \dots}\wedge\Omega$
7	$p \geq 2$	$t_1\omega = \mathrm{Alt}(Jc_J^m\omega)\wedge\ \Omega\wedge\dots\ \dots\wedge\Omega$
8	$p \geq 2$	$t_1\omega = \mathrm{Alt}(J^m c_J^n\omega\otimes\Omega\otimes\dots\ \dots\otimes\Omega)$
9	$p \geq 2$	$t_1\omega = \mathrm{Alt}(J_1^m c_J^n\omega\otimes\Omega\otimes\dots\ \dots\otimes\Omega)$
10	$p \geq 1$	$t_1\omega = (c_{11}\omega)\wedge\Omega$
11	$p \geq 2$	$t_1\omega = (Jc_{11}\omega)\wedge\Omega$
12	$p \geq 3$	$t_1\omega = \mathrm{Alt}(J^m c_{11}\omega\otimes\Omega)$
13	$p \geq 4$	$t_1\omega = (c_J^m c_{11}\omega)\wedge\Omega\wedge\overset{(p-m)/2}{\dots\ \dots}\wedge\Omega$
14	$p \geq 4$	$t_1\omega = (Jc_J^m c_{11}\omega)\wedge\Omega\wedge\overset{(p-m)/2}{\dots\ \dots}\wedge\Omega$
15	$p \geq 5$	$t_1\omega = \mathrm{Alt}(J^n c_J^m c_{11}\omega\otimes\Omega\otimes\overset{(p-m)/2}{\dots\ \dots}\otimes\Omega$

TABLE II

	$t_1 \circ D$
1'	$(t_1 \circ D)\omega = d^D\omega$
2'	$(t_1 \circ D)\omega = Jd^D\omega$
3'	$(t_1 \circ D)\omega = (c_J^p\delta^D\omega)\quad\Omega^{(p+1)/2}$
6'	$(t_1 \circ D)\omega = \mathrm{Alt}(c_J^m D\omega)\wedge\Omega^{(p+1-m)/2}$
7'	$(t_1 \circ D)\omega = \mathrm{Alt}(Jc_J^m D\omega)\wedge\Omega^{(p+1-m)/2}$
10'	$(t_1 \circ D)\omega = \delta^D\omega\wedge\Omega$
11'	$(t_1 \circ D)\omega = J\delta^D\omega\wedge\Omega$
13'	$(t_1 \circ D)\omega = c_J^m\delta^D\omega\wedge\Omega^{(p-m)/2}$
14'	$(t_1 \circ D)\omega = Jc_J^m\delta^D\omega\wedge\Omega^{(p-m)/2}$

TABLE III

$t_0 \quad (p = 1)$
$t_0\omega = \nabla_{\omega}\#\Omega$
$t_0\omega = \omega(Alt_2(\nabla J))$
$t_0\omega = \omega \wedge \delta\Omega$
$t_0\omega = J\omega \wedge \delta\Omega$
$t_0\omega = \omega(\delta J)\Omega$
$t_0\omega = \nabla_{J\omega}\#\Omega$
$t_0 = J^1\nabla_{\omega}\#\Omega$

$t_0 \quad (p = 1)$
$t_0\omega = J^1\nabla_{J\omega}\#\Omega$
$t_0\omega = \omega(Alt(J_1\nabla J))$
$t_0\omega = J\omega(Alt(\nabla J))$
$t_0\omega = (J\omega)(Alt(J_1\nabla J))$
$t_0\omega = \omega \wedge J\delta\Omega$
$t_0\omega = J\omega \wedge J\delta\Omega$
$t_0\omega = \omega(J\delta J)\Omega$

$t_0 \quad (p = 0)$
$t_0 f = f\delta\Omega$

$t_0 \quad (p = 0)$
$t_0 f = fJ\delta\Omega$

TABLE IV

$t_0 \quad (p = 2)$
$t_0\omega = \omega(Alt^2\nabla J)$
$t_0\omega = \omega(\nabla\ \Omega^{\cdot\cdot})$
$t_0\omega = (c^1_J\omega)\delta\Omega$
$t_0 = c^1_1(\omega \otimes \delta J)$
$t_0 = c^1_1(J^1\omega \otimes \delta J)$
$t_0 = J^1\omega(Alt^2(\nabla J))$
$t_0 = J\omega\, Alt^2(\nabla J)$

$t_0 \quad (p = 2)$
$t_0\omega = J^1\omega(\nabla\Omega^{\cdot\cdot})$
$t_0\omega = \omega(J^1\nabla\Omega^{\cdot\cdot})$
$t_0\omega = J^1\omega(J^1\nabla\Omega^{\cdot\cdot})$
$t_0\omega = (c^1_J\omega)J\delta\Omega$
$t_0\omega = c^1_1(\omega \otimes J\delta J)$
$t_0\omega = c^1_1(J^1\omega \otimes J\delta J)$

$t_0 \quad (p = 1)$
$t_0 = \omega(\delta J)$

$t_0 \quad (p = 1)$
$t_0 = \omega(J\delta J)$

REFERENCES

[A-B-P] M.F. Atiyah, R. Bott and K. Patodi, "On the heat equation and the index theorem" Invent. Math. 19 (1973), 279-330.

[C-F-G] L.A. Cordero, M. Fernández and A. Gray, "Symplectic manifolds with no Kaehler structure", to appear in Topology.

[D 1] H. Donnelly, "Invariance theory of Hermitian manifolds" Proc. Amer. Math. Soc. 58 (1976), 229-233.

[D 2] H. Donnelly, "A spectral condition determining the Kaehler property", Proc. Amer. Math. Soc. 47 (1975), 187-195.

[D 3] H. Donnelly, "Heat Equation Asymptotics with Torsion", Indiana Univ. Math. J. Vol. 34 (1985), 105-113.

[E] D.B.A. Epstein, "Natural tensors on Riemannian manifolds", J. Diff. Geom. 10 (1975), 631-635.

[E-K] D.B.A. Epstein and M. Kneser, "Functors between categories of vector spaces", Lecture Notes in Math. Vol. 99, (1969), 154-170.

[F-M] A. Ferrández and V. Miquel, "Hermitian natural tensors", Preprint.

[G 1] P. Gilkey, "curvature and the eigenvalues of the Laplacian for elliptic complexes", Advances in Math. 10 (1973), 344-382.

[G 2] P. Gilkey, "Spectral geometry and the Kaehler condition for complex manifolds", Invent. Math. 26 (1974), 231-258, and "Corrections" , Invent. Math. 29 (1975), 81-82.

[G3] P. Gilkey, "The spectral geometry of real and complex manifolds", Proc. of Sympos. in Pure Math. Vol. 27 (1975), 265-280.

[G 4] P. Gilkey, "Recursion relations and the asymptotic behavior of the eigenvalues of the Laplacian", Compositio Math. 38 (1979), 201-240.

[GR 1] A. Gray, "Nearly Kahler manifolds", J. Diff. Geom. 4 (1970), 283-309.

[GR 2] A. Gray, "The structure of Nearly Kaehler manifolds", Math. Ann. 248 (1976), 233-248.

[G-H] A. Gray and L. M. Hervella, "The sixteen classes of almost hermitian manifolds and their linear invariants", Annali di Mat. pura ed applicata IV, vol. XXIII (1980), 35-58.

[McK-S] H. P. McKean, Jr and I. M. Singer, "Curvature and eigenvalues of the Laplacian", J. Diff. Geom. 1 (1967), 43-69.

[M] V. Miquel, "Volumes of certain small geodesic balls and almost hermitian geometry", Geometriae Dedicata 15 (1984), 261-267.

[P] W. A. Poor, "Differential Geometric Structures", Mc Graw-Hill, Inc., 1981.

[S] P. Stredder, "Natural differential operators on Riemannian manifolds and representations of the orthogonal and special orthogonal groups", J. Diff. Geom. 10 (1975), 647-660.

[T] K. Tsukada, "Hopf manifolds and spectral geometry", Trans. Amer. Math. Soc. 270 (1982), 609-621.

AN EXAMPLE OF AN ALMOST COSYMPLECTIC HOMOGENEOUS MANIFOLD

D. Chinea and C. González
Departamento de Geometría y Topología
Facultad de Matemáticas
Universidad de La Laguna
Islas Canarias. Spain

As is well-known, E. Cartan proved that a connected, complete and simply connected Riemannian manifold is a symmetric space if and only if the curvature is constant under parallel translation. In [1], Ambrose and Singer extended this theory and gave a characterization of homogeneous Riemannian manifolds through a tensor field T of type (1,2) satisfying certain conditions (see §1). K. Sekigawa [9] extended this result and characterized the homogeneous almost Hermitian manifolds.

On the other hand, F. Tricerri and L. Vanhecke [10] gave a classification for the homogeneous Riemannian spaces into eight different classes by properties the T's.

In this paper we characterize the almost contact metric homogeneous manifolds (i.e. almost contact metric manifolds with transitive almost contact isometry groups) and we construct a parametrized family of almost cosymplectic homogeneous manifolds $(H(1,r),g)$, $r>1$, where $H(1,r)$ is a generalized Heisenberg group and g a left invariant metric. Also, we study and classify the homogeneous structures T on these groups.

In §1 we give some results on almost contact metric manifolds and homogeneous structures on Riemannian manifolds.

In §2 we study the almost contact metric manifolds (M,ϕ,ξ,η,g) with transitive isometry group G and such that ϕ is G-invariant.

In §3 we describe the almost cosymplectic structure on the generalized Heisenberg group $H(1,r)$ in the same way as in [3] for $M(1,r)\equiv H(1,r)/\Gamma$.

In §4 we give a detailed study of the homogeneous structures on $(H(1,r),g)$, g being a left invariant metric, and show that is does not admit any homogeneous structure T of type $\mathcal{T}_1\oplus\mathcal{T}_2$ or $\mathcal{T}_1\oplus\mathcal{T}_3$. Finally, we obtain a one parameter family of almost contact homogeneous structures (T_λ,ϕ), with T_λ of type $\mathcal{T}_2\oplus\mathcal{T}_3$. Moreover, we show that in $H(1,2)$ all the homogeneous structures are given by T_λ.

1. PRELIMINARIES

A (2n+1)-dimensional real differentiable manifold M of class C^∞ is said to have a (ϕ,ξ,η)-structure or an almost contact structure if it admits a field ϕ of endomorphisms of the tangent spaces, a vector field ξ, and a 1-form η satisfying

$$(1.1) \qquad \eta(\xi) = 1,$$

$$(1.2) \qquad \phi^2 = -I + \eta\otimes\xi,$$

where I denotes the identity transformation, [2].
Denote by $\chi(M)$ the Lie algebra of C^∞-vector fields on M. Such a (paracompact) manifold M with a (ϕ,ξ,η)-structure admits a Riemannian metric g such that

$$(1.3) \qquad g(\phi X,\phi Y) = g(X,Y) - \eta(X)\eta(Y),$$

where $X,Y\varepsilon\chi(M)$. Then M is said to have a (ϕ,ξ,η,g)-structure or an almost contact metric structure and g is called a compatible metric.
The 2-form Φ on M defined by

$$(1.4) \qquad \Phi(X,Y) = g(X,\phi Y)$$

is called the fundamental 2-form of the almost contact metric structure.
If ∇ is the Riemannian connection of g, then

$$(1.5) \qquad (\nabla_X\eta)Y = g(Y,\nabla_X\xi),$$

$$(1.6) \qquad (\nabla_X\Phi)(Y,Z) = g(Y,(\nabla_X\phi)Z).$$

An almost contact metric structure (ϕ,ξ,η,g) is said to be

(I) normal if $(\nabla_X\phi)Y - (\nabla_{\phi X}\phi)\phi Y + \eta(Y)\nabla_{\phi X}\xi = 0,$

(II) almost cosymplectic if $d\Phi = d\eta = 0,$

(III) cosymplectic if it is (I) and (II), (or $\nabla\phi=0$).

A connected Riemannian manifold (M,g) is said to be homogeneous if there exists a connected Lie group G which acts on (M,g) as a transitive and effective group of isometries.
Ambrose and Singer [1] proved that a connected, complete and simply connected Riemannian manifold (M,g) is homogeneous if and only if there exists a tensor field T of type (1,2) such that

$$(AS)\begin{cases} \text{(i)} & g(T_XY,Z) + g(Y,T_XZ) = 0, \\ \text{(ii)} & (\nabla_XR)_{YZ} = [T_X,R_{YZ}] - R_{T_XYZ} - R_{YT_XZ}, \end{cases}$$

(iii) $(\nabla_X T)_Y = [T_X, T_Y] - T_{T_X Y}$

for $X,Y,Z \in \chi(M)$. Here ∇ denotes the Levi Civita connection and R is the Riemann curvature tensor of M. These conditions are equivalent to

(i) $\tilde{\nabla} g = 0,$

(ii) $\tilde{\nabla} R = 0,$

(iii) $\tilde{\nabla} T = 0,$

where $\tilde{\nabla}$ is the connection determined by

$$\tilde{\nabla} = \nabla - T.$$

A homogeneous (Riemannian) structure on (M,g) is a tensor field T of type (1,2) which is a solution of the system (AS).

F. Tricerri and L. Vanhecke obtained in [10] a classification of the homogeneous structures in eight different classes. These are:

(1) symmetric if $T=0$,

(2) $\mathcal{T}_1$ if $T_{XYZ} = g(X,Y)\psi(Z) - g(X,Z)\psi(Y), \quad \psi \in \Lambda^1(M),$

(3) $\mathcal{T}_2$ if $\underset{X,Y,Z}{\mathfrak{S}} T_{XYZ} = 0$ and $C_{12}(T) = 0,$

(4) $\mathcal{T}_3$ if $T_{XYZ} + T_{YXZ} = 0,$

(5) $\mathcal{T}_1 \oplus \mathcal{T}_2$ if $\underset{X,Y,Z}{\mathfrak{S}} T_{XYZ} = 0,$

(6) $\mathcal{T}_1 \oplus \mathcal{T}_3$ if $T_{XYZ} + T_{YXZ} = 2g(X,Y)\psi(Z) - g(X,Z)\psi(Y) - g(Y,Z)\psi(X)$ with $\psi \in \Lambda^1(M),$

(7) $\mathcal{T}_2 \oplus \mathcal{T}_3$ if $C_{12}(T) = 0,$

(8) $\mathcal{T}_1 \oplus \mathcal{T}_2 \oplus \mathcal{T}_3$ no conditions.

2. ALMOST CONTACT HOMOGENEOUS MANIFOLDS

An almost contact metric manifold (M, ϕ, ξ, η, g) is said to be almost contact homogeneous if (M,g) is homogeneous and ϕ is invariant under the action of the group.

From the results obtained in [6] by Kiričenko on homogeneous Riemannian spaces with invariant tensor structure, we have

THEOREM 2.1 Let (M, ϕ, ξ, η, g) be an almost contact homogeneous manifold. Then, there exists a tensor field T of tupe (1,2) satisfying the conditions (AS), and furthermore

(iv) $\nabla_X \phi = T_X \phi - \phi T_X, \quad$ for all $X \in \chi(M)$.

Conversely, if a connected, simply connected, complete almost contact metric manifold (M, ϕ, ξ, η, g) admits a tensor field T of type (1,2) satisfying (i)-(iv), then (M,ϕ,ξ,η,g) is an almost contact homogeneous manifold.

We shall call (T,ϕ) an almost contact homogeneous structure on (M,ϕ,ξ,η,g) if it satisfies (i)-(iv) in theorem 2.1.

PROPOSITION 2.1 If (M,ϕ,ξ,η,g) is an almost contact homogeneous manifold then ξ, η and Φ are invariant under the action of the group.

Proof. By [6, theorem 1] it is sufficient to find a tensor field T of type (1,2) satisfying the conditions of (AS) and furthermore

$$(2.1) \qquad \nabla_X\xi = T_X\xi ,$$

$$(2.2) \qquad \nabla_X\eta = -\eta T_X ,$$

$$(2.3) \qquad (\nabla_X\Phi)(Y,Z) = -\Phi(T_XY,Z) - \Phi(Y,T_XZ).$$

Now, from theorem 2.1, there exists an almost contact homogeneous structure (T,ϕ) on M. Thus,

$$(\nabla_X\phi)\xi = -\phi T_X\xi ,$$

but,

$$(\nabla_X\phi)\xi = -\phi\nabla_X\xi$$

and

$$g(T_X\xi,\xi) = g(\nabla_X\xi,\xi) = 0,$$

so, we obtain (2.1).

(2.2) is deduced from (2.1) and (1.5). Finally, (2.3) is obtained from (1.4),(1.6) and (iv).

3. ALMOST COSYMPLECTIC STRUCTURE ON H(1,r)

The generalized Heisenberg group $H(1,r)$, $r>1$, is the Lie group of real matrices of the form

$$\begin{pmatrix} 1 & 0 \ldots 0 & a^{r+1} & a^{r+2} \\ 0 & 1 & & \\ \cdot & 0 \quad 1 & a^{r+1}_r & a^{r+2}_r \\ \cdot & 0 \quad 1 & & a^{r+2}_{r+1} \\ 0 & \ldots \quad 0 & & 1 \end{pmatrix}, \qquad a^j_i \in R.$$

This group is a connected and simply connected nilpotent group of dimension $2r+1$, (defined in [4]).

Denote by x_i, z, y_i, $1\leq i\leq r$, the coordinate functions on $H(1,r)$ defined by

$$x_i(A) = a^{r+1}_i , \quad z(A) = a^{r+2}_{r+1} , \quad y_i(A) = a^{r+2}_i , \quad \text{for any } A\in H(1,r).$$

If L_B is the left translation by an element $B\in H(1,r)$, we have

$$L_B^*(dx_i) = dx_i \ , \ L_B^*(dz) = dz, \quad L_B^*(dy_i) = dy_i + b_i^{r+1}dz,$$

$$L_B^*(dy_i - x_i dz) = dy_i - x_i dz,$$

where L_B^* is the induced map on the space of 1-forms on H(1,r). So,

$$\alpha_i = dx_i \ , \quad dz \ , \ \beta_i = dy_i - x_i dz$$

are linearly independent and invariant under the action of H(1,r).
Put,

$$\Phi = - \sum_{i=1}^{r} \alpha_i \wedge \beta_i \ ,$$

$$\eta = dz.$$

Then, Φ is a 2-form on M of maximal rank. Furthermore Φ and η are closed and $\eta \wedge \Phi^n$ is a volume element. In this way (Φ,η) is a cosymplectic structure on H(1,r), (in the sense of Libermann, [8]).
There are many almost contact metric structures on H(1,r) compatible with Φ and η. Next, we write down such an almost contact metric structure on H(1,r).
First we determine the dual fields of the left invariant forms$\{\alpha_i, \eta, \beta_i\}$. They are, respectively,

$$X_i = \frac{\partial}{\partial x_i} \ , \ \xi = \frac{\partial}{\partial z} + \sum_{i=1}^{r} x_i \frac{\partial}{\partial y_i} \ , \quad Y_i = \frac{\partial}{\partial y_i} \ ,$$

and they form an orthonormal frame field with respect to the left invariant metric on H(1,r) defined by

$$g = \sum_{i=1}^{r} (\alpha_i^2 + \beta_i^2) + \eta^2 \ .$$

From $\Phi(X,Y) = g(X,\phi Y)$, it turns out that the tensor field ϕ, defined by

$$\phi X_i = Y_i \ , \ \phi Y_i = -X_i \ , \ \phi\xi = 0,$$

togethed with ξ and η form an almost contact structure compatible with the metric g. Because Φ and η are closed, it follows that H(1,r) is an almost cosymplectic manifold.
The Riemannian connection ∇ of g is given by

$$\nabla_{X_i} Y_i = \nabla_{Y_i} X_i = - \frac{1}{2}\xi \ ,$$

$$\nabla_{X_i} \xi = - \nabla_\xi X_i = \frac{1}{2} Y_i \ ,$$

$$\nabla_{Y_i} \xi = \nabla_\xi Y_i = \frac{1}{2} X_i \ ,$$

the remainder derivatives being zero. Since $\nabla_{X_i} \xi \neq 0$, it follows that the almost contact metric structure is not cosymplectic.

4. HOMOGENEOUS STRUCTURES ON H(1,r)

In this section we give a detailed study of the homogeneous space $(H(1,r), g)$.
First we start with the calculation of the connection forms and the curvature.
The connection forms are given by

$$w_{x_i\xi}(X)=g(\nabla_X X_i,\xi)=-\frac{1}{2}\beta_i(X);\ w_{x_iy_i}=-\frac{1}{2}\eta\ ;\ w_{y_i\xi}=-\frac{1}{2}\alpha_i\ ,$$

for any $X\varepsilon\chi(H(1,r))$, the remainder forms being zero. The curvature tensor is given by

$$R(X_i,X_j,Y_i,Y_j) = -\frac{1}{4}\ ,\quad i\neq j,$$
$$R(X_i,Y_j,X_j,Y_i) = \frac{1}{4}\ ,$$
$$R(X_i,\xi,X_i,\xi) = -\frac{3}{4}\ ,$$
$$R(Y_i,\xi,Y_i,\xi) = \frac{1}{4}\ ,$$

the remainder components being zero.
Next we determine the homogeneous structures on $(H(1,r),g)$.
Let T be a $(0,3)$-tensor field such that

$$T_{XYZ} + T_{XZY} = 0,$$

for all $X,Y,Z\varepsilon\chi(H(1,r))$. By the condition ii) of (AS), that is

$$(\nabla_X R)(Y,Z,W,V) = -R(T_X Y,Z,W,V) - R(Y,T_X Z,W,V) - R(Y,Z,T_X W,V) - R(Y,Z,W,T_X V),$$

and replacing (Y,Z,W,V) by (X_i,X_j,ξ,X_i), (X_i,X_j,ξ,Y_i), (X_i,X_j,X_i,Y_j) and (Y_i,ξ,ξ,X_j) we obtain, respectively,

$$T_{X\xi X_i} = \frac{1}{2}\beta_i(X),\qquad T_{X\xi Y_i} = \frac{1}{2}\alpha_i(X),$$
$$T_{XY_iX_i} = \frac{1}{2}\eta(X),\qquad T_{XX_iY_j} = 0.$$

Put,

$$T_{XX_iX_j} = a_{ij}(X)\quad\text{and}\quad T_{XY_iY_j} = b_{ij}(X).$$

Let $\tilde{\nabla}$ be the connection determined by

$$\tilde{\nabla} = \nabla - T.$$

Then the connection forms of $\tilde{\nabla}$ are given by

$$(4.1)\qquad \tilde{w}_{x_ix_j} = -a_{ij}\ ,\quad \tilde{w}_{y_iy_j} = -b_{ij}\ ,$$

the remainder forms being zero.
By the condition (iii) of (AS):

$$(\tilde{\nabla}_X T)(Y,Z,W) = 0,$$

and replacing Z,W by X_i, X_j and X_i, Y_j we obtain, respectively

$$(4.2)\quad (\tilde{\nabla}_X a_{ij})Y + \sum_{\substack{k=1\\ k\neq i,j}}^{r} (a_{ik}(X)a_{kj}(Y) + a_{jk}(X)a_{ik}(Y)) = 0,$$

$$(4.3)\quad a_{ij} = b_{ij} .$$

We conclude:

<u>THEOREM 4.1</u> All the homogeneous structures T on (H(1,r),g) are given by

$$2T = \sum_{\substack{i,j=1\\ i<j}}^{r} [\beta_i\otimes(\eta\wedge\alpha_i) + \alpha_i\otimes(\eta\wedge\beta_i) + \eta\otimes(\beta_i\wedge\alpha_i) + 2a_{ij}\otimes((\alpha_i\wedge\alpha_j)+(\beta_i\wedge\beta_j))]$$

where the 1-forms a_{ij} satisfy (4.2).

Next, we classify the homogeneous structures on (H(1,r),g) and we shall prove that is a homogeneous manifold of type $\mathcal{T}_2\oplus\mathcal{T}_3$ but it is not $\mathcal{T}_1\oplus\mathcal{T}_2$ nor $\mathcal{T}_1\oplus\mathcal{T}_3$.

<u>THEOREM 4.2</u> (H(1,r),g) does not admit any homogeneous structure T of type $\mathcal{T}_1\oplus\mathcal{T}_2$ or $\mathcal{T}_1\oplus\mathcal{T}_3$.

Proof. From theorem 4.1

$$\mathfrak{S}_{X_i,Y_i,\xi}\, T_{X_iY_i\xi} = -\tfrac{1}{2}.$$

Thus

$$\mathfrak{S}\, T_{XYZ} \neq 0.$$

On the other hand,

$$T_{X_iY_i\xi} + T_{Y_iX_i\xi} = -1.$$

So, T is not $\mathcal{T}_1\oplus\mathcal{T}_2$ nor is it $\mathcal{T}_1\oplus\mathcal{T}_3$.

<u>COROLLARY 4.1</u> (H(1,r),g) does not admit any homogeneous structure of type $\mathcal{T}_1$, $\mathcal{T}_2$ or $\mathcal{T}_3$ (or, equivalently, (H(1,r),g) is not naturally reductive).

<u>LEMMA 4.1</u> A homogeneous structure T on (H(1,r),g) is of type $\mathcal{T}_2\oplus\mathcal{T}_3$ if and only if

$$(4.4)\quad \sum_{\substack{i=1\\ i\neq j}}^{r} a_{ij}(X_i) = \sum_{\substack{i=1\\ i\neq j}}^{r} a_{ij}(Y_i) = 0, \qquad \text{for all } j,\ 1\leq j\leq r.$$

Proof. By definition

$$C_{12}(T)X = \sum_i T_{E_iE_iX}\ , \qquad X\in\chi(H(1,r)),$$

for an arbitrary orthonormal basis $\{E_i\}$ on H(1,r). Thus, using the orthonormal basis $\{X_i, Y_i, \xi\}$, we have

$$C_{12}(T)(X_j) = \sum_{\substack{i=1\\ i\neq j}}^{r} a_{ij}(X_i)\ ,\qquad C_{12}(T)(\xi) = 0\ ,\quad \text{and}$$

$$C_{12}(T)(Y_j) = \sum_{\substack{i=1\\ i\neq j}}^{r} a_{ij}(Y_i).$$

This proves the lemma.

It is possible to obtain examples of homogeneous structures on (H(1,r), g) satisfying (4.4). More specifically, let T_λ be as in theorem 4.1 with

(4.5) $a_{ij} = \lambda\eta \quad i<j$, and $a_{ij} = -\lambda\eta \quad i>j$.

Then, from (4.1) and (4.2) we have

$$\tilde{\nabla}_{X_i}\lambda = \tilde{\nabla}_{Y_i}\lambda = \tilde{\nabla}_{\xi}\lambda = 0.$$

Thus λ is constant.

Trivially, by lemma 4.1 this one-parameter family of homogeneous structures is of type $\mathcal{T}_2\oplus\mathcal{T}_3$. Furthermore, (T_λ, ϕ) is an almost contact homogeneous structure for all λ, where ϕ is the almost contact structure defined in §3.

So, we can conclude:

THEOREM 4.3 (H(1,r),g) is an almost cosymplectic homogeneous manifold of type $\mathcal{T}_2\oplus\mathcal{T}_3$.

Finally, we study the homogeneous structures on H(1,2) and we shall obtain that all the homogeneous structures are of type T_λ given in (4.5). First, from (4.2), the condition $\tilde{\nabla}_X T = 0$ is equivalent to $\tilde{\nabla}_X a_{12} = 0$. Putting,

$$a_{12} = \sum_{i=1}^{2}\{\delta_i\alpha_i + \mu_i\beta_i\} + \lambda\eta,$$

we obtain

$$(4.6)\qquad \begin{aligned} X(\delta_1) &= -\delta_2 a_{12}(X),\\ X(\delta_2) &= \delta_1 a_{12}(X),\\ X(\mu_1) &= -\mu_2 a_{12}(X),\\ X(\mu_2) &= \mu_1 a_{12}(X),\\ X(\lambda) &= 0.\end{aligned}$$

Hence λ is a constant. Moreover, from the definition of the bracket operation on $\chi(H(1,2))$ and (4.6) we have

$$\begin{aligned}\delta_i da_{12} &= 0,\\ \mu_i da_{12} &= 0.\end{aligned} \qquad i = 1,2$$

So we obtain $\delta_i = \mu_i = 0$, $i=1,2$, or $da_{12}=0$. In the last case, since $d\alpha_i = d\eta = 0$ and $d\beta_i = -\alpha_i\wedge\eta$,

$$da_{12} = \sum_{i=1}^{2}\{d\delta_i\wedge\alpha_i + d\mu_i\wedge\beta_i - \mu_i(\alpha_i\wedge\eta)\} = 0,$$

and this also implies that $\delta_i = \mu_i = 0$.

REMARKS.

1.- The Lie algebra $\underline{G}$ of the transitive and effective group G of isometries of (H(1,r),g) associated with the homogeneous structure T_λ is isomorphic to the direct sum $\mathfrak{m}+\mathcal{K}$, where $\mathcal{K}$ is the holonomy algebra of $\tilde{\nabla}_\lambda = \nabla - T_\lambda$ and $\mathfrak{m} = T_pH(1,r)$, $p\in H(1,r)$. From [7, theorem 9.1], $\mathcal{K}$ is generated by the operators $(\tilde{R}_\lambda)_{p\,XY}$, where $X,Y\in\mathfrak{m}$ and $\tilde{R}_\lambda$ is the curvature tensor of $\tilde{\nabla}_\lambda$. By (4.1), it is easy to verify that $\tilde{R}_\lambda = o$. Then $\mathcal{K}$ has dimension zero and G̲ is isomorphic to the Lie algebra H̲(̲1̲,̲r̲)̲ of the generalized Heisenberg group. Thus, the corresponding group of isometries is H(1,r) itself.

2.- In [5] A. Kaplan introduced the notion of groups of Heisenberg type and proved that these spaces are naturally reductive if and only if the center of the corresponding algebra has dimension 1 or 3. Moreover, he proved that for dim(center)=1 these groups are the Heisenberg groups. On the other hand, let H(p,q), p,q⩾1, be the group of matrices of real numbers of the form

$$\begin{pmatrix} I_q & S & T \\ 0 & I_p & Q \\ 0 & 0 & 1 \end{pmatrix}$$

where S is a (qxp)-matrix, T and Q are (qx1) and (px1)-column matrices, respectively. H(p,q) is a connected simply connected two-step nilpotent Lie group of dimension p(q+1)+q which is called a generalized Heisenberg group (see Haraguchi [4]). In general, these groups are not Heisenberg groups. In fact, the dimension of the center of H(1,r), with r>1, is r and so H(1,r) is not a Heisenberg group. However, H(r,1) is an example of a group of Heisenberg type. Furthermore, it is a Heisenberg group and so it is of type $\mathcal{T}_3$.

REFERENCES

[1] AMBROSE, W and SINGER, I.M., "On homogeneous Riemannian manifolds". Duke Math. J., 25 (1958) 647-669.

[2] BLAIR, D.E., "Contact manifolds in Riemannian geometry", Lecture Notes in Math., 509(1976), Springer, Berlin-Heidelberg-New York.

[3] CORDERO, L.A; FERNANDEZ, M. and LEON, M.de, "Examples of compact almost contact manifolds admitting neither Sasakian nor cosymplectic structures", (to appear).

[4] HARAGUCHI, Y., "Sur une généralisation des structures de contact", Thèses, Univ. du Haute Alsace, Mulhouse (1981).

[5] KAPLAN, A., "On the geometry of groups of Heisenberg type", Bull. London Math. Soc. 15(1983) 35-42.

[6] KIRIČENKO, V.F., "On homogeneous Riemannian spaces with invariant tensor structure", Soviet Math. Dokl., vol. 21(1980)nº3,734-737 .

[7] KOBAYASHI, S. and NOMIZU, K., "Foundations of differential geometry I", Inter. Publ. (1963), New York.

[8] LIBERMANN, P., "Sur les automorphismes infinitésimaux des structu-

res symplectiques et des structures de contact", Colloq. Géométrie Différentielle Globale (Bruxelles, 1958), Louvain(1959),37-59.

[9] SEKIGAWA, K., "Notes on homogeneous almost Hermitian manifolds", Hokkaido Math. J., vol. 7 (1978), 206-213.

[10] TRICERRI, F. and VANHECKE, L., "Homogeneous structures on Riemannian manifolds", London Math. Soc. Lecture Note Series 83, Cambridge University Press (1983).

POSITIVE SOLUTIONS OF THE HEAT AND EIGENVALUE EQUATIONS ON RIEMANNIAN MANIFOLDS

Harold Donnelly
Department of Mathematics
Purdue University
West Lafayette, IN 47907

1. Introduction

Let M be a complete Riemannian manifold of dimension n. Suppose that the Ricci curvature of M is bounded below by $-(n-1)c$, $c \geq 0$. The Laplacian of M, acting on functions, will be denoted by Δ.

We consider positive solutions of the heat equation problem:

$$\left(\frac{\partial}{\partial t} + \Delta\right) u(x, t) = 0$$

$$u(x,0) = f(x)$$

Here $u(x,t)$ is a continuous function on $M \times [0,\infty)$. Suppose that $K(x,y,t)$ is the fundamental solution of the heat equation. One has:

Theorem 1.1. If $u(x,t)$ is any non-negative solution of the heat equation, then

$$u(x,t) = \int_M K(x,y,t)f(y)dy$$

In particular, the integral converges and u is uniquely determined by the initial data f.

If M is the real line, then Theorem 1.1 is due to Widder [17]. We follow the outline of his proof. However, an explicit formula for $K(x,y,t)$ is no longer available. One must use appropriate estimates instead. In particular, this provides an interesting use for the lower bound of the heat kernel, an estimate of Cheeger and Yau [1]. An alternative proof of Theorem 1.1 has been given by Li and Yau [13]. A special case was treated by Koranyi and Taylor [11].

The Laplacian Δ of M, acting on smooth compactly supported functions, is essentially self adjoint. Let λ be the infimum of the spectrum of Δ. It was proved in [6] that $\lambda \leq (n-1)^2 c/4$.

Suppose that ϕ is a positive solution of the eigenvalue equation $\Delta\phi = \mu\phi$, for some $\mu \leq \lambda$. Such positive eigenfunctions exist precisely when μ satisfies the given inequality [7]. Let $r(x,y)$ denote the geodesic distance from x to y. Define $\mathrm{Vol}(B_1(y))$ as the volume of a geodesic ball of radius one centered at y. We will prove:

Theorem 1.2. Suppose $\alpha < \mu$ and set $\beta = (n-1)\sqrt{c}/2 + [(n-1)^2 c/4 - \alpha]^{1/2}$. Then one has for $r(x,y) \geq 2$:

$$C_1 \exp(-\beta r(x,y)) \mathrm{Vol}(B_1(y)) \phi(y) \leq \phi(x) \leq C_2 \exp(\beta r(x,y)) \mathrm{Vol}^{-1}(B_1(x)) \phi(y).$$

Here C_1 and C_2 are constants.

Similar bounds, of exponential type, follow from [2, p.351]. However, our method apparently gives a better value for the exponent β. The proof also contains an interesting application of Theorem 1.1.

A function $\psi \in L^2 M$ satisfying $\Delta\psi = \lambda\psi$ is called a ground state for Δ. In general, such square integrable ψ need not exist. For example, when $M = R^n$ then $\lambda = 0$ and there is no L^2 harmonic function on R^n. Suppose that a ground state ψ exists. Then ψ is determined up to multiplication by a constant. We may normalize ψ to be positive. Fixing y, Theorem 1.2 provides an interesting lower bound for the decay of $\psi(x)$. This complements the upper bounds established in [5].

2. Reduction to Zero Initial Data

Let $K(x,y,t)$ be the fundamental solution of the heat equation, as in [4]. Then $K(x,y,t)$ is the positive solution obtained by taking a δ measure, at y, as initial data. Suppose that $u(x,t)$ is any non-negative solution of the heat equation. One has

Lemma 2.1. $u(x,t) \geq \int_M K(x,y,t) u(y,0) dy$. In particular, the integral converges.

Proof. Let D_i be an exhaustion of M by relatively compact domains. Suppose that ϕ_i is a non-negative continuous function of compact support, which is equal to one on D_i. Set $u_i(x,t) = \int_M K(x,y,t)\phi_i(y) u(y,0) dy$. Then u_i satisfies the heat equation since the integral has compact support. Also, $u_i(x,t)$ vanishes at infinity, for fixed t, since the heat semigroup preserves the bounded continuous functions vanishing at infinity, [4, p.713]. Applying the maximum principle of [4, p.705] to the compact domains D_j, $j \geq i$, we obtain $u(x,t) - u_i(x,t) \geq -\epsilon_j$, for $x \in D_j$. Since u_i vanishes at infinity, $\epsilon_j \to 0$ as $j \to \infty$. This gives $u(x,t) \geq u_i(x,t)$ for $x \in M$. Recalling the definition of u_i and applying the monotone convergence theorem [15, p.227] gives Lemma 2.1.

We introduce the notation $\overline{u}(x,t) = \int_M K(x,y,t)u(y,0)dy$. Lemma 2.1 states that $u \geq \overline{u}$. We will eventually prove equality. One first observes:

Lemma 2.2. $\overline{u}(x,t)$ satisfies the heat equation. Moreover, $\overline{u}(x,t)$ is continuous and has initial values $u(x,0)$.

Proof. The functions u_i form a non-decreasing sequence of solutions to the heat equation. Moreover, the local L^1-norms of $u_i(x,t)$, $0 < t_1 < t < t_2$, are uniformly bounded since $u_i(x,t) \leq u(x,t)$. Therefore, one may apply the convergence criterion of [4, p.7ll]. This proves that $\overline{u}$ satisfies the heat equation and is continuous on $M \times (0,\infty)$.

It remains to check that $\overline{u}$ has the required initial values. Suppose that D is a sufficiently small relatively compact domain containing x. Then

$$u(x,0) = \lim_{t \to 0} u(x,t) \geq \lim_{t \to 0} \overline{u}(x,t)$$

and

$$\begin{aligned} \lim_{t \to 0} \overline{u}(x,t) &= \lim_{t \to 0} \int_M K(x,y,t)u(y,0)dy \\ &\geq \lim_{t \to 0} \int_D K(x,y,t)u(y,0)dy. \end{aligned}$$

However, by the local asymptotic expansion of the heat kernel [1, p. 468]:

$$\lim_{t \to 0} \int_D K(x,y,t)u(y,0)dy = u(x,0).$$

Combining the above inequalities gives $u(x,0) = \lim_{t \to 0} \overline{u}(x,t)$. The proof of Lemma 2.2 is complete.

In summary, $w(x,t) = u(x,t) - \overline{u}(x,t)$ is a non-negative solution of the heat equation with zero initial data.

3. Uniqueness of Positive Solutions.

Let $w(x,t)$ be a non-negative solution of the heat equation with $w(x,0) = 0$. We need to show that $w(x,t) = 0$. Define $v(x,t) = \int_0^t w(x,s)ds$. Clearly, it suffices to show that v vanishes identically, since w is non-negative. One begins by observing:

Lemma 3.1. $v_t = -\Delta v = w$. In particular, v is non-negative, satisfies the heat equation, and is subharmonic in x.

Proof. Obviously, $v_t = w$, by the fundamental theorem of calculus. Also $\Delta v = \int_0^t \Delta w(x,s)ds = \int_0^t -w_s(x,s)ds = -w(x,t) + w(x,0) = -w(x,t)$. The differentiation under the integral is justified by local regularity theorems for parabolic equations, [8, p.75].

We now obtain a growth estimate for $v(x,t)$. Suppose that $r(p,x)$ is the geodesic distance from a fixed basepoint p in M. One has

Lemma 3.2. For any $\epsilon > 0$ and $0 \leq t \leq \epsilon$ we may write

$$v(x,t) \leq C_1 \exp(C_2 r^2(p,x)).$$

The constants C_1 and C_2 are independent of t.

Proof. Let B denote the ball centered at x and having radius $r(p,x) + 1$. Suppose that $T > 0$ is arbitrary. Lemma 2.1 gives:

$$v(p,t+T) \geq \int_M K(p,y,T)v(y,t)dy \geq \int_B K(p,y,T)v(y,t)dy.$$

The main result of [1] is a lower bound for the heat kernel, $K(p,y,T) \geq C_3 \exp(-C_4 r^2(p,y))$. However, $y \in B$, so from the triangle inequality $r(p,y) \leq 2r(p,x) + 1$. Substitution yields

$$\int_B v(y,t)dy \leq C_5 \exp(C_6 r^2(p,x))v(p,t+T).$$

The mean value estimate of [12], applied to the non-negative subharmonic function v, gives

$$v(x,t) \leq C_7 \exp(C_8 r(p,x)) \int_B v(y,t)dy.$$

Combining the last two inequalities yields

$$v(x,t) \leq C_9 \exp(C_{10} r^2(p,x))v(p,t+T).$$

As t varies over the interval $0 \leq t \leq \epsilon$, the quantity $v(p,t+T)$ remains uniformly bounded in t. This proves Proposition 3.2.

To complete the proof of Theorem 1.1, we recall the following:

Proposition 3.3. Let $v(x,t)$ be any solution of the heat equation, for $(x,t) \in M \times [0,\epsilon]$, which satisfies

$$|v(x,t)| \leq C_1 e^{C_2 r^2(p,x)}$$

for some C_1 and C_2. If $v(x,0) = 0$, then v is identically zero.

Proof. This follows from the method of [3, pp.1038-1039]. For additional details, and generalizations to weighted L^p-spaces, the reader may consult [10].

By Lemma 3.2 and Proposition 3.3, one has that v is identically zero. Thus $w = u - \bar{u}$ is identically zero. Recalling the definition of u, we have $u(x,t) = \int_M K(x,y,t)f(y)dy$, where $u(y,0) = f(y)$. This completes the proof of Theorem 1.1.

4. Resolvent for Constant Curvature

Let M be a simply connected complete space of constant curvature $-c$, where $c \geq 0$. If n denotes the dimension of M, then the spectrum of Δ consists of the entire half line $[(n-1)^2c/4,\infty)$. This spectrum is purely continuous.

Choose a basepoint $y \in M$. The exponential map $\exp: T_yM \to M$ is a diffeomorphism. Thus, the geodesic distance r from y is smooth away from y. Using geodesic polar coordinates, the metric may be written as

$$(ds)^2 = (dr)^2 + g(r)(d\omega)^2 \tag{4.1}$$

$$g(r) = \begin{cases} (\sinh(\sqrt{c}r)/\sqrt{c})^2 & c > 0 \\ r^2 & c = 0 \end{cases}$$

The associated volume element is $\theta = g^{(n-1)/2}$.

Suppose that $\alpha < (n-1)^2c/4$. The resolvent $(\Delta-\alpha)^{-1}: L^2M \to L^2M$ may be defined via the spectral theorem. Moreover, $(\Delta-\alpha)^{-1}$ is represented by a kernel $R_\alpha(x,y)$, which is smooth outside the diagonal. It follows, from the rotational symmetry of the metric (4.1), that $R_\alpha(x,y)$ depends only upon the geodesic distance r from x to y. We will need the following estimate:

Proposition 4.2. If $r(x,y)$ is large, one has

$$|R_\alpha(x,y)| \geq C_3\exp(-(n-1)\sqrt{c}r/2 - \sqrt{(n-1)^2c/4 - \alpha r})$$

Proof. By definition, $(\Delta-\alpha)R_\alpha = 0$ for $x \neq y$. Since R_α depends only upon $r(x,y)$, we deduce:

$$\frac{-d^2R_\alpha}{dr^2} - \frac{\theta'}{\theta}\frac{dR_\alpha}{dr} - \alpha R_\alpha = 0.$$

Set $S = \theta^{1/2}R_\alpha$. An elementary computation verifies:

$$\frac{-d^2S}{dr^2} + (\omega(r) - \alpha)S = 0$$

where $\omega(r) = (n-1)/4f''(r) + ((n-1)/4)^2(f'(r))^2$ and $g(r) = \exp(f(r))$.

Clearly, ω decays rapidly to $(n-1)^2c/4$ as $r \to \infty$. The proposition now follows from the method of asymptotic integrations [9].

5. Estimates of the Eigenfunctions

Suppose that M is a complete noncompact Riemannian manifold with Ricci curvature bounded from below by $-(n-1)c$, with $c \geq 0$. Here n is the dimension of M. Let λ be the infimum of the spectrum of Δ.

If $\alpha < \lambda$, then R_α will denote the resolvent kernel for $(\Delta - \alpha)^{-1}$ on L^2M. Since λ is the infimum of the spectrum, $(\Delta - \alpha)^{-1}$ is a well defined bounded operator. Note that $\lambda \leq (n-1)^2c/4$ was proved in [6]. Therefore, the resolvent kernel $R_{\alpha,c}$, for the simply connected complete space of constant curvature $-c$, exists by the spectral theorem.

One has

Proposition 5.1. If $r(x,y)$ is the geodesic distance from x to y, then

$$R_\alpha(x,y) \geq R_{\alpha,c}(r(x,y)) > 0.$$

Proof. Let $K(t,x,y)$ be the heat kernel of M. Then $K(t,x,y) \geq K_c(t,r(x,y))$, where K_c is the heat kernel for the simply connected complete space of constant curvature $-c$. This heat kernel lower bound was proved in [1].

By the spectral theorem

$$R_\alpha(x,y) = \int_0^\infty e^{t\alpha}K(t,x,y)dt \geq \int_0^\infty e^{t\alpha}K_c(t,r(x,y))dt = R_{\alpha,c}(r(x,y)).$$

The positivity of the resolvent kernels follows from the well known positivity of the heat kernels [4].

We now proceed to present the proof of Theorem 1.2. Let ϕ be a positive function satisfying $\Delta\phi = \mu\phi$, for some $\mu \leq \lambda$.

By the uniqueness of positive solutions to the heat equation, Theorem 1.1:

$$e^{-\mu t}\phi(x) = \int_M K(t,x,z)\phi(z)dz.$$

In particular, the integral converges.

For $\alpha < \mu$, one has

$$(\mu - \alpha)^{-1}\phi(x) = \int_0^\infty e^{t\alpha} \int_M K(t,x,z)\phi(z)dzdt.$$

Since the integrand is positive, we may apply Fubini's theorem to write

$$(\mu - \alpha)^{-1}\phi(x) = \int_M R_\alpha(x,z)\phi(z)dz.$$

Recall that $r(x,y) \geq 2$. Let B denote a ball of radius one centered at y. Clearly

$$\phi(x) \geq (\mu - \alpha) \int_B R_\alpha(x,z)\phi(z)dz.$$

Since $r(y,z) \leq 1$, it follows from [3] that $C_3\phi(z) \leq \phi(y) \leq C_4\phi(z)$. Thus

$$\phi(x) \geq C_4^{-1}(\mu - \alpha)\phi(y) \int_B R_\alpha(x,z)dz.$$

Now apply Proposition 5.1 to give

$$\phi(x) \geq C_4^{-1}(\mu - \alpha)\phi(y) \int_B R_{\alpha,c}(r(x,z))dz.$$

The lower bound for $R_{\alpha,c}$, from Proposition 4.2, yields the estimate

$$\phi(x) \geq C_1 \exp(-\beta r(x,y))\phi(y)\mathrm{Vol}(B_1(y)).$$

By symmetry in x and y, one immediately deduces

$$\phi(x) \leq C_2 \exp(\beta r(x,y))\phi(y)\mathrm{Vol}^{-1}(B_1(x))$$

with $C_2 = C_1^{-1}$.

This completes the proof of Theorem 1.2.

6. Positivity of the Ground State

Suppose that λ is the infimum of the spectrum of Δ. Let $\psi \in L^2M$ satisfy the equation $\Delta\psi = \lambda\psi$. We normalize ψ to be positive at some point. One has the following results:

Proposition 6.1. The function ψ is positive at every point in M.

Proof. According to the spectral theorem, $e^{-t\Delta}\psi = e^{-t\lambda}\psi$. However, $e^{-t\Delta}$ is positivity improving [4] and thus $e^{-t\Delta}(|\psi| \pm \psi) \geq 0$. So $e^{-t\Delta}|\psi| \geq e^{-t\lambda}|\psi|$. Taking the L^2 inner product with $|\psi|$ gives $e^{-t\lambda}\langle |\psi|, |\psi| \rangle \leq \langle e^{-t\Delta}|\psi|, |\psi| \rangle$.
By the spectral theorem, this implies $e^{-t\Delta}|\psi| = e^{-t\lambda}|\psi|$. Subtraction yields $e^{-t\Delta}(|\psi| - \psi) = e^{-t\lambda}(|\psi| - \psi)$. Since $e^{-t\Delta}$ is positivity improving, $\psi = |\psi|$ is strictly positive.

Proposition 6.2. If $\omega \in L^2M$ and $\Delta\omega = \lambda\omega$, then ω is a constant multiple of ψ.

Proof. Choose a constant b so that $\omega - b\psi$ vanishes at some point. Then $\omega - b\psi$ must vanish identically. Otherwise, Proposition 4.1 shows that $\omega - b\psi$ never vanishes.

Thus, one may apply Theorem 1.2 to give a lower bound for ground state eigenfunctions. Let $\omega \in L^2M$ satisfy $\Delta\omega = \lambda\omega$. We have

Corollary 6.3. Fix a basepoint y, then

$$|\omega(x)| \geq C_4 \exp(-\beta r(x,y)) \mathrm{Vol}(B_1(y)) |\omega(y)|$$

with $\beta = (n-1)\sqrt{c}/2 + \left[\frac{(n-1)^2 c}{4} - \alpha\right]^{1/2}$.

One may easily construct examples of manifolds having a ground state. Let M be obtained by a compactly supported perturbation of the metric on the simply connected space of constant curvature $-c$. By the decomposition principle [6], the essential spectrum of Δ is $[(n-1)^2c/4, \infty)$. Suppose that M contains a sufficiently large Euclidean ball. The minimax principle implies that Δ has discrete spectrum below $(n-1)^2c/4$. In particular, there must be a ground state eigenfunction. One notes that Corollary 6.3 provides a reasonable estimate in this example.

BIBLIOGRAPHY

1 Cheeger, J. and Yau, S. T., A Lower Bound for the Heat Kernel, Communications on Pure and Applied Mathematics, 34 (1981), pp. 465-480.

2 Cheng, S. Y. and Yau, S. T., Differential Equations on Riemannian Manifolds and their Geometric Applications, Communications on Pure and Applied Mathematics, 28 (1975), pp. 333-354.

3 Cheng, S. Y., Li, P., and Yau, S. T., On the Upper Estimate of the Heat Kernel of a Complete Riemannian Manifold, American Journal of Mathematics, 103 (1981), pp. 1021-1063.

4 Dodziuk, J., Maximum Principle for Parabolic Inequalities and the Heat Flow on Open Manifolds, Indiana University Mathematics Journal, 32 (1983), pp. 703-716.

5 Donnelly, H., Eigenforms of the Laplacian on Complete Riemannian Manifolds, Communications in Partial Differential Equations, 9 (1984), pp. 1299-1321.

6 Donnelly, H., On the Essential Spectrum of a Complete Riemannian Manifold, Topology, 20 (1981), pp. 1-14.

7 Fischer-Colbrie, D. and Schoen, R., The Structure of Complete Stable Minimal Surfaces in 3-Manifolds of Non-Negative Scalar Curvature, Communications on Pure and Applied Mathematics, 33 (l980), pp. 199-211.

8 Friedman, A., Partial Differential Equations of Parabolic Type, Prentice Hall, Englewood Cliffs, 1964.

9 Hartman, P., Ordinary Differential Equations, Wiley, New York, 1984.

10 Karp, L. and Li, P., The Heat Equation on Complete Riemannian Manifolds, Preprint.

11 Koranyi, A. and Taylor, J. C., Minimal Solutions of the Heat Equation and Uniqueness of the positive Cauchy problem on Homogeneous Spaces, Proceedings of the American Mathematical Society, 94 (1985), pp. 273-278.

12 Li, P. and Schoen, R., L^p and Mean Value Properties of Subharmonic Functions on Riemannian Manifolds, Acta Math., 153 (1984), pp. 279-302.

13 Li, P. and Yau, S. T., On the Parabolic Kernel of the Schrodinger Operator, Preprint.

14 Reed, M. and Simon, B., Methods of Modern Mathematical Physics IV: Analysis of Operators, Academic Press, New York, 1978.

15 Royden, H., Real Analysis, MacMillan, N.Y., 1968.

16 Strichartz, R., Analysis of the Laplacian on a Complete Riemannian Manifold, J. Func. Anal., 52 (1983), pp. 48-79.

17 Widder, D. V., The Heat Equation, Academic Press, N.Y., San Francisco, and London, 1975.

ETUDE DES ALGEBRES DE LIE RESOLUBLES REELLES QUI ADMETTENT

DES IDEAUX UNIDIMENSIONNELS N'APPARTENANT PAS AU CENTRE

par FRANCISCO JAVIER ECHARTE REULA
Facultad de Matemáticas
41012-SEVILLA -ESPAÑA-

Ce travail a pour objet la définition d'une base spéciale pour les algèbres de Lie resolubles reelles qui admettent des idéaux unidimensionnels.

1-Idéaux unidimensionnels d'une algèbre de Lie-

Soit $\mathcal{L}$ une algèbre de Lie réelle.Considérons ses idéaux unidimensionnels engendrés par les champs X tels que $[X,Z]=0$, $\forall Z \in \mathcal{L}$;ou bien par des champs Y tels que $[Y,Z]=aY$; $\forall Z \in \mathcal{L}$,en étant a dépendant de Y,Z.
Nous désignerons des champs centraux les champs X,et les champs normaux les champs Y.Les champs X constituent le centre de $\mathcal{L}$.
Les champs normaux ont les propiétés définies par les théorèmes suivants:
Thérème 1-a)

Si Y est un champ normal,aY l'est aussi.

Démonstration:

Si $[Y,Z]=bY$,on peut déduire que $[aY,Z]=abY=b(aY)$.

Théorème 1-b

Si Y_1 et Y_2 sont normaux; $[Y_1,Y_2]=0$

Démonstration: $[Y_1,Y_2]=a_1Y_1=a_2Y_2$,par conséquent $[Y_1,Y_2]=0$

Théorème 1-c

Si les champs normaux; $Y_1,\ldots\ldots,Y_r$ sont linéalment indépendants tels que $[Y_i,Z]=b_iY_i$;c'est condition necessaire et suffissante que $b_1=\ldots\ldots=b_r$ pour que toute combinaison lineaire d'entre eux soit aussi normale.

Démonstration:

Si $\sum a_iY_i$ est normal,on deduit que $[\sum a_iY_i,Z]=\sum a_ib_iY_i$ mais on verifie aussi $[\sum a_iY_i,Z]=c(\sum a_iY_i)$,ce qui exige que $b_1=\ldots\ldots\ldots=b_r=c$.
Le reciproque est immédiat.

Définition:

Nous disons que deux champs normaux Y_1,Y_2 sont conjugués,si

$a_1Y_1+a_2Y_2$ est aussi un champ normal. La conjugaison est une relation d'équivalence.

2-Exemples des champs centraux et des champs normaux en algèbres resolubles.

(2-1) Le group $\begin{pmatrix} e^x & y \\ 0 & 1 \end{pmatrix}$; $(x,y) \in R^2$ admet comme base de son algèbre de Lie les champs invariants à gauche:

$Y(0,e^x)$; $Z(1,0)$; où $[Y,Z] = -Y$, Y est champ normal.

(2-2) Le group $\begin{pmatrix} 1 & x & y \\ 0 & 1 & z \\ 0 & 0 & 1 \end{pmatrix}$; $(x,y,z) \in R^3$,admetcomme base de son algèbre de Lie, les champs invariants à gauche:

$X(0,1,0)$; $Z_1(1,0,0)$; $Z_2(0,x,1)$ où $[X,Z_1] = [X,Z_2] = 0$; $[Z_1,Z_2] = X$

X champ central.

(2-3) Le group $\begin{pmatrix} e^x & z \\ 0 & e^y \end{pmatrix}$; $(x,y,z) \in R^3$,admet la base

$X(1,1,z)$; $Y(0,0,e^x)$; $Z(1,0,0)$; où $[X,Y]=0$; $[Y,Z]=-Y$; $[Z,X]=0$

et pourtant X est champ central et Y champ normal.

(2-4)

Groupes de Heisenberg généralisés

$$\begin{pmatrix} 1 & 0 & \ldots 0 & x_1 & z_1 \\ 0 & 1 & \ldots 0 & x_2 & z_2 \\ \ldots & \ldots & \ldots & \ldots & \ldots \\ 0 & 0 & \ldots 1 & x_p & z_p \\ 0 & 0 & \ldots 0 & 1 & y \\ 0 & 0 & \ldots 0 & 0 & 1 \end{pmatrix}$$

admet comme base de son algèbre de Lie la suivante:

$$X_i = \frac{\partial}{\partial x_i} \ ; \quad Y = \frac{\partial}{\partial y} + \sum x_j \frac{\partial}{\partial z_j} \quad ; \quad Z_h = \frac{\partial}{\partial z_h}$$

où:

$[X_i,X_j]=0$; $[Z_h,Z_k]=0$; $[X_m,Z_n]=0$; $[X_l,Y]=Z_l$; $[Z_q,Y]=0$

Les champs $Z_1,\ldots\ldots,Z_p$ sont des champs centraux,et il n'y a pas de champs normaux.Ce sont des groupes nilpotents.

3-Centralisateur d'un champ normal

Pour l'étude des algèbres de Lie resolubles qui admettent des idéaux unidimensionnels,on va choisir une base de la manière suivante:
En premier lieu une base $X_1,\ldots,X_m$ de son centre,suivie du plus grand nombre de champs normaux linéalement indépendants;$Y_1,\ldots,Y_n$;en completant la base totale avec d'autres champs que nous allons étudier à la suite.
Définition: On appelle centralisateur d'un champ U,respecte une algèbre de Lie $\mathcal{L}$,$\mathrm{cen}_{\mathcal{L}}$ U , l'ensemble de champs $Z \in \mathcal{L}$ tels que $[U,Z]=0$.
Théorème 3-a)

Si Y est normal, $\dim \mathrm{cen}_{\mathcal{L}} Y = \dim \mathcal{L} - 1$

Démonstration: Soit $X_1,\ldots\ldots,X_m,Y_1,\ldots,Y_n,Z_1,\ldots\ldots,Z_p$ une base de $\mathcal{L}$,étant X_i,Y_j les champs nommés ci-dessus,et Z_h d'autres champs qui complètent une base de $\mathcal{L}$ .Soit Y un champ normal;X_i ,Y_j appartiennent au centralisateur de Y. Parmi les champs $Z_1,\ldots,Z_p$ on distingue deux classes: $Z_1,\ldots,Z_s$ tels que $[Z_i,Y]=0$ $(i \leq s)$;et les $Z_{s+1},\ldots,Z_p$ tels que: $[Z_j,Y]=a_jY$,$(a_j \neq 0; j=s+1,\ldots\ldots,p)$.Dans ce dernier cas $[Z_j/a_j,Y]=Y$,et par conséquent: $[Y,Z_j/a_j - Z_p/a_p]=0$,ce qui permet de remplacer $Z_j(j=s+1,..,p-1)$, dans la base donnée,par $Z'_j=Z_j/a_j-Z_p/a_p$,ainsi nous considérons comme nouvelle base pour $\mathcal{L}$:

$$X_1,\ldots,X_m,Y_1,\ldots,Y_n,Z_1,\ldots\ldots,Z_s,Z'_{s+1},\ldots\ldots,Z'_{p-1},Z_p$$

et le centralisateur de Y ,$\mathrm{cen}_{\mathcal{L}}$ Y,admet comme base:

$$X_1,\ldots,X_m,Y_1,\ldots,Y_n,Z_1,\ldots,Z_s,Z'_{s+1},\ldots,Z'_{p-1}$$

et par conséquent:

$$\dim \mathrm{cen}_{\mathcal{L}} Y.\dim\mathcal{L} -1$$

Théorème 3-b)

Si deux champs normaux sont conjugués,on vérifie qu'ils ont le même centralisateur.

Démonstration:

Soient Y_1,Y_2 les deux champs normaux,si $Z \in \mathrm{cen}_{\mathcal{L}} Y_1$ on vérifie $[Y_1,Z]=0=0.Y_1$,alors en étant Y_1,Y_2 conjugués $[Y_2,Z]=0.Y_2=0$; par conséquent: $Z \in \mathrm{cen}_{\mathcal{L}} Y_2$ .Le reciproque n'estpas vrai deux champs peuvent ne pas être conjugués et pourtant avoir le même centralisateur,par exemple:

l'algèbre tridimensionnellequi admetcomme base Y_1, Y_2, Z;étant:

$[Y_1, Y_2]=0$; $[Y_1, Z]=aY_1$; $[Y_2, Z]=bY_2$,où $0\neq a\neq b\neq 0$.Les centralisateurs de Y_1, Y_2 sont les mêmes,et malgré tout Y_1, Y_2 ne sont pas conjugués.

La relation $\mathrm{cen}_{\mathcal{L}} Y_1=\mathrm{cen}_{\mathcal{L}} Y_2$ est d'équivalence.

Touts les champs qui ont le même centralisateur qu'un champ normal,ils ne sont pas normaux,car si Y_1, Y_2 sont deux champs normauxqui ont le même centralisateur et qui ne sont pas conjugués,on vérifie que $a_1Y_1+a_2Y_2$,a le même centralisateur et n'est pas normal.

Une classe de champs conjugués est contenue dans une classe de champs avec le même centralisateur.

Si deux champs normaux Y_1, Y_2 d'une algèbre de Lie ont un centralisateur différent,l'intersection des deux centralisateurs respectifs est tel que:

$$\dim(\mathrm{cen}_{\mathcal{L}} Y_1 \cap \mathrm{cen}_{\mathcal{L}} Y_2)=\dim \mathcal{L} -2$$

et,en général ,s'il y a un maximun de <u>h</u> champs normaux avec un centralisateur different deux a deux,on a:

$$\dim (\mathrm{cen}_{\mathcal{L}} Y_1 \cap \mathrm{cen}_{\mathcal{L}} Y_2 \cap \ldots \cap \mathrm{cen}_{\mathcal{L}} Y_h)=\dim \mathcal{L} -h$$

On appelle <u>centralisateur normal</u> l'intersection des centralisateurs de tous les champs normaux. Dans ces conditions on peut prende comme base de $\mathcal{L}$:

$$X_1,\ldots,X_m,Y_1,\ldots,Y_n,Z_1,\ldots,Z_s,Z_{s+1},\ldots,Z_{s+h}$$

où:

$$X_1,\ldots,X_m,Y_1,\ldots,Y_n,Z_1,\ldots,Z_s$$

c'est la base du centralisateur normal,et pour chaque Y_i existe un seul Z_k $(k=s+1,\ldots,s+h)$ tel que $[Y_i, Z_k]\neq 0$

Théorème 3-c)

$[Z_j, Z_k]$ appartient au centralisateur normal $(j,k=1,\ldots,s+h)$

Démonstration:

Soit Y_i un champ normal,Z_j, Z_k deux champs quelconques;en utilisant l'identité de Jacobi on obtient :

$$[[Z_j, Z_k], Y_i]=0$$

d'où $[Z_j, Z_k]\in \mathrm{cen}_{\mathcal{L}} Y_i$,et comme ce resultat est valable $\forall i,j,k$,on démontre le théorème.

Corollaire-

$\mathcal{L}^1=[\mathcal{L}, \mathcal{L}]$ appartient au centralisateur normal.

Exemple 3-1)

Le group:

$$(x,y,z)\ \#(x',y',z')=(x+e^{z}x'+ze^{z}y',y+e^{z}y',z+z')$$

admet comme base de son algèbre de Lie, la suivante:

$$Y(e^{z},0,0)\ ;\ Z(ze^{z},e^{z},0)\ ;U(0,0,-1)$$

où $[Y,Z]=0$; $[Y,U]=Y$; $[Z,U]=Y+Z$

Y champ normal, le centralisateur de Y est l'ideal défini par les champs Y,Z.

4-Types d'algèbres de Lie en relation avec leurs idéaux unidimensionnels-

Une algèbre de Lie qui contient quelque idéal unidimensionnel no-trivial, ne peut pas être semi-simple, car ces algèbres ne contiennent pas d'idéaux abeliens non triviaux.

Théorème 3-a)

Si une algèbre de Lie a des champs normaux ne peut pas être nilpotente.

Démonstration:

Si Y est un champ normal qui appartient à l'algèbre $\mathcal{L}$, il y aura au moins un $Z\varepsilon\mathcal{L}$ tel que $[Y,Z]=aY(a\neq 0)$, par conséquent $Y\varepsilon\mathcal{L}_1=[\mathcal{L},\mathcal{L}]$ et par analogie $Y\varepsilon\mathcal{L}_2,\ldots,Y\varepsilon\mathcal{L}_n$, ainsi que $\mathcal{L}_n\neq\{0\},\forall n$. Et comme toute algèbre nilpotente a un centre non vide, nous pouvons affirmer qu'une algèbre nilpotente a des champs centraux mais elle n'a pas de champs normaux.

Corollaire-

Une algèbre de Lie qui a des champs normaux n'est ni semi-simple ni nilpotente.

REFERENCES

BERNAT-CONZE-DUFLO-LEVY-RAIS-VERGNE-Representations des groupes de Lie resolubles-Dunod-Paris-1972-(pag 1-13)

CHOW-General theory of Lie algebras-vol I-Gordon and Breach-New York-1978-(pag 3-16;35-77)

CORDERO-FERNANDEZ-GRAY-Symplectic Manifolds with no Kahler structure-Santiago-1985-Prepint-

GOODMAN-Lecture Notes nº562-Nilpotent Lie groups-Springer Verlag-1976-(pag 1-32)

PROCEEDINGS NEW BRUNSWICK-NEW JERSEY 1981-Lecture Notes nº 933-Lie Algebras and related topics-Springer Verlag-1982-(pag 111-116)

THE IWASAWA MANIFOLD

Marisa Fernández(*) and Alfred Gray(**)

(*) Departamento de Geometría y Topología. Fac. Matemáticas. Univ. de Santiago de Compostela. España.

(**) University of Maryland. College Park, Maryland 20742. USA.

One of the simplest compact manifolds is the Iwasawa manifold I(3). Let H(3) be the group of 3×3 matrices of the form

$$\begin{pmatrix} 1 & x & z \\ & 1 & y \\ & & 1 \end{pmatrix}$$

where x, y and z are complex numbers. Let $\Gamma(3)$ be the subgroup of H(3) consisting of those whose entries are Gaussian integers. Then I(3) is defined as the quotient space $\Gamma(3)\backslash H(3)$.

In many textbooks (for example [Ch p.4], [GH p.444], [MK p.115]) the Iwasawa manifold is described as a nontrivial complex manifold which is not Kählerian. The usual method to prove this is to exhibit a nonclosed holomorphic 1-form on I(3). In fact it is easy to see that the complex differential forms dx, dy and dz-xdy on H(3) are left invariant; hence they project to holomorphic 1-forms α, β and γ on I(3). But $d\gamma=-\alpha\wedge\beta$ and so γ is a nonclosed holomorphic 1-form. Since the complex Laplacian is 1/2 the ordinary Laplacian on any Kähler manifold, this shows that the natural complex structure on I(3) cannot be Kählerian.

But what is to preclude I(3) from having a different Kähler structure? As a first attempt to establish this stronger assertion let us examine the cohomology of I(3). There are strong cohomological conditions for a compact manifold M to be Kählerian: (a) the even dimensional Betti numbers are nonzero, (b) the odd dimensional Betti numbers are even, and (c) $b_{2i-1}(M)\leq b_{2i+1}(M)$ up the middle dimension of M.

It is very easy to compute the real cohomology of I(3) using differential forms. We have

$$H^1(I(3),R)=\{[\alpha],[\bar\alpha],[\beta],[\bar\beta]\},$$

$$H^2(I(3),R)=\{[\alpha\wedge\bar\alpha],[\beta\wedge\bar\beta],[\alpha\wedge\bar\beta],[\bar\alpha\wedge\beta],[\alpha\wedge\gamma],[\bar\alpha\wedge\bar\gamma],[\beta\wedge\gamma],[\bar\beta\wedge\bar\gamma]\},$$

$$H^3(I(3),R)=\{[\alpha\wedge\bar\alpha\wedge\gamma],[\alpha\wedge\bar\alpha\wedge\bar\gamma],[\beta\wedge\bar\beta\wedge\gamma],[\beta\wedge\bar\beta\wedge\bar\gamma],[\alpha\wedge\beta\wedge\gamma],[\alpha\wedge\bar\beta\wedge\gamma],[\bar\alpha\wedge\beta\wedge\gamma],$$
$$[\bar\alpha\wedge\bar\beta\wedge\bar\gamma],[\bar\alpha\wedge\beta\wedge\bar\gamma],[\alpha\wedge\bar\beta\wedge\bar\gamma]\}.$$

So $b_1(I(3))=b_5(I(3))=4$, $b_2(I(3))=b_4(I(3))=8$ and $b_3(I(3))=10$. Thus the cohomology of I(3) satisfies the Kähler conditions (a) and (b).

THEOREM. The Iwasawa manifold I(3) has the following properties:

(i) I(3) has no Kähler structure;

(ii) I(3) has indefinite Kähler structures;

(iii) I(3) has symplectic forms each of which is Hermitian with respect to a complex structure.

PROOF. To prove (i) we use the Main Theorem of [DGMS]. This theorem states that the minimal model of a compact Kähler manifold is formal. Consequently all Massey products are zero for compact Kähler manifolds. But is easy to exhibit nonzero Massey products on I(3). For example the Massey product $<[\alpha],[\alpha],[\beta]>$ is nonzero because it is represented by the nonexact form $-\alpha\wedge\gamma$.

For (ii) we note that the natural complex structure on I(3) has no indefinite Kähler metric. This is because any 2-dimensional cohomology class $[\Omega]$ of type (1,1) with respect to the natural complex structure must be a linear combination of the classes $[\alpha\wedge\bar{\alpha}],[\beta\wedge\bar{\beta}],[\alpha\wedge\bar{\beta}]$ and $[\alpha\wedge\bar{\beta}]$; hence $[\Omega]^3=0$.

However there are other complex structures on I(3) that possess indefinite Kähler structures. In fact there is a 1-parameter family J_θ of almost complex structures such that each J_θ gives rise to an indefinite Kähler metric. To establish this fact let $\{X_1,X_2,Y_1,Y_2,Z_1,Z_2\}$ be the real vector fields dual to $\{\mathrm{Re}(\alpha),\ \mathrm{Im}(\alpha),\ \mathrm{Re}(\beta),\ \mathrm{Im}(\beta),\ \mathrm{Re}(\gamma),\ \mathrm{Im}(\gamma)\}$. Define J_θ by

$J_\theta X_1=\cos\theta Y_1+\sin\theta Y_2,\ J_\theta X_2=-\sin\theta Y_1+\cos\theta Y_2,\ J_\theta Z_1=Z_2.$

Then it can be checked that the Nijenhuis tensor of J_θ vanishes. Put

$$\lambda=\mathrm{Re}\alpha+\sqrt{-1}(\cos\theta\mathrm{Re}\beta+\sin\theta\mathrm{Im}\beta)$$

$$\mu=\mathrm{Im}\alpha+\sqrt{-1}(-\sin\theta\mathrm{Re}\beta+\cos\theta\mathrm{Im}\beta).$$

Let $\#$ denote the symmetric product and let

$$ds^2_\theta=1/4\{\cos\theta(\lambda\#\bar{\gamma}+\bar{\lambda}\#\gamma)-\sin\theta(\mu\#\bar{\gamma}+\bar{\mu}\#\gamma)$$
$$-\sqrt{-1}(\cos\theta(\lambda\#\bar{\mu}-\bar{\lambda}\#\mu+\mu\#\bar{\gamma}-\bar{\mu}\#\gamma)+\sin\theta(\lambda\#\bar{\mu}-\bar{\lambda}\#\mu+\lambda\#\bar{\gamma}-\bar{\lambda}\#\gamma))\}$$

Then ds^2 is an indefinite Kähler metric for J_θ. The corresponding Kähler form is the symplectic form

$$F_\theta=1/2\{\cos\theta(\lambda\wedge\bar{\mu}+\bar{\lambda}\wedge\mu+\mu\wedge\gamma+\bar{\mu}\wedge\gamma)+\sin\theta(\lambda\wedge\bar{\mu}+\bar{\lambda}\wedge\mu+\lambda\wedge\bar{\gamma}+\bar{\lambda}\wedge\gamma)$$
$$+\sqrt{-1}((\cos\theta(\lambda\wedge\bar{\gamma}-\bar{\lambda}\wedge\gamma)-\sin\theta(\mu\wedge\bar{\gamma}-\bar{\mu}\wedge\gamma))\}.$$

So (iii) is proved as well.

REMARK. Let I denote the natural complex structure of I(3) and let J_θ be the complex structure defined above. Then $h^{p,q}(I(3),J_\theta)$ does not depend on θ and some of the $h^{p,q}(I(3),I)$ are different from the

$h^{p,q}(I(3),J_\theta)$. In fact we have:

$$h^{0,0}(I(3),I)=h^{0,3}(I(3),I)=1$$
$$h^{1,0}(I(3),I)=h^{2,0}(I(3),I)=3$$
$$h^{0,1}(I(3),I)=h^{0,2}(I(3),I)=2$$
$$h^{1,1}(I(3),I)=h^{1,2}(I(3),I)=6$$

but

$$h^{0,0}(I(3),J_\theta)=h^{0,3}(I(3),J_\theta)=1$$
$$h^{1,0}(I(3),J_\theta)=h^{2,0}(I(3),J_\theta)=2$$
$$h^{0,1}(I(3),J_\theta)=h^{0,2}(I(3),J_\theta)=3$$
$$h^{1,1}(I(3),J_\theta)=h^{1,2}(I(3),J_\theta)=6$$

(The other $h^{p,q}$'s can be determined using Serre duality.)

Moreover I(3) is real parallelizable and the complex manifold $(I(3),I)$ is complex parallelizable in the sense of Wang [Wa]. However no $(I(3),J_\theta)$ is complex parallelizable.

REFERENCES

[Br] N. Brotherton: Some parallelizable four manifolds not admitting a complex structure. *Bull. London Math. Soc.* 10, 303-304 (1978).

[Ch] S. S. Chern: *Complex manifolds without potential theory*, Springer-Verlag (1979).

[CFG1] L. A. Cordero, M. Fernández, and A. Gray: Variétés symplectiques sans structures kählériennes. *C. R. Acad. Sci. Paris* 301, 217-218 (1985).

[CFG2] L. A. Cordero, M. Fernández, and A. Gray: Symplectic manifolds without Kähler structure. *Topology* (to appear).

[DGMS] P. Deligne, P. Griffiths, J. Morgan, D. Sullivan: Real homotopy theory of Kähler manifolds. *Invent. Math.* 29, 245-274 (1975).

[FGG] M. Fernández, M. J. Gotay and A. Gray: Four dimensional parallelizable symplectic and complex manifolds (to appear).

[Gr] A. Gray: Minimal varietes and almost Hermitian submanifolds. *Michigan Math. J.* 12, 273-287 (1965).

[GH] P. Griffiths and J. Harris: *Principles of Algebraic Geometry*, John Wiley, New York (1978).

[MK] J. Morrow and K. Kodaira: *Complex manifolds*, Holt Rinehart Winston New York (1971).

[Wa] H. C. Wang: Complex parallelizable manifolds. *Proc. Amer. Math. Soc.* 5, 771-776 (1954).

CONNECTED SUMS AND THE INFIMUM OF THE YAMABE FUNCTIONAL

Olga Gil-Medrano
Departamento de Geometría y Topología
Facultad de Matemáticas
Universidad de Valencia
Burjasot (Valencia), Spain

§ 1.- INTRODUCTION.-

In this paper we show the existence of metrics on the connected sum of two compact Riemannian manifolds with the property that the infimum of the Yamabe functional has upper bounds which depend on the infimum of the Yamabe functional of the two involved manifolds.

The fact that any manifold is diffeomorphic to that obtained by taking its connected sum with a sphere of the same dimension, allows us to prove that on every compact manifold M and for each real number K there exists a Riemannian metric g on M such that the infimum of the Yamabe functional of (M.g) is less than K.

As it is known, Yamabe's functional is deeply related with scalar curvature ([1],[5]) and so from the result above we obtain that on every compact manifold and for each $K \in \mathbb{R}$ there is a Riemannian metric of volume 1 and such that the scalar curvature R is constant and $R < K$. This generalizes a result in [3].

§ 2.- PRELIMINARIES.-

In [5] Yamabe defined the functional

$$J_g(u) = \left(\frac{4(n-1)}{n-2} \int_M |\nabla u|^2 \, dv_g + \int_M R\, u^2 \, dv_g \right) \cdot \left(\int_M u^N \, dv_g \right)^{-2/N}$$

where $N = 2n/(n-2)$, in order to show the existence, for every compact Riemannian manifold (M,g) of a conformal transformation of constant scalar curvature, and took $\mu_g = \operatorname{Inf} J_g(u)$ for $u \in H_1(M)$, $u \geq 0$, $u \not\equiv 0$.

Then, he shows that μ_g is attained for a strictly positive function $u_0 \in C^\infty(M)$ and then $u_0^{4/(n-2)} g$ is the required metric. But his proof was shown

---

Work partially supported by CAICYT, 1985-87 nº 120.

to be incorrect and the result has become the so called Yamabe problem.

As we are not specially concerned with that problem here, we refer the reader to the survey by Aubin in these Proceedings and we only give without proof some properties of the infimum μ_g that will be used in the sequel.

Result 1.- (see [2])

The infimum of the Yamabe functional is a conformal invariant.

Result 2.- (see [2])

For every compact, n-dimensional, Riemannian manifold (M,g)

$\mu_g \leq n(n-1)\omega_n^{2/n}$. If $\mu_g < n(n-1)\omega_n^{2/n}$,

then there is a strictly positive function $u_0 \in C^\infty(M)$ such that $g' = u_0^{4/(n-2)} g$ verifies $vol_{g'}(M) = 1$ and the scalar curvature R' of g' is constant and equal to μ_g. ω_n is the volume of the unit sphere of dimension n.

Result 3.- (see [2])

A compact n-dimensional manifold (n>2) carries a metric whose scalar curvature is a negative constant.

Result 4.- (see [4]).

Let (M,g) a compact Riemannian manifold of dimension $n (n \geq 2)$. For each $p \in M$ and $\varepsilon > 0$, there exists a function $u \in C^\infty(M)$, $u \geq 0$, $u \not\equiv 0$ such that $J(u) < \mu_g + \varepsilon$ and such that u vanishes on a geodesic ball centered at p.

Remark.-

A more general version of the Proposition above, not used here, can be shown by the same method. Namely, a function $u \in C^\infty(M)$ $u \geq 0$, $u \not\equiv 0$ can be exhibited such that $J(u) < \mu_g + \varepsilon$ and such that u vanishes on a tubular neighborhood of a compact submanifold of codimension greater than 2.

§ 3.- EXISTENCE OF NICE METRICS ON THE CONNECTED SUM OF TWO COMPACT MANIFOLDS.-

Let (M_1,g_1) (M_2,g_2) be two compact Riemannian manifolds of dimension n, and let C_i be a n-cell of M_i, i = 1,2. It is known that on the connected sum $M_1 \# M_2$ a metric g can be constructed such that g agrees with g_i on $M_i \sim C_i$=

Theorem 5.-

Let (M_1,g_1), (M_2,g_2) be two compact Riemannian manifolds of dimension n $(n>2)$. Let $\mu_i (i = 1,2)$ be the infimum of the Yamabe functional J_i of (M_i,g_i). For each $\varepsilon > 0$ there is a metric g on $M_1 \# M_2$ such that the infimum μ_g, of the functional J_g, verifies the inequality $\mu_g < \min(\mu_1,\mu_2) + \varepsilon$.

Proof.-

Let us assume $\mu_1 \leq \mu_2$. For a fixed $\varepsilon > 0$, from Result 4, there is a real number $\delta>0$ and a smooth function $u_\delta \in C^\infty(M_1)$ such that u_δ vanishes on a geodesic ball $B_\delta(x_1)$ of radius δ and centered at $x_1 (x_1 \in M_1)$ and such that $J_1(u_\delta) < \mu_1 + \varepsilon$.

Let g be a metric on $M_1 \# M_2$ which agrees with g_1 when restricted to $M_1 \sim B_\delta(x_1)$. We can define the function $\tilde{u}_\delta$ on $M_1 \# M_2$ given by

$$\tilde{u}_\delta = \begin{cases} u_\delta & \text{on } M_1 \sim B_\delta(x_1) \\ 0 & \text{elsewhere} \end{cases}$$

then

$$J_g(\tilde{u}_\delta) = \left(\frac{4(n-1)}{n-2} \int_{M_1 \sim B_\delta(x_1)} |\nabla^g \tilde{u}_\delta|^2 dv_g + \int_{M_1 \sim B_\delta(x_1)} R_g \tilde{u}_\delta^2 \, dv_g \right) \cdot$$

$$\left(\int_{M_1 \sim B_\delta(x_1)} \tilde{u}_\delta^N \, dv_g \right)^{-2/N}$$

But on $M_1 \sim B_\delta(x_1)$ g agrees with g_1 and $\tilde{u}_\delta = u_\delta$ thus

$$J_g(\tilde{u}_\delta) = \left(\frac{4(n-1)}{n-2} \int_{M_1 \sim B_\delta(x_1)} |\nabla^1 u_\delta|^2 \, dv_{g_1} + \int_{M_1 \sim B_\delta(x_1)} R_1 u_\delta^2 \, dv_{g_1} \right) \cdot$$

$$\cdot \left(\int_{M_1 \sim B_\delta(x_1)} u_\delta^N \, dv_{g_1} \right)^{-2/N}$$

Now, u_δ vanishes on $B_\delta(x_1)$ and then

$$J_g(\tilde{u}_\delta) = J_1(u_\delta) < \mu_1 + \varepsilon \,.$$

Consequently, $\mu_g < \mu_1 + \varepsilon = \min(\mu_1, \mu_2) + \varepsilon$.

The same conclusion is obtained if we assume $\mu_2 \leq \mu_1$ by interchanging the roles of M_1 and M_2.

Theorem 6.-

Under the same hypothesis than in Theorem 5, for each $\varepsilon > 0$ there is a metric g on $M_1 \# M_2$ such that μ_g verifies $\mu_g < (\mu_1 + \mu_2) 2^{-2/N} + \varepsilon$, where $N = 2n/(n-2)$.

Proof.-

For a fixed $\varepsilon > 0$, $x_1 \in M_1$, $x_2 \in M_2$ we can find (Result 4) two positive real numbers δ_1, δ_2 and smooth functions u_{δ_1}, u_{δ_2}, $u_{\delta_1} \in C^\infty(M_1)$, $u_{\delta_2} \in C^\infty(M_2)$ with the following properties

1) u_{δ_1} vanishes on $B_{\delta_1}(x_1)$ and $J_1(u_{\delta_1}) < \mu_1 + \varepsilon'$

2) u_{δ_2} vanishes on $B_{\delta_2}(x_2)$ and $J_2(u_{\delta_2}) < \mu_2 + \varepsilon'$

where $\varepsilon' = 2^{-2/n}\varepsilon$. We can assume that in addition

3) $\int_{M_1} u_{\delta_2}^N \, dv_{g_1} = 1 = \int_{M_2} u_{\delta_2}^N \, dv_{g_2}$

because, otherwise, it suffices to consider the new functions

$$u_{\delta_1} / \| u_{\delta_1} \|_{L_N(M_1)} \quad \text{and} \quad u_{\delta_2} / \| u_{\delta_2} \|_{L_N(M_2)}.$$

A metric g on $M_1 \# M_2$ can be constructed such that g agrees with g_i when restricted to $M_i \sim B_{\delta_i}(x_i)$.

Let us consider now the function $\tilde{u}$ on $M_1 \# M_2$ given by

$$\tilde{u} = \begin{cases} u_{\delta_1} & \text{on } M_1 \sim B_{\delta_1}(x_1) \\ u_{\delta_2} & \text{on } M_2 \sim B_{\delta_2}(x_2) \\ 0 & \text{elsewhere,} \end{cases}$$

then

$$J_g(\tilde{u}) = \Big\{ \frac{4(n-1)}{n-2} \int_{M_1 \sim B_{\delta_1}(x_1)} |\nabla^g \tilde{u}|^2 dv_g + \int_{M_1 \sim B_{\delta_1}(x_1)} R_g \tilde{u}^2 \, dv_g \quad +$$

$$+ \frac{4(n-1)}{n-2} \int_{M_1 \sim B_{\delta_1}(x_2)} |\nabla^g \tilde{u}|^2 dv_g + \int_{M_1 \sim B_{\delta_1}(x_1)} R_g \tilde{u}^2 dv_g \Big] \cdot$$

$$\cdot \Big(\int_{M_1 \sim B_{\delta_1}(x_1)} \tilde{u}^N dv_g + \int_{M_2 \sim B_{\delta_2}(x_2)} \tilde{u}^N dv_g \Big)^{-2/N} .$$

Using 1), 2) and the condition on g we obtain

$$J_g(\tilde{u}) = \Big(\frac{4(n-1)}{n-2} \int_{M_1 \sim B_{\delta_1}(x_1)} |\nabla^1 u_{\delta_1}|^2 dv_g + \int_{M_1 \sim B_{\delta_1}(x_1)} R_1 u_{\delta 1}^2 dv_{g_1} +$$

$$+ \frac{4(n-1)}{n-2} \int_{M_2 \sim B_{\delta_2}(x_2)} |\nabla^2 u_{\delta_1}|^2 dv_{g_1} + \int_{M_2 \sim B_{\delta_2}(x_2)} R_2 u_{\delta_1}^2 dv_{g_2} \Big] \cdot$$

$$\cdot \Big(\int_{M_1 \sim B_{\delta_1}(x_1)} u_{\delta_1}^N dv_{g_1} + \int_{M_2 \sim B_{\delta_2}(x_2)} u_{\delta_2}^N dv_{g_2} \Big)^{-2/N} .$$

As $u_{\delta_1}, u_{\delta_2}$ verify conditions 1),2) and 3) we have

$$\int_{M_i \sim B_{\delta_i}(x_i)} u_{\delta_i}^N dv_{g_i} = \int_{M_i} u_{\delta_i}^N dv_{g_i} = 1 \quad \text{for } i = 1,2$$

and

$$\frac{4(n-1)}{n-2} \int_{M_i \sim B_{\delta_i}(x_i)} |\nabla^i u_{\delta_i}|^2 dv_{g_i} + \int_{M_i \sim B_{\delta_i}(x_i)} R_i u_{\delta_i}^2 dv_{g_i} =$$

$$= J_i(u_{\delta_i}) < \mu_i + \varepsilon' .$$

Thus

$$\mu_g = J_g(\tilde{u}) < (\mu_1 + \mu_2 + 2\varepsilon')2^{-2/N} = (\mu_1 + \mu_2)2^{-2/N} + \varepsilon .$$

Remark.-

In general, the two statement above are independent, but if $\max(\mu_1,\mu_2) = \min(\mu_1,\mu_2)\,(2^{2/N} - 1)$ then

$2^{-2/N}(\mu_1,\mu_2) = \min(\mu_1,\mu_2)$, and theorem 5 implies theorem 6.

That condition is accomplished, for instance, if at least one of the two reals μ_1, μ_2 is positive.

On the contrary, if max (μ_1,μ_2) = min (μ_1,μ_2) $(2^{2/N} - 1)$, Theorem 6 is stronger than Theorem 5.

The applications of the two Theorems above will be different. Theorem 5 is related with Aubin's Conjecture which asserts that for every compact n-diemsional Riemannian manifold (M,g) other than those conformal to the sphere, $\mu_g < n(n-1)\omega_n^{2/N}$. In fact, we have

Corollary 7.-

Same hypothesis as in Theorem 5. If μ_1 (resp. μ_2) is strictly less than $n(n-1)\omega_n^{2/n}$ and if C is a conformal class of metrics on $M_1 \# M_2$ wich contains a metric g such that g agrees with g_1 (resp. g_2) on a subset $M_1 \sim C_2$) where C_1 (resp. C_2) is a sufficiently small n-cell of M_1 (resp. M_2) then $\mu_C < n(n-1)\omega_n^{2|n}$.

Proof.-

The proof is similar to the proof of Theorem 5 by taking $\varepsilon > 0$ such that min $(\mu_1,\mu_2) + \varepsilon < n(n-1)\omega_n^{2/n}$.

Remark.-

μ_C makes sense because of the conformal invariance of the infimum of the Yamabe functional

On the other hand, Theorem 6 is going to allow us to prove that on every com-- pact manifold, there is a metric g with μ_g as small as wanted.

It is known that every compact manifold admits a metric such that the infimum of the Yamabe functional is negative. Using that, we are going to show first that the assertion is true for the sphere.

Proposition 8.-

For each real number K there is a metric g on S_n such that $\mu_g < K$.

Proof.-

Let us assume that there is a $K_0 \in \mathbb{R}$ such that for every metric g on S_n $\mu_g = K_0$. Let μ_0 be defined by $\mu_0 = \inf \mu_g$ for $g \in \mathcal{M}(S_n)$ where $\mathcal{M}(S_n)$ is the set of the metrics on S_n. Then $\mu_0 < 0$.

For a given $\varepsilon > 0$ there is a metric $\tilde{g}$ on S_n such that $\mu_{\tilde{g}} < \mu_0 + \varepsilon$. Using Theorem 6 for $M_1 = M_2 = S_n$ and $g_1 = g_2 = \tilde{g}$, a metric g' on $S_n \# S_n$ (which is diffeomorphic to S_n) can be found so that $\mu_{g'} < 2^{-2/N}(\mu_1 + \mu_2) + \varepsilon = 2^{2/n}\mu_{\tilde{g}} + \varepsilon < 2^{2/n}\mu_0 + (2^{2/n} + 1)\varepsilon$, in particular if

$0 < \varepsilon \leq -\mu_0(2^{2/n} - 1)/(2^{2/n} + 1)$, then the metric obtained as above satisfies $\mu_g < \mu_0$ which is impossible.

Theorem 9.-

For every compact manifold M and for each number K there is a metric g on M such that $\mu_g < K$.

Proof.-

For a given $\varepsilon > 0$ there is a metric g_1 on S_n $(n = \dim M)$ such that $\mu_{g_1} < K - \varepsilon$.

Let g_2 be an arbitrary chosen metric on M. Using Theorem 5 for $(M_1, g_1) = (S_n, g_1)$ and $(M_2, g_2) = (M.g_2)$ one proves the existence of a metric g on $M\#S_n$, and consequently on M, such that $\mu_g < \min(\mu_1, \mu_2) + \varepsilon$ and then $\mu_g < \mu_{g_1} + \varepsilon < K$.

Corollary 10.-

For every compact manifold M and for each real number K, there is a metric g on M whose scalar curvature R is constant, $R < K$ and $\mathrm{vol}_g(M) = 1$.

Proof.-

Follows directly from Theorem 9 and Result 2.

Remark.-

A slightly different discussion shows (in [4]) that Theorem 5 can be improved if the manifolds are assumed to be locally conformally flat; namely we have.

Theorem ([4]).-

Let $(M_i,g_i)(i=1,2)$ be an n-dimensional (n> 2), compact, locally conformally flat Riemannian manifold and let $\mu_i(i=1,2)$ be the infimum of the Yamabe functional - for the metric g_i. If $(M_1 \# M_2,\tilde{g})$ is the connected sum, then $\mu_C = \min(\mu_1,\mu_2)$. Where C is the conformal class of $\tilde{g}$.

The proof depends on the fact that, in this case, for each $\delta > 0$, there exists a metric $\tilde{g}'$, in C, such that $\tilde{g}'$ is conformal to g_1 in a set of the form $M_1 \sim B$ where B is a geodesic ball of M_1 of radius δ. This result is not true for generic manifolds.

Example.-

Let S_n be the sphere and let g be a metric on S_n with $\mu_g < 0$. For a given $\varepsilon > 0$ a metric $\tilde{g}$ on $S_n \# S_n = S_n$ constructed by removing small enough balls must verify $\mu_{\tilde{g}} < 2^{2/n}\mu_g + \varepsilon$ and then, when $\delta < \mu_g(1-2^{2/n})$ we must have $\mu_{\tilde{g}} < \mu_g$.

On the other hand, if we take antipodal points $x,y \in S_n$ and the connected sum $S_n \# S_n$ is made by gluing $S_n \sim C(x)$ with $S_n \sim C(y)$ where C(x) (resp. C(y)) is the hemisphere centered at x (resp. at y), the so obtained metric is exactly g. The number μ_g being a conformal invariant, g and $\tilde{g}$ are not in the same conformal class

REFERENCES

[1] AUBIN, T.- "Equations différentielles non linéaires et problème de Yamabe concernant la coubure scalaire" J. Math. pues et appl. 55 (1976) 269-296.

[2] AUBIN, T.- "Nonlinear Analysis on Manifolds. Monge-Ampère Equations".Springer, New-York, 1982.

[3] AVEZ, A.- "Valeur moyenne du scalaire de courbure sur une variété compacte. Applications relativistes". C.R. Acad. Sci. Paris 256(1963),5271-5273.

[4] GIL-MEDRANO, O.- "On the Yamabe Problem concerning the compact locally conformally flat manifolds". To appear in J. of Funct. Anal.

[5] YAMABE, H.- "On the deformation of Riemannian structures on compact manifolds". Osaka Math. J. 12(1960) 21-37.

ISOCLINIC WEBS W(4,2,2) OF MAXIMUM 2-RANK

V. V. Goldberg
Department of Mathematics
New Jersey Institute of Technology
Newark, NJ 01102/USA

0. Statement of Results

The author showed (see [1,2]) that webs $W(4,2,2)$ of maximum 2-rank are exceptional in the sense that they are not necessarily algebraizable while a web $W(d,2,2)$, $d > 4$, is of maximum 2-rank if and only if it is algebraizable. It has been proved in [1,2] that a web $W(4,2,2)$ is of maximum 2-rank if and only if it is almost algebraizable (in this case it is necessarily isoclinic) or it is a non-isoclinic almost Grassmannizable web satisfying certain conditions.

However, the existence of webs $W(4,2,2)$ of maximum 2-rank was not discussed in [1,2]. In the present paper the author proves the existence of isoclinic webs of maximum 2-rank presenting a step-by-step construction of such a web. This construction starts from a given isoclinic three-web $W(3,2,2)$ which eventually will be a three-subweb of the constructed $W(4,2,2)$. For a given isoclinic web $W(3,2,2)$ its extension to a web $W(4,2,2)$ of maximum 2-rank can be unique or can depend on s, $s = 1,2,3$, constants. The construction is realized in three examples.

Note that the last two examples are the first examples of non-algebraizable webs $W(4,2,2)$ of maximum 2-rank.

1. Almost Grassmannizable and Almost Algebraizable Webs

A four-web $W(4,2,2)$ of codimension 2 is given in an open domain D of a differentiable 4-dimensional manifold X^4 by four 2-codimensional foliations X_a, $a = 1,2,3,4$, in D if the tangent 2-planes to the leaves (web surfaces) of X_a passing through any point of D are in general position.

Note that the first number in the notation $W(4,2,2)$ gives the number of foliations, the third one means the codimension and the second number is the ratio of the dimension of the ambient manifold and the codimension.

Two webs $W(4,2,2)$ and $\tilde{W}(4,2,2)$ are equivalent to each other if there exists a local diffeomorphism $\phi: D \to \tilde{D}$ transferring the foliations of W into the foliations of $\tilde{W}$.

The foliations X_a can be given by completely integrable systems of Pfaffian equations $\underset{a}{\omega}^i = 0$, $a = 1,2,3,4$; $i = 1,2$, where the forms $\underset{1}{\omega}^i$ and $\underset{2}{\omega}^i$ are the basis forms of X^4 and

$$\begin{cases} -\underset{3}{\omega}^i = \underset{1}{\omega}^i + \underset{2}{\omega}^i, \quad -\underset{4}{\omega}^i = \lambda^i_j \underset{1}{\omega}^j + \underset{2}{\omega}^i, \\ \det(\lambda^i_j) \neq 0, \quad \det(\delta^i_j - \lambda^i_j) \neq 0 . \end{cases} \quad i,j = 1,2, \tag{1.1}$$

The quantities λ^i_j form a (1,1)-tensor. It is called the <u>basis affinor</u> of W(4,2,2) [3,4].

For $x \in D \subset X^4$ we have

$$dx = \underset{1}{\omega}^i \vec{e}^{\,1}_i + \underset{2}{\omega}^i \vec{e}^{\,2}_i . \tag{1.2}$$

It follows from (1.1) and (1.2) that the vectors $\vec{e}^{\,2}_i$, $\vec{e}^{\,1}_i$, $\vec{e}^{\,2}_i - \vec{e}^{\,1}_i = -\vec{e}^{\,3}_i$, and $\vec{e}^{\,1}_i - \lambda^j_i \vec{e}^{\,2}_j = \vec{e}^{\,4}_i$ are tangent vectors to leaves V_1, V_2, V_3, and V_4 at the point x.

Let V be a 2-dimensional surface in D which is determined by the system $\gamma \underset{1}{\omega}^i + \underset{2}{\omega}^i = 0$ where γ is a function of a point $x \in D$. On the surface V we have $dx = \underset{1}{\omega}^i (\vec{e}^{\,1}_i - \gamma \vec{e}^{\,2}_i)$.

A web W(4,2,2) whose basis affinor λ^i_j is scalar:

$$\lambda^i_j = \delta^i_j \lambda \tag{1.3}$$

is said to be an <u>almost Grassmannizable</u> web. We will denote it by AGW(4,2,2).

The vectors $\vec{\xi}^a = \xi^i \vec{e}^{\,a}_i$, $a = 1,2,3,4$, are tangent to the leaves V_a at the point x. For AGW(4,2,2) they lie in a 2-plane. The bivector $\vec{\xi}^1 \wedge \vec{\xi}^2$ determined by $\vec{\xi}^a$ is said to be a <u>transversal bivector</u> of AGW(4,2,2). Equation (1.2) shows that the tangent plane of V intersects $\vec{\xi}^1 \wedge \vec{\xi}^2$ in the direction of the vector $\vec{\xi} = \xi^i(\vec{e}^{\,1}_i - \gamma \vec{e}^{\,2}_i)$. The anharmonic ratio of $\vec{\xi}$ and $\vec{\xi}^1$, $\vec{\xi}^2$, $\vec{\xi}^3(\vec{\xi}^4)$ is equal to $\gamma(\gamma/\lambda)$ and does not depend on ξ^i. The surface V is called an <u>isoclinic surface</u> of AGW(4,2,2).

A web AGW(4,2,2) is said to be <u>isoclinic</u> if there exists a one-parametric family of isoclinic surfaces through any point $x \in D$.

A web AGW(4,2,2) is said to be <u>transversally geodesic</u> if for any $\vec{\xi}^1 \wedge \vec{\xi}^2$ there exists a two-dimensional surface W tangent to $\vec{\xi}^1 \wedge \vec{\xi}^2$ at x and each $\vec{\xi}^1 \wedge \vec{\xi}^2$ is tangent to one and only one W.

A web AGW(4,2,2) which is isoclinic and transversally geodesic is a <u>Grassmannizable web</u>, i.e., it is equivalent to a <u>Grassmann 4-web</u> formed by 4 foliations of Schubert varieties of codimension 2 on the Grassmannian G(1,3) in a 5-dimensional projective space P^5. Each foliation X_a of Schubert varieties is the image of the bundles of straight lines of a three-dimensional projective space P^3 whose vertices are on the hypersurfaces U_a. If hypersurfaces U_a belong to an algebraic surface V_4^2 of degree 4, the Grassmann web is algebraic. A Grassmannizable web which is equivalent to an algebraic web is said to be algebraizable.

For an almost Grassmannizable web AGW(4,2,2) condition (1.3) allows us to write (1.1) in the form

$$-\underset{3}{\omega}^i = \underset{1}{\omega}^i + \underset{2}{\omega}^i, \quad -\underset{4}{\omega}^i = \lambda\underset{1}{\omega}^i + \underset{2}{\omega}^i, \quad \lambda \neq 0,1. \tag{1.4}$$

In addition, if a web AGW(4,2,2) is isoclinic, we have for it (see [1,2]):

$$\begin{cases} d\underset{1}{\omega}^i = \underset{1}{\omega}^j \wedge \omega_j^i + a_j \underset{1}{\omega}^j \wedge \underset{1}{\omega}^i, \\ d\underset{2}{\omega}^i = \underset{2}{\omega}^j \wedge \omega_j^j - a_j \underset{2}{\omega}^j \wedge \underset{2}{\omega}^i, \end{cases} \tag{1.5}$$

$$d\omega_j^i - \omega_j^k \wedge \omega_k^i = b^i_{jk\ell} \underset{1}{\omega}^k \wedge \underset{1}{\omega}^\ell, \tag{1.6}$$

$$d\lambda = \lambda(b_i - a_i)\underset{1}{\omega}^i + (b_i - \lambda a_i)\underset{2}{\omega}^i, \tag{1.7}$$

$$da_i - a_j\omega_i^j = (\underset{1}{k}_{ij} - \underset{3}{k}_{ij})\underset{1}{\omega}^j + (\underset{2}{k}_{ij} - \underset{3}{k}_{ij})\underset{2}{\omega}^j, \tag{1.8}$$

$$db_i - b_j\omega_i^j = [b_i(b_j - a_j) + \lambda(\underset{1}{k}_{ij} - \underset{4}{k}_{ij})]\underset{1}{\omega}^j + (\underset{2}{k}_{ij} - \underset{4}{k}_{ij})\underset{2}{\omega}^j. \tag{1.9}$$

Here $\underset{a}{k}_{ij}$, $a = 1,2,3,4$; $i,j = 1,2$, are symmetric in i and j. The quantities

$$a^i_{jk} = a_{[j}\delta^i_{k]}, \tag{1.10}$$

$$b^i_{jk\ell} = a^i_{jk\ell} + \underset{1}{k}_{jk}\delta^i_\ell + \underset{2}{k}_{\ell j}\delta^i_k + \underset{3}{k}_{k\ell}\delta^i_j, \tag{1.11}$$

are the torsion and curvature tensors of such W(4,2,2).

A web W(4,2,2) is isoclinic if and only if the Pfaffian derivatives of a_i are symmetric (see [1,2]). In our case these derivatives are:

$$p_{ij} = \underset{1}{k}_{ij} - \underset{3}{k}_{ij}, \quad q_{ij} = \underset{2}{k}_{ij} - \underset{3}{k}_{ij}. \tag{1.12}$$

The quantities $a^i_{jk\ell}$ in (1.11) are symmetric in their lower indices and satisfy the condition (see [5])

$$a^i_{ik\ell} = 0 \tag{1.13}$$

which allows us to find $\underset{3}{k}_{ij}$:

$$\underset{3}{k}_{ij} = \frac{1}{4}b^k_{(kij)} - \frac{1}{3}(p_{ij} + q_{ij}). \tag{1.14}$$

Exterior differentiation of (1.8) and (1.6) gives

$$(\nabla \underset{1}{k}_{ij} - \nabla \underset{3}{k}_{ij}) \wedge \omega^j + (\nabla \underset{2}{k}_{ij} - \nabla \underset{3}{k}_{ij}) \wedge \underset{2}{\omega}^j + a_k b^k_{ijm} \underset{1}{\omega}^j \wedge \underset{2}{\omega}^m$$

$$+ (\underset{1}{k}_{ij} - \underset{3}{k}_{ij})a_m \underset{1}{\omega}^m \wedge \underset{1}{\omega}^j - (\underset{2}{k}_{ij} - \underset{3}{k}_{ij})a_m \underset{2}{\omega}^m \wedge \underset{2}{\omega}^j = 0 \qquad (1.15)$$

$$[\nabla a^i_{sjm} + \nabla \underset{1}{k}_{sj} \delta^i_m + \nabla \underset{2}{k}_{ms} \delta^i_j + \nabla \underset{3}{k}_{jm} \delta^i_s + a_\ell b^i_{sjm}(\underset{1}{\omega}^\ell - \underset{2}{\omega}^\ell)] \underset{1}{\omega}^j \wedge \underset{2}{\omega}^m = 0 \qquad (1.16)$$

where

$$\nabla \underset{\alpha}{k}_{ij} = d\underset{\alpha}{k}_{ij} - \underset{\alpha}{k}_{mj} \omega^m_i - \underset{\alpha}{k}_{im} \omega^m_j, \quad \alpha = 1,2,3,$$

$$\nabla a^i_{jkm} = da^i_{jkm} - a^i_{\ell km} \omega^\ell_j - a^i_{j\ell m}\omega^\ell_k - a^i_{jk\ell}\omega^\ell_m + a^\ell_{jkm}\omega^i_\ell .$$

Contracting (1.16) with respect to i and s and using (1.13), we obtain

$$[\nabla \underset{1}{k}_{jm} + \nabla \underset{2}{k}_{jm} + 2\nabla \underset{3}{k}_{jm} + a_\ell(\underset{1}{k}_{jm} + \underset{2}{k}_{jm} + 2\underset{3}{k}_{jm}) (\underset{1}{\omega}^\ell - \underset{2}{\omega}^\ell)] \wedge \underset{1}{\omega}^j \wedge \underset{2}{\omega}^m = 0. \qquad (1.17)$$

It follows from (1.15) and (1.17) that $\underset{\alpha}{k}_{ij}$ have the form:

$$\nabla \underset{\alpha}{k}_{ij} = \underset{\alpha 1}{k}_{ijm} \omega^m + \underset{\alpha 2}{k}_{ijm} \omega^m, \quad \alpha = 1,2,3, \qquad (1.18)$$

where

$$\underset{11}{k}_{i[jm]} - \underset{31}{k}_{i[jm]} + (\underset{1}{k}_{i[j} - \underset{3}{k}_{i[j})a_{m]} = 0, \qquad (1.19)$$

$$\underset{22}{k}_{i[jm]} - \underset{32}{k}_{i[jm]} - (\underset{2}{k}_{i[j} - \underset{3}{k}_{i[j})a_{m]} = 0, \qquad (1.20)$$

$$a_s b^s_{ijm} = \underset{12}{k}_{ijm} - \underset{32}{k}_{ijm} - \underset{21}{k}_{imj} + \underset{31}{k}_{imj}, \qquad (1.21)$$

$$\underset{11}{k}_{i[jm]} + \underset{21}{k}_{i[jm]} + 2\underset{31}{k}_{i[jm]} + (\underset{1}{k}_{i[j} + \underset{2}{k}_{i[j} + 2\underset{3}{k}_{i[j})a_{m]} = 0, \qquad (1.22)$$

$$\underset{12}{k}_{i[jm]} + \underset{22}{k}_{i[jm]} + 2\underset{32}{k}_{i[jm]} - (\underset{1}{k}_{i[j} + \underset{2}{k}_{i[j} + 2\underset{3}{k}_{i[j})a_{m]} = 0. \qquad (1.23)$$

Alternating (1.21) first with respect to i and k and next with respect to i and m and using (1.11), we obtain correspondingly

$$\underset{21}{k}_{m[ij]} - \underset{31}{k}_{m[ij]} = a_{[i}(\underset{2}{k}_{j]m} - \underset{3}{k}_{j]m}), \qquad (1.24)$$

$$\underset{12}{k}_{j[im]} - \underset{32}{k}_{j[im]} = a_{[i}(-\underset{1}{k}_{m]j} + \underset{3}{k}_{m]j}). \qquad (1.25)$$

Substituting $\underset{1s}{k}_{i[jm]}$ and $\underset{2s}{k}_{i[jm]}$ from (1.19), (1.20), (1.24) and (1.25) into (1.22) and (1.23), we have

$$\underset{31}{k}_{i[jm]} = -\underset{3}{k}_{i[j}a_{m]}, \quad \underset{32}{k}_{i[jm]} = \underset{3}{k}_{i[j}a_{m]}. \qquad (1.26)$$

Equations (1.26), (1.19), (1.20), (1.24), and (1.25) give

$$\underset{\alpha 1}{k}{}_{i[jm]} = -\underset{\alpha}{k}{}_{i[j}a_{m]}\,, \quad \underset{\alpha 2}{k}{}_{i[jm]} = \underset{\alpha}{k}{}_{i[j}a_{m]}\,, \quad \alpha = 1,2,3. \tag{1.27}$$

Now, because of (1.27), all the equations (1.19), (1.20), (1.29), (1.23), (1.24), and (1.25) become the identities.

Equations (1.16), (1.18), and (1.27) imply

$$\nabla a^i_{sjm} = \underset{1}{a}{}^i_{sjm\ell}\,\underset{1}{\omega}{}^\ell + \underset{]}{a}{}^i_{sjm\ell}\,\underset{2}{\omega}{}^\ell \tag{1.28}$$

where

$$\underset{1}{a}{}^i_{sm[j\ell]} + a^i_{sm[j}a_{\ell]} = (\underset{21}{k}{}_{ms[j} + \underset{2}{k}{}_{sm}a_{[j})\delta^i_{\ell]}\,, \tag{1.29}$$

$$\underset{2}{a}{}^i_{sj[m\ell]} - a^i_{sj[m}a_{\ell]} = (\underset{12}{k}{}_{sj[m} - \underset{1}{k}{}_{sj}a_{[m})\delta^i_{\ell]}\,. \tag{1.30}$$

An isoclinic almost Grassmannizable web is transversally geodesic (and therefore Grassmannizable) if and only if (see [1,2])

$$a^i_{jk\ell} = 0\,. \tag{1.31}$$

Equations (1.28) and (1.31) imply

$$\underset{1}{a}{}^i_{jk\ell m} = 0\,, \quad \underset{2}{a}{}^i_{jk\ell m} = 0\,. \tag{1.32}$$

Equations (1.29), (1.30), (1.31), and (1.32) show that for a Grassmannizable web we have

$$(\underset{1}{k}{}_{ij[m} - \underset{1}{k}{}_{ij}a_{[m})\delta^i_{\ell]} = 0\,, \quad (\underset{2}{k}{}_{ij[m} + \underset{2}{k}{}_{ij}a_{[m})\delta^i_{\ell]} = 0\,. \tag{1.33}$$

It follows from (1.33) that for a Grassmannizable web the following identities hold:

$$\underset{1}{k}{}_{ijm} = \underset{1}{k}{}_{ij}a_m\,, \quad \underset{2}{k}{}_{ijm} = -\underset{2}{k}{}_{ij}a_m\,. \tag{1.34}$$

Note that starting from (1.15) we were dealing with the isoclinic three-subweb [1,2,3] of an isoclinic web AGW(4,2,2) formed by the foliations X_1, X_2, and X_3. The same considerations for any isoclinic three-web were done in [5] where slightly different notations were used.

An almost Grassmannizable web AGW(4,2,2) is said to be <u>almost algebraizable</u> if and only if its tensors $\underset{a}{k}{}_{ij}$, $a = 1,2,3,4$, satisfy the condition

$$\sum_{a=1}^{4} \underset{a}{k}{}_{ij} = 0\,. \tag{1.35}$$

We will denote such a web by AAW(4,2,2).

An almost algebraizable web AAW(4,2,2) is algebraizable if and only if it is transversally geodesic, i.e., satisfies (1.31).

The isoclinic three-subweb [1,2,3] of an isoclinic web AGW(4,2,2) is algebraizable if and only if the following condition holds (see [5]):

$$\underset{1}{k}_{ij} + \underset{2}{k}_{ij} + \underset{3}{k}_{ij} = 0 \qquad (1.36)$$

Note that for an algebraizable three-web $W(3,2,r)$, $r \geq 2$, the condition (1.31) of transversal geodesicity can be derived from (1.36) [1,2].

2. Webs W(4,2,2) of Maximum 2-rank

Suppose that the leaves of the foliations X_a of a web $W(4,2,2)$ are level sets $u^i_a(x) = \text{const}$ of functions $u^i_a(x)$, $x \in D$. The functions $u^i_a(x)$ are defined up to a local diffeomorphism in the space of u^i_a.

An exterior 2-equation of the form

$$\sum_{a=1}^{4} f_a(u^j_a)\, du^1_a \wedge du^2_a = 0 \qquad (2.1)$$

is said to be an abelian 2-equation. The number R_2 of linearly independent abelian 2-equations is called the 2-rank of $W(4,2,2)$ (see [6]).

The author has proved that $R_2 \leq 1$ and that a web $W(4,2,2)$ is of maximum 2-rank if and only if it is almost algebraizable or it is a non-isoclinic almost Grassmannizable web satisfying certain conditions [1,2]. Note that in both cases a web $W(4,2,2)$ of maximum 2-rank is almost Grassmannizable.

The webs $W(4,2,2)$ of maximum 2-rank are exceptional because they are not necessarily algebraizable while a web $W(d,2,2)$, $d > 4$, is of maximum 2-rank if and only if it is algebraizable [1,2].

The only abelian 2-equation for a web $W(4,2,2)$ of maximum 2-rank is (see [1,2]):

$$(\lambda-\lambda^2)\sigma\underset{1}{\omega}^1 \wedge \underset{1}{\omega}^2 + (\lambda-1)\sigma\underset{2}{\omega}^1 \wedge \underset{2}{\omega}^2 - \lambda\sigma(\underset{1}{\omega}^1+\underset{2}{\omega}^1) \wedge (\underset{1}{\omega}^2+\underset{2}{\omega}^2)$$
$$+ \sigma(\lambda\underset{1}{\omega}^1 + \underset{2}{\omega}^1) \wedge (\lambda\underset{1}{\omega}^2 + \underset{2}{\omega}^2) = 0, \qquad (2.2)$$

where σ is a solution of the completely integrable equation

$$d\ln[\sigma(\lambda-1)] = \omega^i_i + (a_i - b_i/\lambda)\underset{2}{\omega}^i . \qquad (2.3)$$

Note that (2.2) is an identity, and it is an abelian equation for an isoclinic web $W(4,2,2)$ of maximum 2-rank only under conditions (2.3) and (1.35).

3. Procedure for an Extension of an Isoclinic Web W(3,2,2) to an Isoclinic Web W(4,2,2) of Maximum 2-Rank

The main goal of the present paper is to construct examples of isoclinic webs W(4,2,2) of maximum 2-rank. As we saw earlier, such webs are almost algebraizable webs AAW(4,2,2).

In such a construction we will depart from a given isoclinic three-web W(3,2,2) and extend it to an AAW(4,2,2).

If an isoclinic three-web W(3,2,2) is given, it means that the forms $\underset{1}{\omega}^i$, $\underset{2}{\omega}^i$, $\underset{3}{\omega}^i$, ω_j^i and functions a_i, $\underset{\alpha}{k}_{ij}$, $\alpha=1,2,3$, a^i_{jk}, $b^i_{jk\ell}$, $a^i_{jk\ell}$ satisfying equations (1.4_1), (1.5), (1.6), (1.8), (1.10), (1.11), (1.13), (1.14), (1.18), (1.21), (1.27), (1.28), (1.29), and (1.30) are given.

To construct an AAW(4,2,2), we should find functions λ, b_i, and $\underset{4}{k}_{ij}$ satisfying (1.7), (1.9), and (1.35) and eventually find finite equations of the fourth foliation X_4 integrating the system

$$\lambda \underset{1}{\omega}^i + \underset{2}{\omega}^i = 0 . \tag{3.1}$$

We will suppose that three foliations X_1, X_2, and X_3 of the isoclinic three-web are given as level sets $u^i_\alpha = \text{const.}$, $\alpha=1,2,3$, of the following functions:

$$X_1\colon u^i_1 = x^i \ ; \quad X_2\colon u^i_2 = y^i \ ; \quad X_3\colon u^i_3 = f^i(x^j, y^k),\ i,j,k=1,2. \tag{3.2}$$

Let us indicate now four steps which we will perform to extend the isoclinic three-web (3.2) to an AAW(4,2,2).

Step 1. Find the forms $\underset{\alpha}{\omega}^i$, $\alpha = 1,2,3$, ω^i_j and the functions a^i_{jk}, $b^i_{jk\ell}$, a_i, $a^i_{jk\ell}$, $\underset{\alpha}{k}_{ij}$.

The forms and the functions a^i_{jk} **and** b^i_{jkl} can be found by means of the following formulas (see [7]):

$$\underset{1}{\omega}^i = \bar{f}^i_j\, dx^j \ , \quad \underset{2}{\omega}^i = \tilde{f}^i_j\, dy^j \ , \quad \underset{3}{\omega}^i = -\, dz^i \ , \tag{3.3}$$

where

$$\bar{f}^i_j = \partial f^i/\partial x^j \ , \quad \tilde{f}^i_j = \partial f^i/\partial y^j \ , \quad \det(\bar{f}^i_j) \neq 0 \ , \quad \det(\tilde{f}^i_j) \neq 0,$$

and

$$d\underset{1}{\omega}^i = -\, d\underset{2}{\omega}^i = \Gamma^i_{jk}\, \underset{1}{\omega}^j \wedge \underset{2}{\omega}^k \ , \tag{3.4}$$

$$\Gamma^i_{jk} = (-\partial^2 f^i/\partial x^\ell \partial y^m)\, \bar{g}^\ell_j \tilde{g}^m_k \ , \tag{3.5}$$

$$\omega^i_j = \Gamma^i_{kj} \underset{1}{\omega}^k + \Gamma^i_{jk} \underset{2}{\omega}^k , \tag{3.6}$$

$$a^i_{jk} = \Gamma^i_{[jk]} , \tag{3.7}$$

$$b^i_{jk\ell} = \frac{1}{2} \Big(\frac{\partial\Gamma^i_{k\ell}}{\partial x^m} \bar{g}^m_j + \frac{\partial\Gamma^i_{j\ell}}{\partial x^m} \bar{g}^m_k - \frac{\partial\Gamma^i_{kj}}{\partial y^m} \tilde{g}^m_\ell - \frac{\partial\Gamma^i_{k\ell}}{\partial y^m} \tilde{g}^m_j$$

$$+ \Gamma^m_{j\ell} \Gamma^i_{km} - \Gamma^m_{kj} \Gamma^i_{m\ell} + 2\Gamma^m_{k\ell} a^i_{mj} \Big) . \tag{3.8}$$

As to the functions a_i , $\underset{\alpha}{k}_{ij}$, $\alpha = 1,2,3$, and $a^i_{jk\ell}$, they can be easily calculated using (1.10), (1.8), (1.12), (1.14), and (1.11).

Step 2. Find $\underset{4}{k}_{ij}$, λ, and b_i.

The functions $\underset{4}{k}_{ij}$ can be found from (1.35). In order to find λ and b_i , we will take exterior derivatives of (1.9) using (1.5), (1.6), (1.7), (1.8), (1.9), and (1.18).

Equating to zero coefficients in $\underset{1}{\omega}^j \wedge \underset{1}{\omega}^k$, $\underset{2}{\omega}^j \wedge \underset{2}{\omega}^k$, and $\underset{1}{\omega}^j \wedge \underset{2}{\omega}^k$, by means of (1.27) we get two identities and

$$\lambda[(\underset{1}{k}_{ij} - \underset{4}{k}_{ij})a_\ell - (\underset{12}{k}_{ij\ell} - \underset{42}{k}_{ij\ell})] = -b_m a^m_{ij\ell} - 3b_{(j}\underset{4}{k}_{i\ell)} \tag{3.9}$$

$$-a_j(\underset{2}{k}_{i\ell} - \underset{4}{k}_{i\ell}) - (\underset{21}{k}_{i\ell j} - \underset{41}{k}_{i\ell j}) .$$

In general, equation (3.9) gives a dependence between λ, b_1 , and b_2. Differentiating it by means of (1.7), (1.8), (1.9), (1.28), (1.18) and its prolongations and equating to zero coefficients in linearly independent forms $\underset{1}{\omega}^i$ and $\underset{2}{\omega}^i$, we get new relations between λ, b_1 , and b_2. Some of them may be satisfied identically. Others should be checked on their compatibility among each other and (3.9). If no contradiction exists, the same procedure which has been applied to (3.9) should be applied to these new relations until all λ, b_1 , and b_2 will be found or no new relations between them will appear.

The following cases are possible:

i) The obtained relations between λ, b_1 , and b_2 are not compatible. It means that the given isoclinic three-web W(3,2,2) can not be extended to an AAW(4,2,2).

ii) The relations among λ, b_1 , and b_2 allow us to find s of them, s = 0,1,2,3. In this case other 3-s of these functions should be found by integrating the completely integrable system (1.7), (1.9) of differential equations for those 3-s functions. Its solution will depend on 3-s constants. In particular, if s = 0, all the functions, λ, b_1 , and b_2 will be uniquely determined and the given isoclinic three-web W(3,2,2) can be uniquely extended to an AAW(4,2,2).

<u>Step 3</u>. <u>Find finite equations of the fourth foliation X_4 of an AAW(4,2,2)</u> by integrating the completely integrable system (3.1) where ω^i_1 and ω^i_2 are determined by (3.3) and λ is determined in step 2.

<u>Step 4</u>. <u>Find the only abelian equation of the web AAW(4,2,2)</u>.

For this: (i) find σ by integrating (2.3); (ii) write the abelian equation in the form (2.2) substituting λ from step 2 and σ from (i) into (2.2); and (iii) write the abelian equation in the form (2.1) expressing $\underset{a}{\omega}^1 \wedge \underset{a}{\omega}^2$, a = 1,2,3,4, in terms of $du^1_a \wedge du^2_a$.

4. <u>Examples of Isoclinic Webs W(4,2,2) of Maximum 2-rank.</u>

<u>Example 1</u>. Suppose that a given isoclinic three-web W(3,2,2) is algebraizable. In this case we have (1.31), (1.32), (1.34), and (1.36). Equations (1.36) and (1.35) imply that

$$\underset{4}{k}_{ij} = 0 . \qquad (4.1)$$

It is clear that an extended web AAW(4,2,2) is algebraizable (see our remark after definition of an AAW(4,2,2)). Moreover, (1.36) and (4.1) show that, up to equivalence, it is generated by a surface V^2_4 of degree four in a projective space P^3 which is decomposed into a cubic surface V^2_3 and a plane V^2_1 (see [8]).

Conditions (4.1), (1.31), (1.32), (1.34), and (1.36) show that equation (3.9) has the form $0 \cdot \lambda = 0$. Thus the extended web AAW(4,2,2) of the given isoclinic three-web W(3,2,2) is determined by the completely integrable system (1.7), (1.9). Therefore <u>it depends on three constants</u>. One can consider coefficients of the equation of V^2_1 as these constants.

Thus <u>an algebraizable web W(3,2,2) can be extended to an AAW(4,2,2) and the last one is a particular case of an algebraizable web W(4,2,2)</u>.

<u>Example 2</u>. Let us consider the three-web W(3,2,2) defined by (see [9]):

$$X_1\colon x^1 = \text{const}, \quad x^2 = \text{const}; \quad X_2\colon y^1 = \text{const}, \quad y^2 = \text{const};$$
$$X_3\colon z^1 = x^1 + y^1 = \text{const}, \quad z^2 = (x^2+y^2)(y^1-x^1) = \text{const}. \qquad (4.2)$$

<u>Step 1</u>. Using (3.3)-(3.8), (1.10), (1.11), and (1.14) we have for (4.2):

$$\begin{cases} \Gamma^1_{ij} = \Gamma^2_{22} = 0 , \quad \Gamma^2_{11} = 2(x^2+y^2)/(x^1-y^1), \quad \Gamma^2_{21} = -\Gamma^2_{12} = 1/(x^1-y^1); \\ \omega^1_i = 0, \quad \omega^2_1 = (dx^1+dy^1)(x^2+y^2)/(x^1-y^1) - (dx^2-dy^2), \omega^2_2 = -d\ell n(x^1-y^1); \\ a_1 = 2/(y^1-x^1) , \; a_2 = 0 ; \quad p_{2i} = q_{2i} = 0, \quad p_{11} = -q_{11} = 2/(x^1-y^1)^2; \end{cases}$$

$$\begin{cases} b^1_{ijk} = b^2_{222} = b^2_{211} = b^2_{122} = b^2_{212} = b^2_{211} = 0 \ , \\ b^2_{112} = - b^2_{121} = 2/(x^1-y^1)^2 \ , \\ b^2_{111} = - 8(x^2+y^2)/(x^1-y^1)^2 \ ; \\ \underset{3}{k}_{ij} = 0 \ , \quad \underset{1}{k}_{11} = - \underset{2}{k}_{11} = 2/(x^1-y^1)^2 \ , \\ \underset{1}{k}_{ij} = \underset{2}{k}_{ij} = 0 \ , \quad (i,j) \neq (1,1) \ ; \\ a^1_{ijk} = 0 \ ; \quad a^2_{111} = b^2_{111} \ , \quad a^2_{ijk} = 0 \ , \quad (i,j,k) \neq (1,1,1). \end{cases} \tag{4.3}$$

Equations (4.3) show that the web W(3,2,2) is isoclinic and not transversally geodesic. Since we have for it

$$\underset{1}{k}_{ij} + \underset{2}{k}_{ij} = 0 \ , \quad \underset{3}{k}_{ij} = 0 \ , \tag{4.4}$$

it is natural to call it an almost Bol web (see [10] where algebraic Bol webs are introduced).

Step 2. It follows from (1.35) and (4.4) that

$$\underset{4}{k}_{ij} = 0 \ . \tag{4.5}$$

Equations (4.3) and (4.4) imply

$$\begin{gathered} \underset{3s}{k}{}_{ijm} = \underset{4s}{k}{}_{ijm} = 0 \ , \quad s = 1,2; \quad \underset{1s}{k}{}_{ijk} = \underset{2s}{k}{}_{ijk} = 0 \quad (i,j,k)\neq(1,1,1) \\ \underset{11}{k}{}_{111} = - \underset{12}{k}{}_{111} = - \underset{21}{k}{}_{111} = \underset{22}{k}{}_{111} = 4/(y^1-x^1)^3. \end{gathered} \tag{4.6}$$

By means of (4.3), (4.5), and (4.6), equations (3.9) can be written as

$$\lambda = 1 + b_2 z^2 \ , \tag{4.7}$$

where z^2 is defined by (4.2). Differentiation of (4.7) implies three identities and

$$b_1 = (2+b_2 z^2)/(y^1-x^1) \ . \tag{4.8}$$

Differentiation of (4.8) gives no new relations on λ, b_1, and b_2.

Equation (1.9) for i = 2 can be written by means of (4.3), (4.5), (4.7), and (4.8) in the form

$$d\ell n[b_2(x^1-y^1)] = b_2(y^1-x^1)\ dx^2 \ . \tag{4.9}$$

Integration of (4.9) gives

$$b_2 = 1/[(x^1-y^1)(x^2+ C)] \ , \tag{4.10}$$

where C is a constant. Equations (4.7), (4.8), and (4.10) imply

$$\lambda = (C-y^2)/(x^2+C) \ , \tag{4.11}$$

$$b_1 = (2C+x^2-y^2)/[(x^2+C)(y^1-x^1)] \ . \tag{4.12}$$

We can see that <u>an extension of the isoclinic web W(3,2,2) to an AAW(4,2,2) depend on one constant</u>.

<u>Step 3</u>. Substutiting λ from (4.11) and $\underset{1}{\omega}^i$, $\underset{2}{\omega}^i$ from (3.3) into (3.1), we have

$$-\alpha dx^1 + dy^1 = 0 \ , \quad 2(1+\alpha)\, dy^1 + (y^1-x^1)\, d\alpha = 0 \ , \tag{4.13}$$

where $\alpha = (y^2-C)/(x^2+C)$. It follows from (4.13) that

$$\frac{dx^1}{y^1-x^1} = \frac{dy^1}{\alpha(y^1-x^1)} = \frac{d\alpha}{-2\alpha(1+\alpha)} \ . \tag{4.14}$$

Two independent first integrals of (4.14) give a system defining the foliation X_4 of AAW(4,2,2):

$$\begin{cases} u_4^1 = u^1 = (x^1-y^1)^2(x^2+y^2)^2 / [(x^2+C)(y^2-C)] = \text{const}, \\ u_4^2 = u^2 = x^1 + y^1 + [(y^1-x^1)(x^2+y^2)/\sqrt{(x^2+C)(y^2-C)}\ ] \\ \qquad \cdot \arctan \sqrt{(y^2-C)/(x^2+C)} = \text{const}. \end{cases} \tag{4.15}$$

<u>Step 4</u>. Using (4.3), (4.11), (4.12), and (4.10), we can write equation (2.3) in the form

$$d\ell n[\sigma(x^2+y^2)/(x^2+C)] = -\, d\ell n[(y^1-x^1)(C-y^2)] \ . \tag{4.16}$$

It follows from (4.16) that

$$\sigma = A(x^2+C) / [(x^2+y^2)(y^1-x^1)(C-y^2)] \ , \tag{4.17}$$

where A is a constant. Taking $A = 1$, we get from (4.17)

$$\sigma = (x^2+C)/[(x^2+y^2)(y^1-x^1)(C-y^2)] \ . \tag{4.18}$$

By means of (4.18) and (4.11), the only abelian equation (2.2) for our web AAW(4,2,2) can be written in the form

$$\Omega_1 + \Omega_2 + \Omega_2 + \Omega_4 = 0 \tag{4.19}$$

where

$$\Omega_1 = [1/((y^1-x^1)(x^2+C))]\underset{1}{\omega}^1 \wedge \underset{1}{\omega}^2, \quad \Omega_3 = [((x^1-y^1)(x^2+y^2))]\underset{3}{\omega}^1 \wedge \underset{3}{\omega}^2 \ ,$$

$$\Omega_2 = [1/((y^1-x^1)(y^2-C))]\underset{2}{\omega}^1 \wedge \underset{2}{\omega}^2,$$

$$\Omega_4 = [(x^2+C) / ((y^1-x^1)(x^2+y^2)(C-y^2))]\underset{4}{\omega}^1 \wedge \underset{4}{\omega}^2 \ ,$$

and each of Ω_a, $a = 1,2,3,4$, is a closed 2-form (see [1,2]).

Using (4.2), (4.3), (1.4), and (4.11), we find

$$\begin{cases} \underset{1}{\omega}^1 \wedge \underset{1}{\omega}^2 = (y^1-x^1)\, dx^1 \wedge dx^2 , \quad \underset{3}{\omega}^1 \wedge \underset{3}{\omega}^2 = dz^1 \wedge dz^2 , \\ \underset{2}{\omega}^1 \wedge \underset{2}{\omega}^2 = (y^1-x^1)\, dy^1 \wedge dy^2 , \\ \underset{4}{\omega}^1 \wedge \underset{4}{\omega}^2 = [(y^2-C)^2/(2(y^1-x^1)(x^2+y^2))]\, du^1 \wedge du^2 . \end{cases} \tag{4.20}$$

Equations (4.20) allow us to write the abelian equation (4.20) in the form (2.1):

$$(1/x^2)\, dx^1 \wedge dx^2 + (1/y^2)\, dy^1 \wedge dy^2 - (1/z^2)\, dz^1 \wedge dz^2 - (1/(2u^1))\, du^1 \wedge du^2 = 0 . \tag{4.21}$$

Note that examples 1 and 2 were shortly described in the authors paper [11].

<u>Example 3</u>. Let a three-web $W(3,2,2)$ be given by

$$\begin{cases} \text{I.} \quad x^1 = \text{const}, \quad x^2 = \text{const}; \qquad \text{II.} \quad y^1 = \text{const}, \quad y^2 = \text{const}; \\ \text{III.} \quad z^1 = x^1+y^1 = \text{const}, \quad z^2 = -x^1y^2+x^2y^1 = \text{const}. \end{cases} \tag{4.22}$$

<u>Step 1</u>. By means of (3.3)-(3.8), (1.10), (1.11), and (1.14) we have for the web defined by (4.22):

$$\begin{cases} \Gamma^1_{ij} = \Gamma^2_{22} = 0, \quad \Gamma^2_{11} = x^2/x^1 - y^2/y^1 , \quad \Gamma^2_{12} = -1/x^1 , \quad \Gamma^2_{21} = -1/y^1; \\ a_1 = 1/y^1 - 1/x^1, \quad a_2 = 0, \quad p_{12} = q_{12} = p_{21} = q_{21} = p_{22} = q_{22} = 0 , \\ p_{11} = 1/(x^1)^2, \quad q_{11} = -1/(y^1)^2; \quad b^2_{111} = (1/y^1-1/x^1)\,(x^2/x^1 - y^2/y^1), \\ b^2_{121} = -1/(y^1)^2, \quad b^2_{112} = 1/(x^1)^2, \quad b^1_{ijk} = b^2_{211} = b^2_{122} = b^2_{212} = b^2_{221} = 0, \\ \underset{3}{k}_{11} = (1/(y^1)^2 - 1/(x^1)^2)/4, \quad \underset{1}{k}_{11} = (3/(x^1)^2 + 1/y^1)^2)4 , \\ \underset{2}{k}_{11} = -(1/(x^1)^2+3/(y^1)^2)/4, \quad \underset{\alpha}{k}_{ij} = 0, \quad (i,j) \neq (1,1); \ b^2_{222} = 0, \\ a^1_{111} = - a^2_{211} = - a^2_{111} = - a^2_{112} = (1/(y^1)^2 - 1/(x^1)^2)/4 , \\ a^2_{111} = b^2_{111} , \quad a^1_{ijk} = a^2_{122} = a^2_{212} = a^2_{221} = a^2_{222} = 0, \quad (i,j,k) \neq (1,1,1). \end{cases} \tag{4.23}$$

It follows from (4.23) that the web (4.22) is isoclinic and not transversally geodesic. Moreover, since $b^i_{(jk\ell)} \neq 0$, it is not a hexagonal web.

<u>Step 2</u>. Equations (1.35) and (4.23) imply that

$$\underset{4}{k}_{11} = \underset{3}{k}_{11} , \quad \underset{4}{k}_{ij} = 0 , \quad (i,j) \neq (1,1) \tag{4.24}$$

Equations (4.23) and (4.24) give

$$\begin{cases} \underset{21}{k}{}^{111} = \underset{31}{k}{}^{111} = \underset{41}{k}{}^{111} = -\frac{1}{3}\underset{11}{k}{}^{111} = 1/(2(x^1)^3) , \\ \underset{22}{k}{}^{111} = \underset{32}{k}{}^{111} = \underset{42}{k}{}^{111} = -\frac{1}{3}\underset{12}{k}{}^{111} = 1/(2(y^1)^3) , \\ \underset{as}{k}{}^{ijk} = 0 , \quad a = 1,2,3,4; \quad s = 1,2; \quad (i,j,k) \neq (1,1,1). \end{cases} \tag{4.25}$$

By virtue of (4.23), (4.24), and (4.25), equations (3.9) can be written in the form

$$\lambda = (x^1)^2[1/(y^1)^2 - b_1(1/x^1 + 1/y^1) - b_2(x^2/x^1 - y^2/y^1)] . \tag{4.26}$$

Differentiation of (4.26) leads to identities.

Therefore an extended web AAW(4,2,2) of the given isoclinic three-web (4.22) is defined by completely integrable system (1.7), (1.9). Equation (4.26) gives that this system contains only two independent equations (1.9).

Thus <u>an extended web AAW(4,2,2) depends on 2 constants</u>.

We will integrate the system (1.7), (1.9), (4.26) and find λ, b_1, and b_2 explicitly. For this, using (4.23) and (4.26), we write (1.9) in the form

$$db_1 = [1/(y^1)^2 + b_1(b_1 - 2/y^1) - (b_1 - 1/y^1)b_2y^2]\, dx^1$$
$$+ (b_1-1/y^1)\, b_2y^1\, dx^2 - (b_2y^2+1/y^1)/y^1\, dy^1 + b_2\, dy_2 , \tag{4.27}$$

$$db_2 = b_2(b_1 - b_2y^2 - 1/y^1)\, dx^1 + b_2^2y^1\, dx^2 - b_2/y^1\, dy^1 . \tag{4.28}$$

It follows from (4.27) and (4.28) that

$$d[(b_1 - 1/y^1) / (b_2y^1)] = d(y^2/y^1) . \tag{4.29}$$

Equation (4.29) gives

$$b_1 = b_2(y^2 + C_1y^1) + 1/y^1 \tag{4.30}$$

where C_1 is a constant.

Substituting b_1 from (4.30) into (4.28), we easily obtain

$$d(b_2y^1) / (b_2y^1)^2 = C_1\, dx^1 + dx^2 . \tag{4.31}$$

It follows from (4.31) that

$$b_2 = -(x^2 + C_1x^1 + C_2)^{-1} / y^1 \tag{4.32}$$

where C_2 is a constant.

Equation (4.32) allows us to express b_1 determined by (4.30) in the form

$$b_1 = [C_2 + C_1(x^1 - y^1) + x^2 - y^2]\ (x^2+C_1x^1+C_2)^{-1}/y^1 . \quad (4.33)$$

Equations (4.26), (4.32), (4.33) give the following expression for λ:

$$\lambda = x^1(y^2 + C_1y^1 - C_2)\ (x^2 + C_1x^1 + C_2)^{-1}/y^1 . \quad (4.34)$$

If we take $C_1 = C_2 = 0$, equations (4.32), (4.33), and (4.34) become

$$b_2 = -1/(x^2y^1)\ , \quad b_1 = (1-y^2/x^2)/y^1\ , \quad \lambda = x^1y^2/(y^1x^2)\ . \quad (4.35)$$

Step 3. Let us denote

$$\alpha = (y^2+C_1y^1-C_2)/(x^2+C_1x^1+C_2)\ . \quad (4.36)$$

Then $\lambda = \alpha x^1/y^1$, and using (4.23), we can write (3.1) in the form

$$\begin{cases} (\alpha x^1/y^1)\ dx^1 + dy^1 = 0\ , \\ (1+\alpha)\ dy^1 - x^1\ d\alpha\ = 0\ . \end{cases} \quad (4.37)$$

Equations (4.37) can be also written in the form

$$\frac{dx^1}{-y^1} = \frac{dy^1}{x^1\alpha} = \frac{d\alpha}{\alpha(1+\alpha)}\ . \quad (4.38)$$

If we denote the common value of the expressions in (4.38) by dt/t, then we find

$$t = \alpha/(1+\alpha) \quad (4.39)$$

and

$$\frac{dx^1}{dt} = -\frac{y^1}{t}\ , \qquad \frac{dy^1}{dt} = \frac{x^1}{1-t}\ . \quad (4.40)$$

Eliminating y^1, we get from (4.40)

$$t(t-1)\ \frac{d^2x^1}{dt^2} + (t-1)\ \frac{dx^1}{dt} - x^1 = 0\ , \quad (4.41)$$

or

$$\frac{d}{dt}\ [t(t-1)\ \frac{dx^1}{dt} - tx^1] = 0\ . \quad (4.42)$$

Equations (4.40), (4.42), and (4.39) give

$$\begin{cases} x^1 = -A\left(1 + \frac{1}{1+\alpha}\ \ln|\alpha|\right) - B\ \frac{1}{1+\alpha}\ , \\ y^1 = \ \ A\left(1 - \frac{\alpha}{1+\alpha}\ \ln|\alpha|\right) - B\ \frac{\alpha}{1+\alpha}\ , \end{cases} \quad (4.43)$$

where A and B are arbitrary constants.

Solving (4.43) for A and B, we obtain two independent first integrals of (4.37) defining the foliation X_4 of the web AAW(4,2,2):

$$\begin{cases} u_4^1 = u^1 = (z^2 + C_1 z^1)/(x^2+y^2+C_1 z^2) = \text{const}, \\ u_4^2 = u^2 = -u^1 \ln|(y^2+C_1 y^1-C_2)/(x^2+C_1 x^1+C_2)| - z^1 = \text{const}. \end{cases} \tag{4.44}$$

Step 4. Equations (4.23), (4.32), (4.33), and (4.34) allow us to write equation (2.3) in the form

$$d\ln[x^1 y^1(\lambda-1)\sigma] = d\ln[y^1/(y^2 + C_1 y^1 - C_2)] . \tag{4.45}$$

Integrating (4.45) and taking the appropriate constant of integration, we get

$$\sigma = 1/[x^1(\lambda-1)(y^2 + C_1 y^1 - C_2)] . \tag{4.46}$$

By means of (4.46) and (4.34), the only abelian 2-equation (2.2) for our web AAW(4,2,2) has the form

$$\Omega_1 + \Omega_2 + \Omega_3 + \Omega_4 = 0 \tag{4.47}$$

where

$$\Omega_1 = -[1/(y^1(x^2+C_1 x^1+C_2))] \underset{1}{\omega}^1 \wedge \underset{1}{\omega}^2, \quad \Omega_3 = (z^2+C_2 z^1)^{-1} \underset{3}{\omega}^1 \wedge \underset{3}{\omega}^2,$$

$$\Omega_2 = [1/(x^1(y^2+C_1 y^1-C_2))] \underset{2}{\omega}^1 \wedge \underset{2}{\omega}^2, \quad \Omega_4 = -(z^2+C_2 z^1)^{-1}\lambda^{-1} \underset{4}{\omega}^1 \wedge \underset{4}{\omega}^2,$$

and each of Ω_a, $a = 1,2,3,4$, is a closed 2-form (see [1,2]).

Using (4.22), (4.23), (1.4), (4.34), and (4.44), we find

$$\begin{aligned} &\underset{1}{\omega}^1 \wedge \underset{1}{\omega}^2 = y^1\, dx^1 \wedge dx^2, \quad &&\underset{3}{\omega}^1 \wedge \underset{3}{\omega}^2 = dz^1 \wedge dz^2, \\ &\underset{2}{\omega}^1 \wedge \underset{2}{\omega}^2 = -x^1\, dy^1 \wedge dy^2, \quad &&\underset{4}{\omega}^1 \wedge \underset{4}{\omega}^2 = \lambda(x^2+y^2+C_1 z^1)\, du^1 \wedge du^2 . \end{aligned} \tag{4.48}$$

Equations (4.48) allow us to write the abelian 2-equation (4.47) in the form of (2.1):

$$-\frac{1}{x^2+C_1 x^1+C_2} dx^1 \wedge dx^2 - \frac{1}{y^2+C_1 y^1-C_2} dy^1 \wedge dy^2$$

$$+\frac{1}{z^2+C_2 z^1} dz^1 \wedge dz^2 - \frac{1}{u^1} du^1 \wedge du^2 = 0 . \tag{4.49}$$

Note in conclusion that the four-webs constructed in examples 2 and 3 are the first examples of non-algebraizable webs W(4,2,2) of maximum 2-rank.

References

[1] V. V. Goldberg, Tissus de codimension r et de r-rang maximum, C. R. Acad. Sci., Paris, Sér. I, 297 (1983), pp. 339-342.

[2] V. V. Goldberg, r-Rank problems for a web W(d,2,r), submitted.

[3] V. V. Goldberg, On the theory of four-webs of multidimensional surfaces on a differentiable manifold X_{2r} (in Russian), Izv. Vyssh. Uchebn. Zaved., Mat. 21 (1977), No. 11, pp. 118-121. English translation: Soviet Mathematics (Iz. VUZ) 21 (1977), No. 11, pp. 97-100.

[4] V. V. Goldberg, A theory of multidimensional surfaces on a differentiable manifold X_{2r} (in Russian), Serdica 6 (1980), No. 2, pp. 105-119.

[5] M. A. Akivis, On isoclinic three-webs and their interpretation in a ruled space of projective connection (Russian), Sib. Mat. Zh. 15 (1974), No. 1, pp. 3-15. English translation: Sib. Math. J. 15 (1974), No. 1, pp. 1-9.

[6] P. A. Griffiths, On Abel's differential equations. Algebraic Geometry, J. J. Sylvester Sympos., Johns Hopkins Univ., Baltimore, Md., 1976, pp. 26-51. Johns Hopkins Univ. Press, Baltimore, Md., 1977.

[7] M. A. Akivis and A. M. Shelekhov, On the computation of the curvature and torsion tensors of a multidimensional three-web and of the associator of the local quasigroup connected with it (in Russian). Sib. Mat. Zh. 12 (1971), No. 5, pp. 953-960. English translation: Sib. Math. J. 12 (1971), No. 5, pp. 685-689.

[8] V. V. Goldberg, Grassmann and algebraic four-webs in a projective space, Tensor, New Ser. 38 (1982), pp. 179-197.

[9] G. Bol, Über Dreigewebe in vierdimensionalen Raum. Math. Ann. 110 (1935), pp. 431-463.

[10] M. A. Akivis, The local differentiable quasigroups and three-webs that are determined by a triple of hypersurfaces (in Russian) Sib. Mat. Zh. 14 (1973), No. 3, pp. 467-474. English translation: Sib. Math. J. 14 (1973), No. 3, pp. 319-324.

[11] V. V. Goldberg, 4-tissus isoclines exceptionnels de codimension deux et de 2-rang maximum, C. R. Acad. Sci., Paris, Sér. 1 301 (1985), pp. 593-596.

ALMOST TRANSVERSALLY SYMMETRIC FOLIATIONS

F.W. Kamber[1], E.A. Ruh[2] and Ph. Tondeur[1]

([1]) Department of Mathematics, University of Illinois at Urbana Champaign Urbana, IL 61801

([2]) Department of Mathematics, Ohio State University, Columbus, Ohio 43210

1. In this paper we compare Riemannian foliations with transversally homogeneous foliations, where the model transverse structure is of the type of a compact symmetric space G/K. The datum needed for comparison is a connection in the normal bundle, having similar properties as the canonical connection in the case of a transversally symmetric foliation. This similarity is most conveniently formulated in terms of the corresponding Cartan connections. For the symmetric model case the curvature of the Cartan connection vanishes. An almost transversally symmetric foliation is one where this curvature is small in an appropriate norm. In the spirit of Rauch's comparison theorem [RA], and more specifically the comparison theorem of Min-Oo and Ruh [MR], we wish to conclude that this assumption already implies the existence of a transversally symmetric structure of type G/K. We succeed in doing so for harmonic Riemannian foliations, i.e. foliations where all leaves are minimal submanifolds [KT 2].

The precise result is as follows.

THEOREM. <u>Let</u> F <u>be a transversally oriented harmonic Riemannian foliation of codimension</u> $q > 2$ <u>on the compact oriented manifold</u> (M,g_M). <u>Let</u> G/K <u>be an irreducible compact symmetric space. Then there exists a constant</u> $A > 0$ <u>with the following property. If</u> η <u>is a basic</u> $\mathfrak{k}$<u>-connection in a foliated</u> K<u>-reduction of the normal frame bundle with Cartan</u> $\mathfrak{g}$<u>-connection</u> ω <u>and curvature</u> Ω_ω, <u>then</u> $||\Omega_\omega|| < A$ <u>implies that</u> F <u>is transversally symmetric of type</u> G/K.

2. First we explain the terminology in more detail. Let M be a manifold and F a foliation of codimension q on M. Let G/K be a symmetric space of compact type with G and K connected, and with $\dim G/K = q$. F is transversally homogeneous of type G/K if F is given on an atlas of distinguished charts $\mathcal{U} = \{U_\alpha\}$ by local submersions $f_\alpha : U_\alpha \to G/K$, which on $U_\alpha \cap U_\beta$ are related by the left action of an element $g \in G : f_\alpha = gf_\beta$ (see e.g. Blumenthal [B 1]). This can be expressed in terms of the frame bundle $F(Q)$ of the normal bundle Q of F as follows. The isotropy representation of G/K shows that $K \subset SO(q)$. Then there is a K-reduction $K \longrightarrow P \xrightarrow{\pi} M$ of $F(Q)$ with a foliated bundle structure [KT 1]. A

$\underline{\underline{k}}$-valued adapted connection η in P gives rise to $\underline{\underline{g}}$-valued Cartan connection

$$\omega = \eta + \theta .$$

Here θ is the canonical $\mathbb{R}^q$-valued (solder) 1-form on P defined by

$$\theta(X) = u^{-1}(\pi(X)) \quad \text{for} \quad X \in T_u P$$

where the frame u of Q at $\pi(u)$ is considered as a linear map $\mathbb{R}^q \to Q_{\pi(u)}$. The curvature

$$\Omega_\omega = d\omega + \frac{1}{2}[\omega,\omega]$$

is then expressed in terms of the curvature $\Omega_\eta = d\eta + \frac{1}{2}[\eta,\eta]$ and torsion $\Sigma_\eta = d\theta + [\eta,\theta]$ by

$$\Omega_\omega = \Omega_\eta + \frac{1}{2}[\theta,\theta] + \Sigma_\eta .$$

For the unique torsion-free connection η the symmetric space structure implies $\Omega_\eta = -\frac{1}{2}[\theta,\theta]$ and thus $\Omega_\omega = 0$.

The foliation F lifts canonically to a K-invariant foliation $\tilde{F}$ on P. A tangent vectorfield $X \in \Gamma L$ lift to a tangent vectorfield $\tilde{X} \in \Gamma\tilde{L}$. An adapted $\underline{\underline{k}}$-connection η in P is basic if $i_{\tilde{X}}\Omega_\eta = 0$ for all $\tilde{X} \in \Gamma\tilde{L}$ [KT 1]. This condition is equivalent to $i_{\tilde{X}}\Omega_\omega = 0$ for all $\tilde{X} \in \Gamma\tilde{L}$, and the curvature Ω_ω of the corresponding Cartan connection.

A foliation F on M is said to be <u>almost transversally symmetric of type</u> G/K, if there exists a foliated K-reduction P of $F(Q)$ and a basic K-connection η with small curvature Ω_ω for the corresponding basic Cartan connection ω. The norm $||\Omega_\omega||$ is measured in terms of a Riemannian metric g_M on M and a biinvariant metric on G. Since $K \subset SO(q)$, these foliations are necessarily Riemannian. It is therefore no restriction to assume g_M to be a bundle-like metric [RE].

A Riemannian foliation F is harmonic, if all leaves are minimal submanifolds of (M,g_M) [KT 2]. The result stated in the theorem above is then that a harmonic Riemannian foliation which is almost transversally symmetric of type G/K, is necessarily transversally symmetric.

For the foliation of M by points the theorem reduces to the result proved by Min-Oo and Ruh in [MR].

A consequence of the theorem is the existence of a developing map $M \xrightarrow{f} \widetilde{G/K}$ to the universal cover of G/K for the lift $\tilde{F}$ of a harmonic almost transversally symmetric foliation F to the universal cover $\tilde{M}$ (see Haefliger [H] and Blumenthal [B 1]). If $h : \pi_1 M \to G$ denotes the holonomy homomorphism of F with image group Γ, then f is an h-equivariant submersion. This implies by [B 2] that the cohomology $H_B(F)$ of basic forms of F is isomorphic to the De Rham cohomology of $\widetilde{G/K}$: $H_B(F) \cong H_{DR}(\widetilde{G/K})$.

3. An outline of the proof is as follows. We start with a basic Cartan connection $\omega = \eta + \theta$ in the bundle P of oriented orthonormal frames of the normal bundle

$Q = TM/L$ of the given foliation F. The assumption is that $\|\Omega_\omega\| < A$ for a certain constant $A > 0$. We want to construct a new basic Cartan connection

$$(3.1) \qquad \bar{\omega} = \omega + \alpha .$$

whose curvature vanishes. $\bar{\omega}$ will be basic iff for all $\tilde{X} \in \Gamma\tilde{L}$ we have

$$i_{\tilde{X}}\alpha = 0, \; L_{\tilde{X}}\alpha = 0.$$

These conditions on $\underline{\underline{g}}$-valued forms on P define a complex $\Omega_B^{\bullet}(\tilde{F},\underline{\underline{g}})$ of basic forms.

The vanishing of the curvature of $\bar{\omega}$ is equivalent to the differential equation

$$(3.2) \qquad D^{\omega}\alpha + \tfrac{1}{2}[\alpha,\alpha] = -\Omega_{\omega}$$

where $D^{\omega}\alpha = d\alpha + [\omega,\alpha]$ denotes the covariant exterior derivative of α with respect to ω.

The idea of the proof is to construct a convergent sequence ω_i of basic Cartan connections such that the curvatures $\Omega_i \equiv \Omega_{\omega_i}$ converge to 0. Then $\bar{\omega} = \lim_{i\to\infty} \omega_i$ will have the desired properties. Let $\omega = \omega_0$. Then at each step

$$(3.3) \qquad \omega_{i+1} = \omega_i + \alpha_i \quad i = 0, 1,\ldots$$

is to be constructed by a convergent iteration scheme implying $\Omega_i \to 0$ as a consequence of $\|\Omega_0\| < A$. Now

$$(3.4) \qquad \Omega_{i+1} = \Omega_i + D^{\omega_i}\Omega_i + \tfrac{1}{2}[\alpha_i,\alpha_i]$$

and we end up with

$$(3.5) \qquad \bar{\omega} = \omega + \sum_{i=0}^{\infty} \alpha_i \quad \text{with} \quad \bar{\Omega} = 0.$$

For simplicity of notation we drop the subscripts, and examine one step of the iteration scheme

$$(3.6) \qquad \begin{aligned} &\omega \to \omega + \alpha \\ &\Omega \to \Omega + D^{\omega}\alpha + \tfrac{1}{2}[\alpha,\alpha]. \end{aligned}$$

Instead of (3.2) we try to solve the linearized equation

$$(3.7) \qquad D^{\omega}\alpha = -\Omega .$$

The following "Ansatz" will be modified below by changing D^{ω}.

Before going into this refinement, consider the eqation

$$(3.8) \qquad \Delta^{\omega}\beta = -\Omega$$

for $\beta \in \Omega_B^2(\tilde{F},\underline{g})$, where the Laplacian is defined as usual in terms of D^ω and its adjoint $D^{\omega*}$ by

$$\Delta^\omega = D^\omega D^{\omega*} + D^{\omega*}D^\omega.$$

Because of the harmonicity condition on the foliation, the operator $D^{\omega*}$ is up to sign the conjugate of D^ω with respect to the *-operator in the basic complex. (In the presence of a non-trivial mean curvature, the correct adjoint is obtained by twisting the *-conjugate by means of the mean curvature form as in [KT 3]). Using De Rham-Hodge Theory in $\Omega_B^{\bullet}(\tilde{F},\underline{g})$, and assuming the absence of non-trivial harmonic 2-forms, one could conclude that there is a unique solution to (3.8). As a consequence of the Bianchi-identity $D^\omega\Omega = 0$ one verifies $D^\omega\beta = 0$ for a solution of (3.8). Now let

$$(3.9) \qquad \alpha = D^{\omega*}\beta.$$

Then

$$D^\omega\alpha = D^\omega D^{\omega*}\beta = \Delta^\omega\beta = -\Omega$$

and α is a solution of the linearized equation (3.7).

To make this idea work one has to modify (3.8) to a context where the necessary uniqueness theorem for β can be established, at least for $||\Omega|| < A$.

A new differential D' replacing Δ^ω is defined by replacing the bracket $[X_j,X_k]$ in the defining formula by

$$\{X_j,X_k\} = \omega^{-1}([\omega(X_j),\omega(X_k)]).$$

This is applied to the vectorfields of a transverse parallelization of $\tilde{F}$ (see Molino [M]). Since ω has a kernel, the right-hand side is not well-defined. But this ambiguity is irrelevant to the evaluation of a basic form $\alpha \in \Omega_B(\tilde{F},\underline{g})$ on such a vectorfield.

The crucial estimate for the convergence of the curvatures Ω_i to zero in the iteration scheme is that for $||\Omega|| < A$

$$(3.10) \qquad ||D' - D^\omega|| < c||\Omega||$$

for a constant c. As a consequence of the Bianchi identity $D^\omega\Omega = 0$ this yields

$$(3.11) \qquad ||D'\Omega|| < c||\Omega||^2.$$

Let D'^* denote the adjoint of D', and Δ' the corresponding Laplacian in the basic complex $\Omega_B(\tilde{F},\underline{g})$. Then (3.8) is replaced by the equation

$$(3.12) \qquad \Delta'\beta_i = -\Omega_i$$

and α_i is defined by

$$\alpha_i = D'^{*}\beta_i.$$

The curvature of $\omega_i + \alpha_i$ is then

$$(3.13)\qquad \Omega_{i+1} = \Omega_i + D^{\omega_i}\alpha_i + \frac{1}{2}[\alpha_i,\alpha_i].$$

This construction is well-defined since for small $\|\Omega_i\|$ the solution to (3.12) can now be shown to be unique. This is a consequence of the following two facts.

(A) For the basic forms of a transversally oriented Riemannian foliation on a compact oriented Riemannian foliation there is a De Rham-Hodge decomposition [KT 3]. The essential point is that the relevant Laplacian operator on the complex of basic forms is the restriction of an elliptic operator on the total space corrected by a lower order operator. Therefore ellipticity is preserved, and the De Rham-Hodge decomposition follows from general facts about coercive bilinear forms [E].

(B) For small $\|\Omega_i\|$ the operator Δ' is positive definite on 2-forms. This is proved by a Bochner-Weitzenböck formula. It implies the uniqueness of the solution β_i of (3.12) and yields furthermore an estimate of $\|\beta_i\|$ in terms of $\|\Omega_i\|$.

This estimate together with (3.10) implies that the sequence of Cartan connections ω_i converges, with the curvatures Ω_i having limit zero. Thus $\bar{\omega} = \lim_{i\to\infty} \omega_i$ has the desired properties. The smoothness of $\bar{\omega}$ follows from the fact that it solves the differential equation $d\bar{\omega} + \frac{1}{2}[\bar{\omega},\bar{\omega}] = 0$ and the regularity theorem. The necessary estimates in Sobolev norms for the applicability of the regularity theorem are a consequence of (B) above.

REFERENCES

[B1] R. A. Blumenthal, Transversely homogeneous foliations, Annales de l'Institut Fourier (Grenoble) 29(1979), 143-158.

[B2] R. A. Blumenthal, The base-like cohomology of a class of transversely homogeneous foliations, Bulletin des Sciences Mathematiques 104(1980), 301-303.

[E] J. Eells, Elliptic operators on manifolds, Complex Analysis and its Applications, Trieste 1975, Volume I, 95-152.

[H] A. Haefliger, Structures feuilletées et cohomologie à valeurs dans un faisceau de groupoides, Comm. Math. Helv. 32(1958), 248-359.

[KT1] F. W. Kamber and Ph. Tondeur, Foliated bundles and characteristic classes, Springer Lecture Notes 493(1975), 1-208.

[KT2] F. W. Kamber and Ph. Tondeur, Harmonic foliations, Proc. NSF Conference on Harmonic Maps, Tulane University (1980), Springer Lecture Notes 949(1982), 87-121.

[KT3] F. W. Kamber and Ph. Tondeur, Foliations and metrics, Proc. of the 1981-82 year in Differential Geometry, University of Maryland, Birkhäuser, Progress in Math. 32(1983), 103-152.

[MR] Min-Oo and E. A. Ruh, Comparison theorems for compact symmetric spaces, Ann. Scient. Ec. Norm. Sup. 4e série, t. 12(1979), 335-353.

[M] P. Molino, Géometrie globale des feuilletages riemanniens, Proc. Kon. Nederland Akad., Ser. A. 1, 85(1982), 45-76.

[RA] H. E. Rauch, Geodesics, symmetric spaces, and differential geometry in the large, Comment. Math. Helv. 27(1953), 294-320.

[RE] B. L. Reinhart, Foliated manifolds with bundle-like metrics, Ann. Math. 29(1959), 119-132.

This work was supported in part by a grant from the National Science Foundation.

UNIFORMIZATION OF GEOMETRIC STRUCTURES WITH APLICATIONS TO CONFORMAL GEOMETRY

Ravi S. Kulkarni and Ulrich Pinkall**

Max-Planck-Institut für Mathematik,

Gottfried-Claren-Straße 26

5300 Bonn 3, Germany.

§1.- Introduction.

(1.1) The classical uniformization theory of Riemann surfaces is an outstanding meeting place of the classical function theory and topology. There are diverse aspects of this theory which extend in other set-ups in different ways, cf. [8], [9], [10]. In this paper we shall consider it in the context of "geometric structures" as defined below. This is a direct generalization of the uniformization of Riemann surfaces via Fuchsian and Kleinian groups.

(1.2) Let X be a topological space and G a group of homeomorphisms of X, satisfying the "uniformization condition" (U) : each $g \in G$ is uniquely determined by its action on any nonempty open subset. The pair (X,G) is to be thought of as a model space. An (X,G)-structure on a topological space M is given by a covering of M by open sets $\{U_\alpha\}_{\alpha \in A}$ and homeomorphisms $S_\alpha : U_\alpha \to X$ s.t. for all pairs α,β in A with $U_\alpha \cap U_\beta \neq \emptyset$ the mapping $S_\alpha \circ S_\beta^{-1}|_{S_\beta(U_\alpha \cap U_\beta)}$ is a restriction of an element of G. For example, if X is the standard sphere S^n and G is the full group of Möbius transformations $\mathcal{M}(n)$ then by Liouville's theorem for $n \geq 3$ an $(S^n, \mathcal{M}(n))$-structure on an n-dimensinal manifold M^n is the same as a conformal class of locally conformally Euclidean metrics. The case n=2, with $\mathcal{M}(2)$ replaced by its identity component $\mathcal{M}_0(2) \approx PSL_2(\mathbf{C})$, plays a central

---

(*) Both authors were supported by the Max-Planck-Institut für Mathematik, Bonn, Germany. The first author was also partially supported by an NSF Grant.

role in the uniformization theory of Riemann surfaces via the Kleinian groups. In Gunning's terminology an $(S^2, \mathcal{M}_0(2))$-structure is a $\mathbf{CP}^1$-structure on a Riemann surface. As an another example of geometric interest consider X = real (resp. complex) projective space and G = the full group of real (resp. complex) projective transformations.

(1.3) A nice class of (X,G)-structures arises as follows. Let Ω be an open subset of X and Γ a subgroup of G which leaves Ω invariant and acts freely and properly discontinuously there. Then $\Gamma\backslash\Omega$ clearly admits an (X,G)-structure. We shall call an (X,G)-structure on M <u>Kleinian</u> if $M \approx \Gamma\backslash\Omega$ as described above.

Of course X has a distiguished (X,G)-structure σ_0. An (X,G)-structure σ on a simply connected M is always of the form $\delta^*\sigma_0$ where $\delta : M \to X$ is a local homeomorphism. (This is essentially a precise formulation of the "monodromy principle".) Moreover if $\mathrm{Aut}(M,\sigma)$ denotes the automorphism group of this structure then δ determines a homomorphism $\rho : \mathrm{Aut}(M,\sigma) \to G$, and δ is ρ-equivariant i.e. for all $\alpha \in \mathrm{Aut}(M,\sigma)$ and $x \in M$, $\delta(\alpha\ x) = \rho(\alpha)\,\delta(x)$. The map δ is unique up to a left-composition by an element of G, and correspondingly ρ is unique up to a conjugation by an element of G.

If M has an (X,G)-structure σ but M is not necessarily simply connected then assuming that it has a universal cover $\tilde{M}$ we see that $\tilde{M}$ has an induced (X,G)-structure $\tilde{\sigma}$ and the deck-transformation group $\Delta \approx \pi_1(M)$ is clearly a subgroup of $\mathrm{Aut}(\tilde{M},\tilde{\sigma})$. Let $\delta: \tilde{M} \to X$ be a local homeomorphism s.t. $\tilde{\sigma} = \delta^*\sigma_0$. Then δ is called a <u>development</u> of (M,σ). If $\rho : \mathrm{Aut}(\tilde{M},\tilde{\sigma}) \to G$ is the corresponding homomorphism then $\rho|_\Delta$ is called the <u>holonomy representation</u> of (M,σ).

It is obvious that if we are in a category where the covering space theory is valid then an (X,G)-structure σ on M is <u>Kleinian</u> iff $\delta: \tilde{M} \to \delta(\tilde{M})$ is a covering map and $\rho(\Delta) = \Gamma$ acts freely and properly discontinuously on $\delta(\tilde{M})$. We shall say that (M,σ) is <u>almost Kleinian</u> if only $\delta : \tilde{M} \to \delta(\tilde{M})$ is a covering map.

(1.4) A problem of basic geometric interest is to find criteria for an (X,G)-structure to be Kleinian or almost Kleinian. For the case of $\mathbf{CP}^1$-structures, cf. (1.1), Gunning provided a nice criterion, cf. [9] theorem 7, and (1.5) below. This was proved by another method by Kra [12]. Both proofs use facts special to Riemann surfaces. In this paper we shall re-examine this theorem in the context of general geometric structures. In §§ 2 and 3 we develop the notions of limit sets and domains of properness for an arbitrary subgroup $\Gamma \leq G$ acting on X and prove the following general

(1.4.1) Uniformization theorem Let M be a compact space with an (X,G)-structure with $\delta : \tilde{M} \to X$ a development map, $\rho : \pi_1(M) \to G$ the holonomy representation and $\Gamma = \operatorname{im} \rho$. Let $\mathcal{N}_0$ be the union of those components of the domain of normality of Γ which intersect im δ. Then $\delta|_{\delta^{-1}(\mathcal{N}_0)} : \delta^{-1}(\mathcal{N}_0) \to \mathcal{N}_0$ is a covering map.

(1.5) This theorem combined with a theorem of Fried [5] implies a direct extension of Gunning's theorem, cf. (5.3). A compact manifold with a Möbius structure such that the development map is not surjective is almost Kleinian. Conversely of course, except for the manifolds conformal to the spherical space-forms, an almost Kleinian manifold with a Möbius structure has development onto a proper subset of S^n.

Here is another quite different criterion, cf. (5.4). A compact manifold with a Möbius structure so that the domain of properness of its holonomy group is connected and has finitely generated π_1 is almost Kleinian. It may be remarked that in the proofs of Gunning or Kra the domain of properness plays no direct role.

In [13] it was proved that a connected sum of manifolds with Möbius structures admits a Möbius structure. A convenient source of Kleinian examples is a partial refinement of this statement, cf. (5.6). A connected sum of Kleinian manifolds with a Möbius structure admits a Kleinian Möbius structure. This is an analogue on the "space"-level of the famous Klein-Maskit "combination theorems" cf. [17] which are statements on the "group"-level. This result has been known for some time, cf. Goldman [6] § 5, but no proof is print.

Perhaps it should be pointed out that not every manifold with a Möbius structure is Kleinian or even almost Kleinian. There are some very interesting examples illustrating various phenomena, cf. (5.7). Moreover the above-mentioned results are valid in a much greater generality as pointed out in (5.8). In fact the "ideal boundary" of an arbitrary connected, simply connected, complete Riemannian manifold of curvature $\leq -a<0$ admits many features of the standard conformal geometry of S^n.

The hypothesis of compactness of the space with a geometric strucutre in Gunning's theorem and also in the theorems proved here is admittedly adhoc. It excludes some geometrically interesting cases, e.g. the noncompact hyperbolic manifolds with finite volume. It is easy to see that the statements of the theorems are no longer valid if compactness is simply dropped. However in replacing compactness by appropriate hypotheses on development, limit sets, etc. would bring forth the "geometry" in a more transparent way. This entails some entirely new ideas which so far we have only partially carried out. We shall present these extensions in a subsequent publication.

We wish to thank P. Pansu for explaining to us his ideas on a "coarse conformal geometry", cf. [19]. This significantly extended the validity of our results.

§ 2. Wandering points, twins, and polars.

(2.1) The study of dynamics of the holonomy group is an important part of the study of a geometric structure. With this in view we shall develop appropriate notions in a sufficient general set-up, which were motivated by the notions of the limit set and the domain of discontinuity of a Kleinian group in the classical theory. This discussion also extends that in [14] § 1.

(2.2) Let X be a locally compact, Hausdorff space which has a countable base for topology, and which is locally simply connected, and locally path-connected. Let G be a closed group of homeomorphisms of X with respect to the compact-open topology. The pair (X,G) is to be thought of as "a model space" in the sense described in (1.2). For pairs of spaces X,Y let $\mathcal{C}$(X,Y) denote the space of continuous functions from X to Y again equipped with the compact-open topology. For $A \subset X$ let

(2.2.1) $$G|_A = \{g|_A \mid g \in G\}$$

considered as a subset of $\mathcal{C}(A,X)$. Let A^- denote the closure of A in X and

(2.2.2) $$G(A) = \bigcup_{g \in G} gA$$

A point $x \in X$ is said to be <u>a recurrent point of the G-orbit of</u> A if for every neighborhood V of x the subset $\{g \in G \mid gA \cap V \neq \emptyset\}$ has a noncompact closure in G. We set

(2.2.3) $G(A)'$ = the set of recurrent points of A.

Clearly this set is a closed G-invariant set. We also set

(2.2.4) $$\mathcal{Z}(A) = \{g \in G \mid gA \cap A \neq \emptyset\}.$$

(2.3) A point $x \in X$ is said to be <u>wandering</u> (with respect to G) if it has a compact neighborhood U_x such that $\mathcal{Z}(U_x)$ is compact. We set

(2.3.1) L_0 = the set of non-wandering points.

(2.4) Let $p : X \to G\backslash X$ be the orbit-space projection so that $G\backslash X$ has a quotient topology. We say that $x,y \in X$ are <u>twins</u> (with respect to G) if $p(x),p(y)$ have no disjoint neighborhoods. This means that for every neighborhood U of x and V of y there exists $g \in G$ such that $gU \cap V \neq \emptyset$. Let

(2.4.1) $\tau(x)$ = the set of twins of x.

Clearly $y \in \tau(x)$ iff $x \in \tau(y)$, and $\tau(x)$ is a closed G-invariant subset.

(2.5) We say that $y \in X$ is a <u>polar</u> of $x \in X$ if y is a recurrent point of every neighborhood of x. Write

(2.5.1) $\mathcal{P}(x)$ = the set of polars of x.

Clearly $\mathcal{P}(x)$ is a closed G-invariant set and $\mathcal{P}(x) \subset \tau(x)$.

(2.6) <u>Proposition</u> Let (X,G) be as above and $x \in X$.

Then a) $\tau(x) = \bigcap_U G(U)^-$, $\mathcal{P}(x) = \bigcap_U G(U)'$, where U runs over neighborhoods of x,

b) $G(x)^- \subset \tau(x)$,

c) $G(x)' \subset L_0 \cap \mathcal{P}(x)$,

d) If x is a wandering point then $\tau(x) = G(x) \cup \mathcal{P}(x)$, and moreover $\mathcal{P}(x) \neq G(x)'$ if x is a recurrent point of a compact subset of $X - L_0$

<u>Proof</u>. The parts a), b) are clear from definitions. In c) it is again clear from a) that $G(x)' \subset \mathcal{P}(x)$. We now show $G(x)' \subset L_0$. Let $y \in G(x)'$ so there exist $g_n \in G$ such that $g_n x \to y$, and g_n is a divergent sequence in G. Let V be any neighborhood of y. So whenever $g_n x$, $g_m x \in V$ we see that $g_n g_m^{-1} V \cap V \neq \emptyset$. It is clear that $\mathcal{Z}(V)$ is not compact. Since this holds for every neighborhood of y it follows that $y \in L_0$.

Now we prove d). It follows from a) and b) that for any x we have $\tau(x) \supseteq G(x) \cup \mathcal{P}(x)$. Now assume x to be a wandering point and $y \in \tau(x)$. If $y \in G(x)^-$ then $y \in G(x)$ or $y \in G(x)'$ and $G(x)' \subset \mathcal{P}(x)$ so $y \in G(x) \cup \mathcal{P}(x)$. Suppose $y \notin G(x)^-$. Then since x is a wandering point we see that for small neighborhoods V of x we have $y \notin G(V)$. But by a) we see that $y \in \mathcal{P}(x)$. This proves the first part of d). Now suppose that x is a recurrent point of a compact subset K of $X - L_0$ Let U_n be a decreasing sequence of neighborhoods of x converging to x. There exist $g_n \in G$ and $k_n \in K$ so that $g_n \cdot k_n \in U_n$, or $k_n \in g_n^{-1} U_n$ Let k_0 be a cluster point of k_n It is clear that $k_0 \in \mathcal{P}(x) \cap K$. But $k_0 \notin G(x)'$ since otherwise, by c), k_0 would belong to L_0, but we chose K to lie in $X - L_0$. This finishes the proof.

q.e.d.

§ 3. Limit sets, properness - and normality - domains

(3.1) Let (X,G) be as in (2.2). We shall use the notations in § 2. We now assume

(U_1) If for a non-empty open subset V of X and $g_1, g_2 \in G$ we have $g_1|_V = g_2|_V$ then $g_1 = g_2$.

(U_2) For a non-empty open subset V of X if $G|_V$ has a cluster point g_0 in $\mathcal{C}(V,X)$ so that g_0 is injective then there exists $g \in G$ such that $g|_V = g_0$.

These assumptions of course hold for Möbius or projective structures or for the geometric structures defined by an integrable G-structure of finite type, cf. [13] §2. The assumption (U_1) is the same as the assumption (U) of [13] § 1.

(3.2) We say that G acts locally properly on X if every point $x \in X$ is a wandering point. More stringently, G is said to act properly on X if for every compact subset $K \subset X$, we have $\mathcal{Z}(K)$ compact. The set

(3.2.1) Ω_{loc} = the set of wandering points = $X - L_0$ is called the domain of local properness of G. The set L_0 is called the 0-limit set of G. Now let

(3.2.2) $L_1 = \{x \in X \mid x$ is a recurrent point of a compact subset of $\Omega_{loc}\}$. This set is called the 1-limit set of G, and

(3.2.3) $\Lambda = L_0 \cup L_1$

is called simply the limit set of G. Correspondingly

(3.2.4) $\Omega = X - L$

is called the domain of properness of G. The proof of the following proposition may be left to the reader.

Proposition G acts locally properly on Ω_{loc} and properly on Ω. Moreover Ω_{loc} is the largest open subset of X on which G acts locally properly.

It should be remarked that in general Ω need not be a maximal domain on which G acts properly. In fact it may happen that Ω can be extended to more than one maximal open subsets of X on which G acts properly, cf. [14], § 1.

(3.3) A point $x \in X$ is called a point of normality (with respect to G) if it has a neighborhood U_x such that $G|_{U_x}$ is a relatively compact subset in $\mathcal{C}(U_x,X)$. Then

(3.3.1) $\mathcal{N}$ = the set of points of normality

is called the normality domain of G.

(3.4) Theorem $\mathcal{N} \subset \Omega$.

Proof Let $x \in \mathcal{N}$. We first show that x is wandering. There exists a neighborhood U_x of x so that $G|_{U_x}$ is relatively compact in $\mathcal{C}(U_x,X)$. We claim that $\mathcal{Z}(U_x)$ is a relatively compact subset of G. Let $g_n \in \mathcal{Z}(U_x)$. Passing to a subsequence we may assume that

$g_n|_{U_x}$ and $g_n^{-1}|_{U_x}$ converge to g_0 and h_0 respectively in $\mathcal{C}(U_x,X)$. By the continuity of the composition and the fact that $g_n \cdot g_n^{-1} = 1$ we see that g_0 and h_0 are injective. So by the hypothesis (U_2), cf. (3.1), we have elements $g,h \in G$ such that $g|_{U_x} = g_0$ and $h|_{U_x} = h_0$. So $\mathcal{Z}(U_x)$ is relatively compact in G. So x is wandering. Thus $\mathcal{N} \cap L_0 = \emptyset$. Now suppose that we have a sequence $g_n \in G$ so that $g_n|_{U_x} \to g_0$ in $\mathcal{C}(U_x,X)$. In particular $g_n x \to g_0 x = y$, say.

Moreover since g_n converges to g_0 uniformly on a compact subset of U_x we see that for every neighborhood U_y of y there is a neighborhood $V_x \subset U_x$ so that

$g_n(V_x) \subset U_y$ for n sufficiently large. It follows that $\bigcap_{V_x} (G_n V_x)' = y$, where V_x runs over all neighborhoods of x. In the notation of § 2 we see that the polar set $\mathcal{P}(x)$ of x coincides with $G(x)'$. So by (2.6), part d), $x \notin L_1$. So $\mathcal{N} \cap L_1 = \emptyset$, and hence $\mathcal{N} \subset \Omega$.

q.e.d.

(3.5) Remark In general $\mathcal{N} \neq \Omega$. It is easy to construct examples when X is non-compact. But in general $\mathcal{N} \neq \Omega$ even when X is compact. Here is an example in dimension 3. Consider the group of projective transformations of $\mathbf{RP}^3$ generated by

$$g : (x,y,z,w) \to (2x,4y,z,w)$$

$$h : (x,y,z,w) \to (x,y,z+w,w)$$

where (x,y,z,w) are the homogeneous coordinates in $\mathbf{RP}^3$ Here g fixes the line $\lambda : x=0=y$ pointwise whereas h fixes the hyperplane $\pi : w = 0$ pointwise. It is easy to see that all points in $\mathbf{RP}^3 - \{\lambda \cup \pi\}$ are wandering. So $L_0 = \lambda \cup \pi$. The recurrent points of any compact set in $\mathbf{RP}^3 - L_0$ are easily seen to lie in L_0. So $L_1 \subset L_0$. Hence $\Lambda = \lambda \cup \pi$ and $\Omega_{loc} = \Omega = \mathbf{RP}^3 - \{\lambda \cup \pi\}$. However looking at the restriction of $\langle g \rangle$ on the line $\mu : y=0=z$

we see that the line μ does not lie in the normality domain N. So $N \subsetneq \Omega$.

(3.6) Remark We have defined the notions of Ω_{loc}, Ω, N, etc. with respect to a closed subgroup of the group of homeomorphisms of X. If G is not closed - as indeed may happen when G = the image of the holonomy of a geometric structure - we define Ω_{loc}, etc. of G to be that of $\bar{G}$.

§ 4. A uniformization theorem

(4.1) Let (X,G) be a model space satisfying the conditions of (2.2) and the assumptions (U_1) and (U_2) of (3.1). Let M be a topological space with an (X,G)-structure, $p: \tilde{M} \to M$ the universal covering projection with deck-transformation group $\Delta \approx \pi_1(M)$. $\delta: \tilde{M} \to X$ a development map and $\rho: \Delta \to G$ a corresponding holonomy representation. Set Ω_M = im δ and Γ = im ρ. Let N_0 be the union of the components of the domain of normality of Γ which have a non-empty intersection with Ω_M. We shall use these notations throughout this section.

(4.2) Theorem Suppose M is a compact space with an (X,G)-structure. Let $\tilde{N} = \delta^{-1}(N_0)$. Then $\delta|_{\tilde{N}} : \tilde{N} \to N_0$ is a covering map.

Proof Since δ is a local homeomorphism it suffices to show that $\delta|_{\tilde{N}}$ has a path-lifting property. Fix a point y_0 in im $\delta|_{\tilde{N}}$, and a path $\beta : [0,1] \to N_0$ with $\beta(0) = y_0$. Let $\tilde{x}_0$ be a point in $\tilde{N}$ with $\delta(\tilde{x}_0) = y_0$, and $\tilde{\alpha}$ a maximal lift of β beginning at $\tilde{x}_0$. By way of contradiction assume that β does not lift entirely. Then shrinking the domain of β if necessary and reparametrizing we may assume that $\tilde{\alpha}$ is defined on [0,1). We will show that $\tilde{\alpha}$ has a continuous extension at 1, and so indeed $\tilde{\alpha}$ is not maximal. Let $\alpha = p \circ \tilde{\alpha}$:

$[0,1) \to M$ be the projected path in M. Since M is compact we may choose an increasing sequence $t_n \in [0,1)$ so that $t_n \to 1$ and $x_n = \alpha(t_n) \to z_0$. Let $\tilde{z}_0 \in \tilde{M}$ be a point lying over z_0, and $\tilde{x}_n = \tilde{\alpha}(t_n)$. Then there exist $g_n \in \Delta$ so that $\tilde{z}_n = g_n \tilde{x}_n \to \tilde{z}_0$. Notice that since $\mathcal{N}_0$ is Γ-invariant, we have $\tilde{N}$ Δ-invariant, so $\tilde{z}_n \in \tilde{N}$. Write $\rho(g_n) = \gamma_n$, $y_n = \delta(\tilde{x}_n)$, $w_n = \delta(\tilde{z}_n)$, $w_0 = \delta(\tilde{z}_0)$, and note that

$$y_n \to y_0 \text{ and}$$

$$w_n = \delta(\tilde{z}_n) = \delta(g_n \tilde{x}_n) = \gamma_n \delta(\tilde{x}_n) = \gamma_n y_n \to \delta(z_0) = w_0.$$

Let $\tilde{V}$ be a neighborhood of $\tilde{z}_0$ and V of w_0 so that $\delta|_{\tilde{V}} : \tilde{V} \to V$ is a homeomorphism.

Now choose a compact neighborhood U_{y_0} of y_0 so that $\Gamma|U_{y_0}$ is a relatively compact subset of $\mathcal{C}(U_{y_0},X)$. For n sufficiently large $\beta([t_n,1]) \subset U_{y_0}$ and by passing to a subsequence if necessary we may assume that $\gamma_n \to \gamma_0 \in \mathcal{C}(U_{y_0},X)$. Since this convergence is uniform we have for n sufficiently large, $\gamma_n(\beta([t_n,1])) \subset V$. Hence the path $\gamma_n \circ \beta|_{[t_n,1]}$ has a lift. This lift would coincide with $g_n \circ \tilde{\alpha}$ on $[t_n,1)$. It is now clear that $\tilde{\alpha}$ itself has a continuation at 1, and the proof is finished.

q.e.d.

§ 5 Applications to conformal geometry

(5.1) We consider the model space $(S^n, \mathcal{M}(n))$. An important feature of this structure, called the Möbius structure, is contained in the following proposition

Proposition Let G be a subgroup of $\mathcal{M}(n)$, and consider the limit sets, etc. w.r.t. G. Then

$$\Lambda = L_0 = L_1 \,, \quad \text{and} \quad \Omega_{loc} = \Omega = \mathcal{N}.$$

Proof If $\Omega_{loc} = \emptyset$ there is nothing to prove. Otherwise, let $x \in \Omega_{loc}$ and U_x a small round ball around x contained in Ω_{loc}. Let g_n be a discrete sequence in G. By passing to a subsequence we may assume that $g_n x \to y$, so that $y \in L_0$. Also, since Ω_{loc} is G-invariant, we may assume that all $g_n(U_x)$ are pairwise disjoint. It is then obvious that in the spherical metric the radius of the round balls $g_n(U_x)$ goes to zero. So indeed $g_n|_{U_x}$ tends to a constant map c_y in $\mathcal{C}(U_x,X)$. We have shown that $G|_{U_x}$ has a compact closure in $\mathcal{C}(U_x,X)$. So $x \in \mathcal{N}$. Since we always have $\mathcal{N} \subset \Omega \subset \Omega_{loc}$, cf. § 4, it follows that $\Omega_{loc} = \Omega = \mathcal{N}$.

q.e.d.

(5.2) Remark It is easy to see that Λ as in (5.1) can be identified with the limit set as defined in the classical situation, cf. [1], [7]. In fact in that case Λ may be identified with $G(x)'$ for any $x \in S^n$, except in the easily analyzed case when Λ = {2 points}, each fixed by G and x coincides with one ot the fixed points. This may be proved in our set-up by a slight extension of the argument in (5.1), - and in fact the argument applies to any group of quasi-conformal transformations, cf. also (5.8) below. In the following we shall use the well-known properties of the limit set as described e.g. in [7].

(5.3) The following is a direct extension of Gunning's theorem 7 in [9].

Theorem Let M^n be a compact manifold with a Möbius structure, $\delta : \tilde{M}^n \to S^n$ its development, $\rho : \Delta \approx \pi_1(M) \to \mathcal{M}(n)$ the corresponding holonomy, and $\Gamma = \rho(\Delta)$. Let Ω be the domain of properness of Γ. Suppose δ is not surjective. Then δ is a covering onto the unique component of Ω which intersects im δ. In particular M is almost Kleinian.

Proof Write im $\delta = \Omega_M$. Then $\partial\Omega_M$ is Γ-invariant. If $\partial\Omega_M$ = {a point} then regarding this point as ∞, Γ may be considered as a group of similarity transformations in $\mathbf{E}^n = S^n - \{\infty\}$. In this case the result follows from a remarkable theorem of D. Fried [5] (which in fact asserts that M^n has a finite covering which is conformal to a flat space-form or else to a Hopf manifold.) Now supose that $\partial\Omega_M$ contains at least two points. Then as is well-known $\partial\Omega_M \supseteq \Lambda$. So $\Omega_M \subset \Omega$. Since Ω_M is connected it is contained in exactly one component Ω_0 of Ω. Since by (5.1) $\Omega = \mathcal{N}$ it follows from the uniformization theorem (4.1) that δ is a covering onto Ω_0.

q.e.d.

(5.4) We now prove another criterion for a Möbius structure to be Kleinian or almost Kleinian.

Theorem Let M^n, δ, ρ, G, Ω be as in (5.2), except that instead of an assumption about δ, we now assume that Ω is connected and $\pi_1(\Omega)$ finitely generated. In particular if $\pi_1(\Omega) = e$, then M is almost Kleinian.

Proof Let Ω_M = im δ. If $\Omega_M \neq S^n$ then the result follows by (5.2). So suppose if possible that $\Omega_M = S^n$

Case 1 Assume $\pi_1\Omega = e$. By (5.1) we know $\Omega = \mathcal{N}$, and if $\tilde{\mathcal{N}} = \delta^{-1}(\Omega)$, by (4.1)

$\delta|_{\tilde{N}} : \tilde{N} \approx \Omega$. But Λ and hence $\delta^{-1}(\Lambda)$ have no interior, and $\delta : \tilde{M} \to S^n$ is a local homeomorphism. Under these conditions it is an easy point-set topological fact that δ itself is a homeomorphism, and in fact M is conformal to a spherical space-form.

Case 2 Assume $\pi_1\Omega \neq e$, but is finitely generated. Let $\tilde{N} = \delta^{-1}(\Omega)$, $\tilde{L} = \delta^{-1}(\Lambda)$. By (4.1), $\delta|_{\tilde{N}} : \tilde{N} \to \Omega$ is a covering. Let $y \in \Lambda$ be an attracting fixed point of an element $g \in \Gamma$. Let $\tilde{x} \in \tilde{M}$ be such that $\delta(\tilde{x}) = y$, and $\tilde{U}$ a neighborhood of $\tilde{x}$ and U a neighborhood of y so that $\delta|_{\tilde{U}}$ is a homeomorphism of $\tilde{U}$ onto U. Since $\pi_1(\Omega)$ is assumed to be finitely generated there is a compact subset $A \subset \Omega$ which carries $\pi_1(\Omega)$. Moreover for n sufficiently large $g^n(A) \subset U - \Lambda$. So $U - \Lambda$ carries $\pi_1(\Omega)$. But $\delta|_{\tilde{U}-\tilde{L}}$ is a homeomorphism. So the inclusion map $U - \Lambda \to \Omega$ which is surjective on π_1 lifts to $\tilde{N}$. It follows that $\delta|_{\tilde{N}}$ must be a homeomorphism. But then again as in case 1 it would follow that δ itself is a homeomorphism, and $\Omega_M = \Omega = S^n$. To summarize: if we assume $\pi_1(\Omega) \neq e$ but finitely generated we must have $\Omega_M \neq S^n$ and so by (5.2) M must be almost Kleinian.

q.e.d.

(5.5) Remark In case a compact manifold M^n with a Möbius structure is not almost Kleinian the development map exhibits a rather quaint behavior reminiscent of the behavior of the holomorphic map near an essential singularity. More precisely in the above notation assume that M is not almost Kleinian, so $\Omega_M = S^n$ and suppose Λ has more than 2 points so it is perfect set. Let $p : \tilde{M} \to M$ be the covering projection and $L = p(\delta^{-1}\Lambda)$. (Notice that $\delta^{-1}\Lambda$ is a closed subset invariant under the deck-transformation group so L is a closed subset of M.)

Assertion For any open subset U such that $U \cap L \neq \emptyset$ we have $\delta(p^{-1}U) = S^n$.

Proof Indeed let $V = \delta(p^{-1}U)$ and $\Gamma = \rho(\Delta)$. Clearly V is Γ-invariant and contains a small disk D which contains a repelling fixed point of a hyperbolic element

$g \in \Gamma$. So V contains $\bigcup_{n=1}^{\infty}(g^n D) = S^n - \{y\}$ where y is the attracting fixed point of g. But D also contains a repelling fixed point of a hyperbolic element $g_1 \in \Gamma$ such that g_1 does not fix y. It is now clear that $V = S^n$. q.e.d.

(5.6) We now point out a convenient construction of Kleinian Möbius structures. It is also useful in another constructions in conformal geometry.

Theorem A connected sum of Kleinian manifolds with a Möbius structure also admits a Kleinian Möbius structure.

Proof Let $M_i = \Gamma_i \backslash \Omega_i$, i = 1,2 be two Kleinian manifolds with a Möbius structure so that Ω_i are two open connected nonempty subsets of S^n and Γ_i are subgroups of $\mathcal{M}(n)$ leaving Ω_i invariant and acting freely and properly discontinuously there. We shall show that the abstractly defined free product $\Gamma = \Gamma_1 * \Gamma_2$ acts freely and properly discontinuously on a certain region on $\Omega \subset S^n$ so that $\Gamma \leq \mathcal{M}(n)$ and $\Gamma \backslash \Omega$ is diffeomorphic to a connected sum M of M_1 and M_2. Indeed let $\Delta_i = \pi_1(M_i)$, i = 1,2. We have projections $p_i : \Delta_i \to \Gamma_i$ Let $\Phi_i = \ker p_i$. Let $\{\gamma^i{}_j\}_{j \in J_i}$ be based loops in M_i so that their homotopy classes $[\gamma^i{}_j]$ normally generate Φ_i. So Ω_i is a connected covering of M_i which is universal with respect to the property that each lift of $\gamma^i{}_j$ is a loop. Consider the complex

$$A = (M_1 \cup M_2 \cup \mathbf{I}) / \sim$$

where $\mathbf{I} = [0,1]$ and 0 is identified with a point in M_1 and 1 with a point in M_2. Then $\pi_1(A) \approx \Delta_1 * \Delta_2$ and we have a canonical projection $f : \pi_1(A) \to \Gamma$ so that $\Phi = \ker p$ is normally generated by $\{\gamma^1{}_j\}_{j \in J1} \cup \{\gamma^2{}_j\}_{j \in J2}$. Take the covering B of A corresponding to

Φ. This is constructed out of the $|\Gamma/\Gamma_i|$ copies of Ω_i (i.e. to say in 1-1 correspondence with Γ/Γ_i), i = 1,2 and $|\Gamma|$ copies of **I**. The copies of **I** may be considered as the "connecting bonds" between the copies of Ω_1 and those of Ω_2. The main point is that (*) each copy of Ω_i is attached with $|\Gamma_i|$ connecting bonds, i = 1,2, and no copy of Ω_1 is joined to a copy of Ω_2 by two connecting bonds.

We now thicken **I** in A to remove the interior so as to obtain the connected sum M of M_1 and M_2 by the process described in [13] so that M has a Möbius structure which restricts to the prescribed Möbius structures on parts of M_i, i = 1,2 which lie in M. We do the corresponding thickenings etc. in B so as to obtain a manifold Ω with a Möbius structure which covers M with the covering group Γ. We shall now embed Ω into S^n preserving the Möbius structure. In the process of obtaining Ω from B, from each copy of Ω_i in B, $|\Gamma_i|$ round disks are removed. We now embed one copy of Ω_1 with $|\Gamma_1|$ round disks removed in S^n preserving the Möbius structure. In each hole of this copy we can insert a copy of Ω_2 (with $|\Gamma_2|$ holes) to which it is attached in Ω. In each hole of a copy of Ω_2 we can insert a copy of Ω_1 (with $|\Gamma_1|$ holes) to which it is attached in Ω. Now the fact (*) we mentioned above implies that we can continue this process to obtain a Möbius structure-preserving embedding of Ω into S^n. We have also used here the existence of inversions and the fact that all round S^{n-1} in S^n are equivalent under $\mathcal{M}(n)$. Now the group Γ acts on the image of Ω into S^n preserving the Möbius structure. Since every Möbius transformation defined on a connected open subset of S^n is a restriction of an element of $\mathcal{M}(n)$, we can regard Γ as a subgroup of $\mathcal{M}(n)$. This finishes the proof.

q.e.d.

(5.7) It should be pointed out that the hypotheses in (5.3) and (5.4) cannot be entirely dropped. In fact the method of proof of (5.6) shows that a connected sum of two compact manifolds with a Möbius structure, one of which is almost Kleinian but non-Kleinian and the other $\not\approx S^n$ admits a Möbius structure with surjective development map, so this structure is non-almost Kleinian.

Another class of very interesting examples is obtained by conformally deforming a neighborhood of a totally geodesic hypersurface in a compact hyperbolic manifold. For $\mathbf{CP}^1$-structures they were noticed by Maskit [16] and Hejhal [11] and in a quite different context by Faltings [4]. The non-trivial infinitesimal deformations of the corresponding groups were noticed by Lafontaine [15] and Millson [18]. Looking at their development they were named "Mickey Mouse" examples by Thurston [20], and their geometry (and also their projective analogues) have been beautifully explained by Goldman [6]. We remark that similar deformations can be obtained also for $M^n \times S^p$ by deforming a neighborhood of $N^{n-1} \times S^p$ where M^n is a compact hyperbolic manifold and N^{n-1} is a totally geodesic hypersurface in M^n. Among these we find examples of non-almost Kleinian M^n whose domain of properness is disconnected, although each component is simply connected - thus showing that the hypothesis of connectedness of the domain of properness in (5.4) cannot be dropped. On the other hand a connected sum of non-Kleinian almost Kleinian compact manifolds with limit set $\approx S^{n-2}$ gives an example of a non-almost Kleinian manifold with a connected domain of properness with non-finitely generated π_1 - thus showing that the hypothesis of finite generation of π_1 in (5.4) also cannot be dropped.

(5.8) <u>Further generalizations</u> In this section we have formulated the theorems for the sake of simplicity only in the case of the standard Möbius structures. But it is apparent from the proofs that the strict "angle-preserving" property is not really used in any crucial way and so the results are valid in much more general set-ups. In fact let H^{n+1} be an (n+1)-dimensional complete, connected, simply connected Riemannian manifold with sectional curvatures $\leq -\varepsilon < 0$. In a well-known manner we can attach to it an ideal boundary Σ^n made up of classes of asymptotic geodesic rays. Then Σ^n is homeomorphic to the n-sphere and $H^{n+1} \cup \Sigma^n$ is homeomorphic to the closed disk. Moreover the isometries of H^{n+1} are classified into elliptic, hyperbolic and parabolic types depending on whether there is a fixed point in H^{n+1}, or exactly two fixed points on Σ^n, or exactly one fixed point on Σ^n : For $p \in H^{n+1}$ the images of round sub-spheres of disks in the unit n-sphere in $T_p(\mathcal{H}^{n+1})$ via the exponential map may be considered as "round" sub-spheres or disks of Σ^n. (Here p is allowed to vary in $\mathcal{H}^{n+1}$.) This defines a kind of "conformal geometry" in Σ^n, cf. [19]. The full group of isometries of $\mathcal{H}^{n+1}$ extends to Σ^n and serves as the "Möbius group" of Σ^n, and we may consider the structures based on (Σ^n,G). The boundaries of rank-1 non-compact symmetric spaces provide interesting examples of this set-up. The proposition (5.1) is valid for (Σ^n,G)-structures. As of this writing we do not know the validity of Fried's theorem quoted in the proof of the theorem (5.3). Otherwise, if we assume in (5.3) that im δ misses two points then (5.3) is valid for (Σ^n,G)-structures. Similarly (5.4) with an appropriate modification goes through. For the validity of connected sums we need to assume

1) there exists $g \in G$ which leaves invariant a tame (n-1)-sphere Σ^{n-1} in Σ^n (e.g. a round (n-1)-sphere in Σ^n) and interchanges the two components of $\Sigma^n - \Sigma^{n-1}$, and

2) G does not fix a point in Σ^n and the fixed points of hyperbolic isometries are dense in Σ^n.

For example 2) holds if $\mathcal{H}^{n+1}$ covers a manifold $\mathcal{M}^{n+1}$ of finite volume; and furthermore 1) holds if M^n admits an isometry $\bar{g}$ such that $\bar{g}^2 = 1$ and the fixed point set is a totally geodesic hypersurface. The conditions 1) and 2) also hold for the boundaries of rank-1 symmetric spaces. The conditions 1) and 2) ensure that there are inversions through sufficiently small spheres, and moreover given any two non-empty open sets U,V of Σ^n there exists $g \in G$, an (n-1)-sphere $\Sigma^{n-1} \subset U$ and $g(\Sigma^{n-1}) \subset V$ so that there exists an inversion σ through Σ^{n-1}. (Then $g \circ \sigma \circ g^{-1}$ is an inversion through $g \cdot \Sigma^{n-1}$). This suffices to perform the connected sums of manifolds with a (Σ,G)-structure. For the case of the boundary of the complex hyperbolic space this fact was observed by Burns and Schnider [3]. Finally if 1) and 2) hold then the theorem (5.6) is valid for (Σ,G)-structures.

REFERENCES

[1] BEARDON, A.F.: "The geometry of discrete groups", Graduate Texts in Math. 91. Springer-Verlag, 1983.

[2] BERS, L.: "Uniformization, moduli, and Kleinian groups", Bull. London Math. Soc. 4 (1972), 257-300.

[3] BURNS, D. Jr.; SCHNIDER, S.: "Spherical hypersurfaces in complex manifolds", Inv. Math. 33, 223-246.

[4] FALTINGS, G.: "Real projective structures on Riemann surfaces", Comp. Math. 48 (1983), 223-269.

[5] FRIED, D.: "Closed similarity manifolds", Comment. Math. Helv. 55 (1980), 576-582.

[6] GOLDMAN, W.: "Projective structures with fuchsian holonomy", preprint MSRI 07215-85.

[7] GREENBERG, L.: "Discrete subgroups of the Lorentz groups", Math. Scand. 10 (1962).

[8] GRIFFITHS, Phillip A.: "Complex analytic properties of certain Zariski open sets on algebraic varieties", Ann. of Math. 94 (1971), 21-51.

[9] GUNNING, R.C.: "Special co-ordinate coverings of Riemann surfaces", Math. Ann. 170 (1967), 67-86.

[10] GUNNING, R.C.: "On uniformization of complex manifolds, the role of connections", Mathematical Notes, Princeton University Press (1978).

[11] HEJHAL, D.: "Monodromy groups and linearly polymorphic functions", Acta Math. 135 (1975), 1-55.

[12] KRA, I.: "Deformation of fuchsian groups", Duke Math. J. 36 (1969), 537-546.

[13] KULKARNI, R.S.: "On the principle of uniformization", Jour. of Diff. Geom. 13 (1978), 109-138.

[14] KULKARNI, R.S.: "Groups with domains of discontinuity", Math. Ann. 237 (1978), 253-272.

[15] LAFONTAINE, J.: "Modules de structures conformes plates et cohomologie de groupes directs", C.R. Acad. Sci. Paris 297 (1983), 655-658.

[16] MASKIT, B.: "On a class of Kleinian groups", Ann. Acad. Sci. Fenn. Ser. A I 442 (1969).

[17] MASKIT, B.: "On Klein's combination theorem", Trans. A.M.S. 120 (1965), 499-509, part II ibid. 131 (1968), 32-39.

[18] MILLSON, J.: "A remark on Raghunathan's vanishing theorem", Topology 24 (1985), 495-498.

[19] PANSU, P.: "Geometrie conforme grossiere", (preprint).

[20] THURSTON, W.: "Geometry and Topology of 3-manifolds", Princeton Lecture Notes (1978).

REPRESENTATION COADJOINTE QUOTIENT ET ESPACES HOMOGENES DE CONTACT

André Lichnerowicz
Collège de France
Rue des Ecoles.
75005 Paris. (France).

On sait l'intérêt multiple que représente la représentation coadjointe d'une algèbre de Lie Ω , intérêt mis en évidence par les travaux de Kirïllov, Kostant, Souriau et leurs successeurs. Sur l'éspace dual Ω^* , le tenseur de structure se traduit par une structure de Poisson dont les feuilles sont les orbites d'un groupe connexe G, d'algèbre de Lie Ω , opérant sur Ω^* par représentation coadjointe. En particulier Souriau a montré que tout espace homogène symplectique exact est revêtement d'une de ces orbites, ce qui fournit un modèle universel, à un revêtement près, pour de tels éspaces.

Par quotient par le groupe des homothéties de Ω^* , la structure de Poisson $\tilde{\Lambda}$ définit sur une sphère de Ω^* une structure de Jacobi (Λ,E) . Ses feuilles sont des orbites de G opérant sur la sphère parce que nous nommons (par abus de langage) <u>la représentation coadjointe quotient</u>, intéressante en elle-même. Ces orbites se trouvent munies soit d'une structure d'<u>espace homogène de contact</u>, si elles sont de dimension impaire, soit d'une structure d'<u>espace homogène localement conformément symplectique</u> (l.c.s.) <u>hamiltonien</u> si elles sont de dimension paire.

Dans le premier cas, la feuille $\tilde{S}$ de $\tilde{\Lambda}$, engendrée par des rayons de Ω^* , au dessus d'une feuille S de (Λ,E) est espace homogène symplectique hamiltonien et S fournit un modèle universel, à un revêtement près, pour les espaces homogènes de contact dit <u>propres</u>

Dans le second cas, les feuilles $\tilde{S}$ de $\tilde{\Lambda}$ au dessus de S sont transverses aux rayons et S est un espace homogène l.c.s. hamiltonien qui fournit encore un modèle universel, à un revêtement près de tels espaces.

Le troisieme cas est celui des espaces homogènes de contact non propres qui correspondent à un revêtement près, aux espaces homogènes pfaffiens (à 1-forme globale de contact invariante). Par contactisation de revêtements de feuilles symplectiques $\tilde{S}$ convenables, on peut donner une construction universelle à un revêtement près, de tels espaces.

1.- Structure de Jacobi et structure conforme de Jacobi:

Soit W une variété différentielle connexe, paracompacte de dimension m et classe C^∞. Tous les éléments introduits ici sont C^∞. On pose $N = N(W) = C^\infty(W,\mathbb{R})$.

a) J'ai introduit en 1977, la notion de structure de Jacobi. Une telle structure est définie sur W par un 2.tenseur contravariant antisymétrique Λ et un vecteur E tels que, si i(.) désigne le produit intérieur, la crochet

(1.1) $[u,v] = i(\Lambda)\, du \wedge dv + u\; i(E)\, dv - v\; i(E)\, du$ $(u,v \in N)$

définisse sur N une algèbre de Lie. Pour qu'il soit ainsi, il faut et il suffit que l'on ait, en termes de crochets de Schouten:

(1.2) $[\Lambda,\Lambda] = 2\, E \wedge \Lambda$ $[E,\Lambda] = \mathcal{T}(E)\Lambda = 0$

où $\mathcal{T}(.)$ est l'opérateur de dérivation de Lie. D'après Kirïllov, toute algèbre de Lie locale (N, [,]) a un crochet de la forme (1-1). Soit h un élement de N partout $\neq 0$; nous posons $f_h: u \in N \to uh \in N$. Deux algèbres de Lie locales (N, [,]) et (N, [,]') sont équivalentes si l'on a

(1.3) $f_h\, [u,v]' = [f_h u, f_h v]$ $(u,v \in N)$

Les structures de Jacobi correspondantes sont reliées par:

(1.4) $\Lambda' = h\Lambda$ $E' = hE + |\Lambda,h|$

La classe d'équivalence correspodant à (1.4) est dite structure conforme de Jacobi. Nous notons L l'algèbre de Lie des automorphismes infinitessimaux de la structure conforme de Jacobi de (W, Λ, E). A tout élément $u \in N$, on fait correspondre le champ dit hamiltonien $X_u = uE + |\Lambda,u|$ et on note L* l'algèbre de Lie des champs hamiltoniens de (W, Λ, E); L* est un ideal de L et $\alpha : u \to X_u$ est un homomorphisme de (N, | , |) sur L* dont le noyau est le centre de (N, [,]). On a pour X_u:

(1.5) $\mathcal{T}(X_u)\, \Lambda = a\, \Lambda$ $\mathcal{T}(X_u) E = aE + [\Lambda, a]$ $(a = -i(E)du)$

Si E = 0, c'est-à-dire si Λ vérifie $[\Lambda.\Lambda] = 0$, Λ définit sur W une structure de Poisson.

b) Si $x \in W$, P_x est le plan engendré par les valeurs en x des vecteurs hamiltoniens de (W,Λ,E). Le champ P des plans P_x est le champ caractéristique de la structure conforme. Si dim P_x = const. la structure est dite régulière, si $P_x = T_x W$, elle est dite transitive.

A partir d'une extension contraviante du théorème de Frobenius (Süssmann P Stefan) et en montrant que P est invariant par tout champ de vecteurs hamiltonien , on établit que P définit su (W,Λ,E) un feuilletage généralisé; une feuille (de dimension variable) est par définition une variété intégrale maximale de P. Sur chaque feuille, les restrictions de Λ et E définissent une structure de Jacobi transitive.

Soit S une feuille de (W,Λ,E). Si dim S est impaire, $(\Lambda|_S,E|_S)$ définit sur une structure pfaffienne (donnée par une 1-forme globale de contact ω_S). Si dim S est paire $(\Lambda|_S,E|_S)$ définit sur S une structure l.c.s. qui peut être équivalente à une structure symplectique (g.c.s.) ou non. Dans ce dernier cas on a une structure l.c.s. vraie et une étude autonome fine de ces variétés a été faite en collaboration avec Mme. Guédira. Pour une variété de Poisson, toute feuille est symplectique.

2.- Variété de Poisson associée a une variété de Jacobi:

a) Considerons la variété $\tilde{W} = W \times R$ et notons z la variable canonique de R, $\tilde{Z}$ le champ de $\tilde{W}$ correspondant à $\partial/\partial z$, π la projection de $\tilde{W}$ sur W; Λ et E définissent sur $\tilde{W}$ des éléments invariants par $\tilde{Z}$ désignés par la même notation. Introduisons sur W les 2-tenseurs $\tilde{\Gamma} = \Lambda + \tilde{Z} \wedge E$, $\tilde{\Lambda} = e^{-z}\,\tilde{\Gamma}$. On vérifie immédiatement que l on a:

$$[\tilde{\Lambda}, \tilde{\Lambda}] = 0 \qquad\qquad T(\tilde{Z})\tilde{\Lambda} = -\tilde{\Lambda} \tag{2.1}$$

On établit:

Proposition A toute structure de Jacobi (Λ,E) sur W est associée canoniquement sur $\tilde{W}$ une structure de Poisson homogène de degré -1 par rapport a $\tilde{Z}$ et inversement.

Soit $\{x^j\}$ $(i,j...=1,...,m)$ une carte de W de domaine U, $\{x^A\} = \{x^0 = z, x^j\}$ la carte correspondante de $\tilde{W}$ au dessus de U. A toute fonction $u \in N$ associons la fonction $\tilde{u} = e^z \pi^* u$ définie sur $\tilde{W}$. On obtient ainsi un isomorphisme entre N et l'espace $\tilde{N}_1$ des fonctions homogènes et de degré 1 sur $\tilde{W}$: Le champ hamiltonien $\tilde{X}_{\tilde{u}} = [\tilde{\Lambda},\tilde{u}]$ associé à $\tilde{u} \in \tilde{N}_1$ a pour composantes $\tilde{X}^i_{\tilde{u}} = \tilde{X}^i_u$ et $\tilde{X}^0_{\tilde{u}} = -i(E)$ du. Soit $\tilde{L}_1$ l'algèbre de Lie des champs hamiltoniens de $(\tilde{W},\tilde{\Lambda})$ associés aux éléments de $\tilde{N}_1$, $\{\ ,\ \}^{\sim}$ le crochet de Poisson sur $(\tilde{W},\tilde{\Lambda})$. On déduit des considérations précédentes:

Proposition. Les algèbres de Lie $(\tilde{N}_1\ \{,\}^{\sim})$ et $(N, [\ ,\ ])$ sont naturellement isomorphes; π définit un isomorphisme de $\tilde{L}_1$ sur l'algèbre L* des champs hamiltoniens de (W,Λ,E).

b) Soit x_0 un point de W, $\tilde{x}_0$ un point de $\tilde{W}$ tel que $\pi\tilde{x}_0 = x_0$. Soit S (resp. $\tilde{S}$) la feuille de (W,Λ, E) (resp. $(\tilde{W},\tilde{\Lambda})$) passant par x_0(resp. $\tilde{x}_0$). L'analyse de ces feuilles montre que:

Proposition. On a $\pi\tilde{S} = S$.

1° Si dim. S est paire, la restriction de π à $\tilde{S}$ définit $\tilde{S}$ comme revêtement de S.

2° Si dim. S est impaire, $\tilde{S}$ coïncide avec $S \times \mathbb{R}$

S'il en est ainsi, S est munie d'une structure pfaffienne ω_S et $\tilde{S} = S \times \mathbb{R}$ admet une structure de variété symplectique exacte donnée par $\tilde{\omega}_{\tilde{S}} = e^z \pi^* \omega_S$. La restriction de $\tilde{Z}$ à $\tilde{S}$ définit un champ de vecteurs $\tilde{Z}_{\tilde{S}}$ de $\tilde{S}$. Si $\tilde{F}_{\tilde{S}}$ est la 2-forme symplectique de S, on a

$$\tilde{F}_{\tilde{S}} = d\tilde{\omega}_{\tilde{S}} \qquad \tilde{\omega}_{\tilde{S}} = i(\tilde{\tilde{Z}}_{\tilde{S}})\tilde{F}_{\tilde{S}}$$

3.- Représentation coadjointe et variété de Poisson:

Soit Ω une algèbre de Lie réelle de dimension n, Ω^* l'espace dual, $< \ , \ >$ la forme bilinéaire de dualité. On note $\Xi : \xi \in \Omega^* \to \xi \in \Omega^*$ le champ de vecteurs générateur des homothéties de Ω^*. On introduit sur Ω^* le 2-tenseur $\tilde{\Lambda}$ donné par:

(3.1) $$i(\tilde{\Lambda})(X \wedge Y) = <\Xi , [X,Y]> \qquad (X,Y \in \Omega)$$

Le 2-tenseur $\tilde{\Lambda}$ vérifie :

(3.2) $$[\tilde{\Lambda}, \tilde{\Lambda}] = 0 \qquad T(\Xi)\tilde{\Lambda} = -\tilde{\Lambda}$$

et définit donc sur Ω^* une structure de Poisson, homogène de degrè -1 par rapport à Ξ. A. $X \in \Omega$ associons la fonction $\tilde{u}_x = -<\Xi , X>$ sur Ω^* . Le vecteur $\tilde{X}$ image de X dans la représentation coadjointe de Ω peut s'écrire

$$\tilde{X} = [\tilde{\Lambda}, \tilde{u}_x]$$

Si X, Y $\in \Omega$, on a $[\tilde{X},\tilde{Y}] = -[X,Y]^{\sim}$; l'application $X \in \Omega \to \tilde{X} = Ad^*X$ est un homomorphisme de l'algèbre de Lie opposée à Ω sur l'algèbre de Lie $Ad^*\Omega$ dont le noyau est le centre de Ω ; $\tilde{\Lambda}$ et Ξ sont invariants par $Ad^*\Omega$ et $\tilde{\Lambda}$ induit sur les feuilles de $(\Omega^*, \tilde{\Lambda})$ des structures symplectiques invariantes.

Si $\{e^R\}$ (R,S,...=1,...n) est une base de Ω, $\{\lambda_R\}$ la base duale de Ω^*, on a $[e^R,e^S] = C^{RS}_T e$ où $C = \{C^{RS}_T\}$ est le tenseur de structure de Ω; $\tilde{\Lambda}$ et $\tilde{X}$ ont pour composantes :

$$\tilde{\Lambda}^{RS} = C^{RS}_T \Xi^T \qquad \tilde{X}^S = -\tilde{\Lambda}^{RS} X_R$$

Nous posons dans la base $\{\lambda_R\}$: $r^2(\xi) = \sum_{R=1}^{n} (\xi^R)^2 \quad (r \geqslant 0)$

4.- Représentation coadjointe quotient :

a) Considérons $\Omega^*_0 = \Omega^* - \{0\}$, munie du champ Ξ des homothéties. Le quotient de Ω^*_0 par le groupe des homothéties peut être représenté par la sphère S^{n-1} de Ω^*_0 de centre o et rayon $r(\xi) = 1$; Ω^*_0 se projette sur S^{n-1} selon p: $\xi \in \Omega^*_0 \to r^{-1}(\xi).\ \xi \in S^{n-1}$.

Soit $\{z^\alpha\}(\alpha,\beta,\ldots =-1,\ldots,n-1)$ une carte locale de S^{n-1}; $\{z^R\} = \{z^o = \log r(\xi), z^\alpha\}$ est une carte de Ω^*_0 par laquelle Ξ , d'après la définition de r, a pour composantes $\Xi^0 = 1$, $\Xi^\alpha = 0$.

La variété Ω^*_0 étant difféomorphe à $S^{n-1} \times \mathbb{R}$, il résulte de (2.2) que S^{n-1} se trouve munie d'une structure de Jacobi (Λ, E).

b) A X $\in \Omega$, nous avons associé la fonction $\tilde{u}_x = -<\Xi, X>$ homogène et de degre 1 par rapport à Ξ. On en déduit sur S^{n-1} une fonction u_x définie par $p^* u_x = r^{-1}\tilde{u}_x = e^{-z^o}\tilde{u}_x$; le champ hamiltonien $\tilde{X} = [\tilde{\Lambda}, \tilde{u}_x]$ de $(\Omega^*_0, \tilde{\Lambda})$ se projette par p selon le champ hamiltonien $\hat{X} = u_x E + [\Lambda , u_x]$ de (S^{n-1}, Λ, E). On obtient ainsi un isomorphisme d'algèbres de Lie entre Ad* Ω et $\hat{\Omega}$, algèbre engendrée par les $\hat{X}$; ceux-ci sont des a.i. de la structure conforme de Jacobi:

$$(4.1) \qquad T(\hat{X})\, \Lambda = a_x \Lambda \qquad T(\hat{X})E = a_x E + [\Lambda , a_x] \qquad (a_x = -i(E)du_x)$$

Soit G un groupe de Lie connexe d'algèbre de Lie Ω . Il opère sur Ω^*_0 par Ad* et sur S^{n-1} par une action ϕ telle que:

$$(4.2) \qquad p\mathrm{Ad}^*\, g^{-1} = \Phi_g p \qquad (g \in G)$$

L'action ϕ de G sur S^{n-1} est dite <u>la représentaton coadjointe quotient.</u> Les feuilles de $(\Omega^*_0 , \tilde{\Lambda})$ sont les orbites de l'action de G par Ad* ; les feuilles de la variété de Jacobi (S^{n-1}, Λ, E) sont les orbites de l'action de G par ϕ , la structure conforme de Jacobi induite par (Λ , E) est invariante par ϕ .

c) En particulier une feuille S de S^{n-1} <u>de dimension impaire</u> admet une structure de contact invariante, induite par la structure pfaffienne ω_S déterminée par (Λ, E) (sur S , $(T(\hat{X}) + a_x)\, \omega_S = 0$) ; S est alors <u>un espace homogène de contact</u> de groupe G. De plus une feuille $\tilde{S}$ de $(\Omega^*_0, \tilde{\Lambda})$ au dessus de S est un espace homogène symplectique exact de 1-forme invariante $\tilde{\omega}_{\tilde{S}} = i(\Xi_{\tilde{S}})\tilde{F}_{\tilde{S}} = e^{z^\circ} p^* \omega_S$ où $\tilde{F}_{\tilde{S}}$ est la 2-forme symplectique de $\tilde{S}$.

Une feuille S de S^{n-1} de <u>dimension paire</u> admet une structure l.c.s., invariante conforme par G qui opère transitivement ; S est dit <u>un espace homogène l.c.s. hamiltonien.</u> Une feuille $\tilde{S}$ au dessus de S est transverse à Ξ ......
et est un espace homogène symplectique hamiltonien de 2-forme $\tilde{F}_{\tilde{S}}$. On montre que pour que S soit <u>l.c.s. vraie</u>, il faut et il suffit que $\tilde{S}$ soit un revêtement non trivial.

<u>5.- Les deux cas d'espaces homogènes de contact</u>:

Soit (W,ω_W) une variété pfaffienne de dimension $m = 2p+1 \geqslant 3$. On suppose que cette variété admet une structure G/H <u>d'espace homogène de contact</u>, où G connexe opère transitivement et quasi-effectivement sur W par transformations de contact, le groupe d'isotropie H en $x_0 \in W$ étant <u>non discret</u>. Par action quasi-effective, on entend que tout sous-groupe invariant de G contenu dans H est discret. On a dim.G=n > m. La structure d'espace homogène de contact est définie au changement $\omega_W \to b\,\omega_W$ près ($b \in N$, $b \neq 0$ partout).

On note (Λ_W, E_W) les éléments contravariants correspondants à ω_W. Avec les notations du §2, $\tilde{W} = W \times \mathbb{R}$ admet une structure symplectique exacte de 1-forme $\tilde{\omega}_{\tilde{W}} = e^z \pi^* \omega_W$ telle que $\tilde{F}_{\tilde{W}} = d\tilde{\omega}_{\tilde{W}}$ correspond au 2-tenseur $\tilde{\Lambda}_{\tilde{W}}$ déduit de (Λ_W, E_W). En modifiant éventuellement G par revêtement, on peut supposer que ce groupe opère sur $\tilde{W}$ par des symplectomorphismes laissant $\tilde{\omega}_{\tilde{W}}$ et $\tilde{Z}$ invariants. Pour cette action, au champ X_W, champ hamiltonien associé à $u_W = i(X_W)\omega_W$ correspond sur $\tilde{W}$ le champ $\tilde{X}_{\tilde{W}} = [\tilde{\Lambda}_{\tilde{W}} , \tilde{u}_{\tilde{W}}]$, où $\tilde{u}_{\tilde{W}} = e^z \pi^* u_W$, qui se projette sur X_W et est tel que $\tilde{X}^{\circ}_{\tilde{W}} = -i(E_W)du_W$.

On a la proposition suivante:

Propositon. Soit Ψ l'algèbre de Lie de H, $\tilde{x}_0$ un point de $\tilde{W}$ au dessus de x_0.

1º S'il existe $X \in \Psi$ tel que $(i(E_w)du_w)(x_0) \neq 0$, G opère transitivement sur W.

2º Si, pour tout $X \in \Psi$, on a $(i(E_w)du_w)(x_0 = 0$, l'orbite $\tilde{W}_0$ de $\tilde{x}_0$ par G est de dimension $2p+1$.

En effet, dans le premier cas, le vecteur $\tilde{X}_{\tilde{W}}$ correspondant à X a pour composantes en $\tilde{x}_0$, dans la carte $\{x^A\}$: $\tilde{X}^i(\tilde{x}_0) = 0$, $\tilde{X}^0(\tilde{x}_0) = -(i(E_w)du_w)(x_0) \neq 0$. Il en résulte que l'orbite de $\tilde{x}_0$ dans $\tilde{W}$ par le sous-groupe à un parametre de H engendré par X est la droite facteur issue de $\tilde{x}_0$. La conclusion s'en déduit immédiatement.

Dans le second cas, soit H_0 la composante connexe de H ; $\tilde{x}_0$ est invariant par H_0; le plus grand sous-groupe $\tilde{H}$ de G laissant $\tilde{x}_0$ invariant est tel que $H_0 \subset \tilde{H} \subset H$ et $\tilde{W}_0$ est isomorphe à $G/\tilde{H}$.

Nous disons dans le premier cas que l'espace homogène de contact est propre, dans le second non propre.

6.- L'application moment ρ et son passage au quotient:

a) Dans le cas général, l'action hamiltonienne de G sur $\tilde{W}$ admet une application moment $\rho : \tilde{x} \in \tilde{W} \longrightarrow \rho(\tilde{x})$ Ω^* donnée par :

(6.1) $\qquad <\rho(\tilde{x}), X> = -\tilde{\omega}_{\tilde{W}}(\tilde{X}_{\tilde{W}}(\tilde{x})) = -\tilde{u}_{\tilde{W}}(\tilde{x}) \qquad (\forall X \in \Omega , \tilde{x} \in \tilde{W})$

Pour $\tilde{x} = (x,z)$, (6.1) peut s'écrire :

(6.2) $\qquad <\rho(\tilde{x}), X> = -e^z\, \omega_W(X_W(x)) = -e^z u_W(x)$

Cette application moment est Ad*-équivariante :

(6.3) $\qquad \rho(g\tilde{x}) = (Ad^* g^{-1})\rho(\tilde{x}) \qquad (g \quad G)$

Supposons que pour un point $\tilde{x}_1$ de $\tilde{W}$ (avec $\pi\tilde{x}_1 = x_1$) on ait $\rho(\tilde{x}_1) = 0$. On déduit de (6.3) que $\rho(g\tilde{x}_1) = 0$ pour tout $g \in G$. Il résulte de (6.2) que $u_W(gx_1) = 0$ pour $\forall X \in \Omega$, $g \in G$. Ainsi $u_W = i(X_W)\omega_W$ est identiquement nulle et $X_W = 0$ pour tout $X \in \Omega$ ce qui est absurde. On voit que l'application moment ρ est à valeur dans Ω_0^*.

b) Par dérivation de (6.2) en z on voit que $(d\rho / dz)(\tilde{x})=\rho(\tilde{x})$.Il en résulte

(6.4) $\qquad \Xi(\rho(\tilde{x})) = \rho_*(\tilde{Z}(\tilde{x}))$

On a de plus :

$$dr\,[\rho(x, z)] / dz = r\,[\rho(x, z)]$$

c'est-à-dire

$$r\,[\rho(x, z)] = r\,[\rho(x,0)]\, e^z$$

En changeant ω_W en $b\omega_W$ ($b \neq 0$ partout) on peut supposer $r[\rho(x,0)] = 1$. On a en effet $\tilde{\omega}_{\tilde{W}} \to (\pi^*.b)\tilde{\omega}_{\tilde{W}}$ et $\rho(\tilde{x}) \to \rho'(\tilde{x}) = b(x)\rho(\tilde{x})$ pour $\tilde{x} = (x,z)$. Il vient alors $r[\rho'(x,z)] = b(x)\, r\,[\rho(x,0)]\, e^z$ et en prenant b telle que $b(x) = r[\rho(x,0)]^{-1}$, on peut supposer dans la suite

(6.5) $$r\,[\rho(x,z)] = e^z$$

On a ainsi $\rho^* z^0 = z$ et les variables z et z^0 peuvent être identifiées.

On démontre aisément que l'on a pour $\tilde{x} \in \tilde{W}$

(6.6) $$\rho_*(\Lambda_{\tilde{w}}(\tilde{x})) = \tilde{\Lambda}(\rho(\tilde{x})) \qquad \rho_*(\tilde{X}_{\tilde{w}}(\tilde{x})) = \tilde{X}(\rho(\tilde{x}))$$

c) Considérons l'application de $\tilde{W}$ dans S^{n-1} donnée par $p\rho(\tilde{x}) = e^{-z}\,\rho(\tilde{x})$. Il resulte de (6-2) que:

$$\langle p\,\rho(\tilde{x}),\, X\rangle = -\,\omega_w(X_w(x)) = -\,u_w(x)$$

Ainsi $p\rho$ définit en fait une application $\sigma: x \in W \to \rho(x) \in S^{n-1}$ donnée par:

(6.7) $$\langle \sigma(x),\, X\rangle = -\omega_w(X_w(x)) = -\,u_w(x)$$

D'après (4.2) et (6.3), on a :

(6.8) $$\sigma(gx) = \phi_g\,\sigma(x)$$

et σ est une application de W sur la feuille S de (S^{n-1}, Λ, E) passant par $\sigma(x_0)$. On a le diagramme commutatif :

$$\begin{array}{ccc} \tilde{W} & \xrightarrow{\;p\;} & \Omega_0^* \\ \pi\downarrow & & \downarrow p \\ W & \xrightarrow{\;\sigma\;} & S^{n-1} \end{array}$$

Pour $X \in \Omega$, on déduit de (6.6) par la projection p:

(6.9) $$\sigma_*(X_w) = \hat{X}|_S$$

Si l'on se ramène par quotient à un groupe G opérant effectivement sur W, on vérifie immédiatement que (6.8) est encore valable. On a :

Théorème 1. - Soit $(W,\omega_w) = G/H$ un espace homogène de contact tel que G connexe opère transitivement et effectivement sur W par transformations de contact, H étant non discret; (6-7) définit une application σ de W sur une feuille de (S^{n-1}, Λ, E) équivariante par rapport à la representation coadjointe quotient.

7.- Le cas des espaces homogènes de contact propres :

a) Soit $\tilde{S}$ la feuille de (Ω^*_0, Λ) passant par $\xi_0 = \rho(\tilde{x}_0)$ qui se projette par p selon S. On a dim S $\leq$ 2p+1. Nous allons établir :

Proposition - Pour que l'espace homogène de contact $(W, \omega_w) = G/H$ soit propre, il faut et il suffit que la feuille $\tilde{S}$ soit engendrée par des trajectoires de Ξ dans Ω_0^* (ou rayons).

En effet supposons $\tilde{S}$ engendrée par des rayons ; S admet alors d après § 4, une structure pfaffienne ω_S et $\tilde{S}$ une structure symplectique exacte donnée par

$\tilde{\omega}_{\bar{S}} = i(\Xi_{\tilde{S}_*})\ \tilde{F}_{\tilde{S}} = e^{z}\rho^*\omega_S$, où $\tilde{F}_{\tilde{S}}$ est la 2-forme symplectique de $\tilde{S}$. Evaluons $i(\tilde{X}|_{\tilde{S}})\omega_{\tilde{S}}$ où $X \varepsilon\ Ad^*\ \Omega$. Un calcul direct évident donne pour $\xi \varepsilon \tilde{S}$

$$(i(\tilde{X}|_{\tilde{S}})\tilde{\omega}_{\tilde{S}})(\xi) = -<\xi, X> \qquad (X \varepsilon \Omega)$$

et par suite :

$$\tilde{\omega}_{\tilde{S}}(\tilde{X}(\rho(\tilde{x}))) = -<\rho(\tilde{x}), X> = \tilde{\omega}_{\tilde{W}}(\tilde{X}_{\tilde{W}}(\tilde{x}))$$

Il vient par produit par e^{-z} :

$$(7.1) \qquad \omega_S(\hat{X}(\sigma(x))) = \omega_W(X_W(x))$$

où, d'après (6.9), $\hat{X}(\sigma(x)) = \sigma_*(X_W(x))$. Il en résulte :

$$(7.2) \qquad \omega_W = \sigma^*\omega_S$$

Comme $\omega_W \wedge (d\omega_W)^p = \sigma^*(\omega_S \wedge (d\omega_S)^p) \neq 0$ partout, on a dim S = 2p + 1, dim $\tilde{S}$ = 2p + 2 et pour l'orbite $\tilde{W}_0$ de $\tilde{x}_0$ par G, dim $\tilde{W}_0$ = 2p + 2.

Inversement supposons que G opère transitivement sur $\tilde{W}$. On a $\rho(\tilde{W}) = \tilde{S}$ et ρ définit $\tilde{W}$ comme revêtement de $\tilde{S}$ qui est de dimension 2p + 2. D'après (6.4), $\tilde{S}$ est engendrée par des rayons de Ω_0^*, ce qui démontre la proposition.

b) Ainsi si (W, ω_W) = G/H est un espace homogène de contact propre de dimension 2p +1, S de dimension 2p + 1 admet une structure de contact invariante donnée par ω_S. Ainsi (S, ω_S) est un espace homogène de contact propre et σ définit W comme revêtement de S, préservant d'après (7.2) les 1-formes de contact choisies.

S étant engendrée par des rayons de Ω_0^*, il existe $X \varepsilon \Omega$ tel qu'en $\xi_0 = \rho(\tilde{x}_0)$, on ait

$$\tilde{X}(\xi_0) = \lambda\ \xi_0 \qquad (\lambda \varepsilon R\ ; \lambda \neq 0)$$

Il en résulte qu'il existe $Y \varepsilon \Omega$ tel que:

$$Ad(X).Y = -\lambda\ Y$$

Ainsi G est non compact et Ω contient l'algèbre de Lie résoluble non abélienne de dim 2 engendrée par (X,Y) et non contenue dans Ψ . La réciproque est immédiate. On a :

Théoreme 2 - Soit (W, ω_W) = G/H un espace homogène de contact propre, où G connexe opère transitivement et effectivement sur W par transformations de contact, H étant non discret. L'application σ définit $(W, \omega\ )$ comme revêtement d'une feuille (S, ω_S) de la sphère S^{n-1} et fait correspondre les structures de contact invariantes de W et S.

Le groupe G est non compact et Ω admet une sous algèbre résoluble de dimension 2, engendrée par $X, Y \varepsilon \Omega$ tels que $[X, Y] = Y$. Inversement, s'il en est ainsi, une feuille S de dimension impaire admet une structure naturelle d'espace homogène de contact propre de groupe G.

(S,ω_S) est un modèle universel, à un revêtement près, pour les espaces homogènes de contact propres.

8.- Feuille $\tilde{S}$ non engendrée par des rayons :

Nous allons maintenant étudier les espaces homogènes de contact non propres. Si $\tilde{S}$ est la feuille de $(\Omega_0^*, \tilde{\Lambda})$ passant par $\xi_0 = \rho(\tilde{x}_0)$, cette feuille n'est pas engendrée par des rayons; $(\tilde{S}, \tilde{\Lambda}|_{\tilde{S}})$ est un espace homogène symplectique $G/\tilde{K}$ .Nous notons $\Sigma \subset \Omega$ l'algèbre de Lie de $\tilde{K}$.

On établit aisément :

Proposition - Pour que $\tilde{S}$ ne soit pas engendrée par des rayons de Ω_0^*, il faut et il suffit qu'il existe $X \in \Sigma$ tel que $\tilde{u}_X(\xi_0) = -<\xi_0, X>$ soit $\neq 0$.

S'il en est ainsi tout élément Y de Σ se décompose d'une manière unique selon $Y = T + cX$ (avec $\tilde{u}_T(\xi_0) = 0$). Le sous-espace Σ' de Σ engendré par les éléments T est un idéal de Σ contenant $[\Sigma, \Sigma]$; Σ/Σ' est une algèbre de Lie abélienne de dimension 1.

9.- Le cas d'un espace de contact non propre :

Soit $(W,\omega_W) = G/H$ un espace homogène de contact de dimension $2p + 1$ non propre, H pouvant être discret.

a) Sur la variété symplectique exacte $(\tilde{W},\tilde{\omega}_{\tilde{W}})$ sur laquelle G opère en laissant $\tilde{\omega}_{\tilde{W}}$ invariante, on établit aisément qu'en $\tilde{x}_0 = (x_0, z_0)$, on a :

(9.1) $$(\pi^*\omega_W \wedge \tilde{F}^p_{\tilde{W}})(\tilde{x}_0) = e^{pz_0}(\pi^*(\omega_W \quad (d_W^p)))(\tilde{x}_0) \neq 0$$

Soit $\tilde{W}_0$ l'orbite de $\tilde{x}_0 \in \tilde{W}$; elle est de dimension $2p + 1$. Nous notons $\tilde{\omega}^{(0)}_{\tilde{W}}$ et $\tilde{F}_{\tilde{W}}$ les restrictions ã $\tilde{W}_0$ de $\tilde{\omega}_{\tilde{W}}$ et $\tilde{F}_{\tilde{W}}$. Il résulte de (9-1) :

(9.2) $$(\tilde{\omega}^{(0)}_{\tilde{W}} \wedge (d\tilde{\omega}^{(0)}_{\tilde{W}})^p)(\tilde{x}_0) \neq 0$$

$\tilde{\omega}^{(0)}_{\tilde{W}}$, invariante par l'action de G sur $\tilde{W}_0$ est une 1-forme de contact. Il vient :

Proposition - Un espace homogène de contact non propre $(W,\omega_W) = G/H$ admet comme revêtement l'espace homogène pfaffien $(\tilde{W}_0, \tilde{\omega}^{(0)}_{\tilde{W}})$.

b) On a $\rho(\tilde{W}_0) = \tilde{S}$, où $\tilde{S}$ est la feuille de $(\Omega_0^*, \tilde{\Lambda})$ passant par $\xi_0 = \rho(\tilde{x}_0)$. On a donc dim $\tilde{S} \leqslant 2p$ et S n'est pas engendrée par des rayons. Il résulte immédiatement de (6.6) que, si $\rho^{(0)}$ est la restriction de ρ à $\tilde{W}_0$, on a :

(9.3) $$\tilde{F}^{(0)}_{\tilde{W}} = \rho^{(0)*}\tilde{F}_{\tilde{S}}$$

D'après (9.2), on a $\tilde{F}^{(0)p}_{\tilde{W}} \neq 0$ partout, donc $\tilde{F}^p_{\tilde{S}} \neq 0$ et dim $\tilde{S} = 2p$; $\tilde{S}$ étant tranverse à (Ξ, p)définit $\tilde{S}$ comme revêtement de S, feuille de S^{n-1} avec dim $S = 2p$.

Ainsi $W = G/H$ de dimension $2p + 1$ est appliquée par σ sur S de dimension $2p$ munie d'une structure l.c.s. conformément invariante par l'action de ϕ

Si $Y \in \Psi$, on a $\tilde{Y}_{\tilde{w}}(\tilde{x}_0) = 0$ et $<\xi_0, Y> = \omega_{\tilde{w}}(\tilde{Y} \sim (\tilde{x}_0)) = 0$. La fonction $\tilde{u}_Y = -<\Xi, Y>$ s'annule en ξ_0 ; Ψ est ainsi un sous espace de Σ' de même dimension (dim K − 1) que lui. Ainsi $\Sigma' = \Psi$ et Ψ est un idéal de Σ tel que Σ/Ψ soit une algèbre abélienne de dimension 1. On a :

<u>Théorème 3</u> − Soit $(W, \omega_w) = G/H$ un espace homogène de contact non propre, de dimension $2p + 1$, tel que G opère transitivement et effectivement sur W par transformations de contact. L'orbite $\tilde{W}_0$, revêtement de W est munie d'une structure d'espace homogène pfaffien.

σ applique (W, ω_w) sur une feuille S de (S^{n-1}, Λ, E) de dimension $2p$, munie d'une structure l.c.s. conformément invariante par ϕ. Si $S = G/K$, l'algèbre de Lie Ψ est un idéal de l'algèbre de Lie Σ de K tel que Σ/Ψ soit une algèbre de Lie abélienne de dimension 1.

<u>10.− Contactisation et espace homogène pfaffien</u> :

a) Un étude classique de la contactisation (voir par exemple Arnold) conduit à la proposition suivante :

<u>Proposition</u> − Soit (M, F) une variété symplectique. Pour qu'un fibré $q : \hat{M} \to M$ en cercles (resp. en droites réelles) admette une structure pfaffienne $\hat{\alpha}$ dont le vecteur fondamental du fibré est le vecteur de Reeb et telle que $d\hat{\alpha}$ soit l'image de F, il faut et if suffit que F soit à cohomologie entière (resp. exacte), $(\hat{M}, \hat{\alpha})$ est dit contactisée de (M, F).

b) Soit (W, ω_w) un espace homogène pfaffien de dimension $2p + 1$. Il existe une application Ad^*-équivariant encore notée ρ (utiliser $\tilde{W}_0$) de ces espaces sur une feuille $\tilde{S}$ de $(\Omega_0^*, \tilde{\Lambda})$ de dimension $2p$. On déduit de la proposition précédente :

<u>Théorème 4</u> − Soit (W, ω_w) un espace homogène pfaffien de dimension $2p + 1$ tel que G opère transitivement et effectivement sur W, en laissant ω_w invariante. L'application ρ applique W sur un espace homogène symplectique hamiltonien $(\tilde{S}, \tilde{F}\tilde{s}) = G/\tilde{K}$ feuille de $(\Omega_0^*, \tilde{\Lambda})$, de dimension $2p$ tel que pour son revêtement $(S', F'_{S'}) = G/K_0$, où K_0 est la composante connexe de $\tilde{K}$, $F'_{S'}$ ait une classe de cohomologie entière.

Inversement soit $(\tilde{S}, \tilde{F}_{\tilde{S}}) = G/\tilde{K}$ une feuille de $(\Omega_0^*, \tilde{\Lambda})$ de dimension $2p$. où $\tilde{K}$ est le groupe d'isotropie de $\xi_0 \in S$, qui vérifie les conditions suivantes :

1º. Il existe $X \in \Sigma$ tel que $<\Xi, X>$ ne s'annule pas en ξ_0.

2º. Le revêtement $(S', F'_{S'}) = G/K_0$ de $\tilde{S}$ est tel que $F'_{S'}$ soit à classe de cohomologie entière .

Il existe alors une contactifiée de $(S',F_{S'})$ soit $(W',\omega_W) = G/H_0$ qui est un espace homogène pfaffien de dimension 2p + 1.

A un revêtement près, tout espace homogène pfaffien et tout espace homogène de contact non propre peut-être obtenu ainsi:

11.- Les espaces l.c.s. hamiltoniens:

Des raisonnements analogues à ceux qui ont été développés permettent d'établir:

Théoreme 5 - Soit $(W,\Lambda_W, E_W) = G/H$ un espace homogène l.c.s. hamiltonien de dimension $2p \geq 4$, tel que G connexe opère transitivement et effectivement sur W par transformations telles que (Λ_W,F_W) se transforment de manière conforme, H étant non discret. Il existe une application σ de W sur une feuille (S,Λ_S,E_S) de S^{n-1}, équivariante par rapport à ϕ, qui définit W comme revêtement de S de dimension 2p et fait correspondre les structures l.c.s. de W et S.

(S,Λ_S,E_S) est un modèle universel, à un revêtement près, pour les espaces homogènes l.c.s. hamiltoniens. Si (W,Λ_W,E_W) est l.c.s. vraie, (S,Λ_S,E_S) est l.c.s. vraie.

12.- Cas où G est compact:

Soit W = G/H un espace homogène à groupe G compact. Il existe sur Ω^* une forme quadratique définie positive invariante par Ad* G. En définissant $r(\xi)$ à partir de cette forme, on voit que la variété de Jacobi portée par S^{n-1} se reduit à une variété de Poisson. Les feuilles de (S^{n-1},Λ) sont toutes symplectiques et les revêtements: $\tilde{S} \longrightarrow S$ triviaux.

Ainsi si G est compact, tout espace homogène (W,Λ_W,E_W) l.c.s. hamiltonien est symplectique hamiltonien.

Les orbites $\tilde{S}$ étant des espaces homogènes symplectiques par rapport à Ad* G semi-simple compact sont des espaces homogènes kähleriens simplement connexes (A. Borel) et il est de même pour les feuilles S, donc pour les espaces l.c.s.. De plus:

Proposition - Tout espace homogène pfaffien G/H à groupe G compact est contactisé d'une orbite $(\tilde{S},\tilde{F}_{\tilde{S}})$ de Ad*, orbite pour laquelle $\tilde{F}_{\tilde{S}}$ est à cohomologie entière.

BIBLIOGRAPHIE

[1] A. Lichnerowicz. Comptes rendus t 285, 1977, 455-461; J. de Maths Pures et Appl.t 57 1978, 453-488; Comptes rendus t 296 I, 1983, 915-920 et 297 I, 1983, 261-266. Les notations sont celles de ces notes.

[2] A.A. Kirillov. Russ.Math. Surveys 31, 1976, p. 55.

[3] F. Guédira et A. Lichnerowicz. Géométrie des algèbres de Lie locales de Kirillov; J. Math. Pures et Appl. t. 63, 1984, p. 407-484.

[4] R. Abraham et J.E. Marsden. Foundations of Mechanis. Benjamin-Cummings 1978, p. 276-282.

[5] A. Lichnerowicz. C.R. Académie Sciences Paris, t. 299 I, 1984, p. 685-689.

ON A GEOMETRIC GENERALIZATION OF THE NOETHER THEOREM

V. MARINO and A. PRASTARO
Dipartimento di Matematica
Università della Calabria
87036 Arcavacata di Rende (Cs) Italy

ABSTRACT

Task of this paper is to compare some geometric approaches to obtain conservation laws associated to partial differential equations. More precisely we intend to consider the methods developed by A. Prastaro and A. M. Vinogradov.

The differential equations are considered from a geometric point of view; namely they are submanifolds of Jet-derivative spaces on fiber bundles. To the symmetries of these submanifolds are associated conservation laws that are not necessarily of Noetherian type.

1. INTRODUCTION

A modern line of research in the context of the geometric characterization of differential equations, is that intended to obtain conservation laws directly associated to the geometric structure giving a differential equation. It is important therefore to study from a side the formal properties of partial differential equations (PDE) and to another to characterize the symmetries of these structures. Really the well known Noether's theorem is unable to reproduce conservation laws for any PDE as we have serious obstructions to represent any PDE in the variational context [10,12]. Furthermore, as a symmetry of a variational PDE is not necessarily a symmetry of the corresponding Lagrangian, we can imagine conservation laws directly associated to the symmetry of the differential equation in the variational context also.

Generalizations in these directions have been studied by numerous authors; (some references are given in ref. [10 , 13 , 14 , 15])

In general conservation laws that generalize Noether's theorem are obtained for variational equations only. But for non variational PDE we know only two generalizations given by Prastaro [2,4,5] and Vinogradov [10].

In sections 3 and 4 we resume these fundamental results.

Scope of this paper is to compare these two methods and see in which sense they can be considered as generalizations of the classical Noether's Theorem.

2. NOTATIONS

In this note we will always consider C^∞ manifolds and C^∞ maps. We will use the following notations:

M = n-dimensional differentiable manifold

$\pi : W \to M$ = a fiber bundle over M

TM = tangent bundle of M

T^*M = cotangent bundle of M

η = a volume form over M

Df = derivative of any $f: M \to W$, i.e. $Df: M \to \mathcal{D}(W) \subset \mathcal{D}(M,W) = T^*M \otimes TW$

where $\mathcal{D}(W) = \bigcup_{p\in M} \left\{ \bigcup_{p\in\pi^{-1}(p)} T^*_pM \otimes T_qW \right\}$

$D^hf = D(D^{h-1}f)$.

$C^\infty(W)$ = the set of C^∞ sections of π over M

$J\mathcal{D}^k(W) = \{ u \in \mathcal{D}^k(W) \mid \exists f \in C^\infty(W): D^kf(\pi_k(u)) = u;$

$\pi_k: J\mathcal{D}^k(W) \to M$ is the canonical projection $\}$

Observe that $J\mathcal{D}^k(W)$ is a subspace of $\mathcal{D}^k(W)$; it is called k-th Jet derivative over W [1]

vTW = vertical tangent bundle of W

$\overset{\bullet\,\circ}{\Lambda}_p W$ = fiber bundle of horizontal p-forms over $\pi : W \to M$, i.e. the p-forms which are zero on the vertical vectors of W

$\mathcal{P}[s]$ = differential operator associated to a first order Lagrangian and section s as given in ref. [6]

$\sigma_K(A)$ = symbol of a k-order differential operator A

$\mathcal{L}_X$ = Lie derivative with respect to a vector field X.

3. A RECENT GENERALIZATION OF NOETHER THEOREM [2,4,5]

Let

$$L : J\mathcal{D}^k(W) \to \mathbb{R} \tag{1}$$

be a k-order Lagrangian on the fiber bundle $W = (W, M, \pi)$. A system of k-order partial differential equations on the fiber bundle W is a sub-bundle E_k of $J\mathcal{D}^k(W)$ and a solution s of E_k is a section of W, $s: M \to W$, such that $D^k s: M \to E_k$ (s can be eventually only locally defi ned). At the first order (k=1), a well known proceeding (see e.g. refs. [3,6]) allows us to canonically associate to L a second order differential operator ε called Euler-Lagrange operator.

Then, the Noether theorem enables us to find for the relating Euler-Lagrange equation some conservation law directly related to the sym metry properties of the Lagrangian. More precisely we have:

Theorem 1 (Noether theorem) [6]:

Let s be an extremal for the Lagrangian L. Let L be invariant with respect to a 1-parameter group ϕ_t of transformations of the fi ber bundle $W = (W, M, \pi)$. Let $\tilde{\phi}_t$ be the corresponding 1-parameter group on M. Then the following (n-1)-form β on M

$$\beta \equiv \left[(L \circ Ds)\tilde{\xi} + (\sigma_1(\mathcal{P}[s]) \cdot \partial \tilde{s}) \right] \rfloor \eta$$

is closed, i.e. $d\beta = 0$, where $\tilde{\xi} \equiv \partial\tilde{\phi}$ and $\tilde{s}$ is the deformation of s by means of ϕ_t. ∂ denotes the infinitesimal variation [1].

The above theorem can be generalized for a k-order Lagrangian. We must previously recall that to any k-order Lagrangian L is associated a horizontal n-form Ω on $\pi_k : J\mathcal{D}^k(W) \to M$ called "Lagrangian density of order k" which is obtained as a section of the fiber bundle of the horizontal n-form on $\pi_k : J\mathcal{D}^k(W) \to M$ (see ref. [2] for details), that

is $\Omega \in C^\infty(\overset{\bullet}{\Lambda}{}^o_n J\mathcal{D}^k(W))$ and $\Omega = L \cdot \pi^*_k \eta$.
We also remind that the Euler-Lagrange equation (EL) $\subset J\mathcal{D}^{2k}(W)$ associated to Ω is the ker of the differential operator (Euler-Lagrange operator)

$$\varepsilon : C^\infty(W) \to C^\infty(\Lambda^o_n M \otimes vTW)$$

that can be intrinsically obtained by some geometric methods as proven for example in refs. [2,4,7,8,10,11,12,13,14,15] .

Furthermore, an infinitesimal symmetry for Ω (and then for L) is a π-related vector field X on W such that $(D^k\gamma)^* \mathcal{L}_{\overset{\kappa}{X}} \Omega = 0$ for any extremal γ of Ω, where $\overset{\kappa}{X}$ is the k-th prolongation of X on $J\mathcal{D}^k(W)$ induced by the functor $J\mathcal{D}^k$.

Then we recognize that a conservation law for Ω is a (n-1)-form

$$\beta \in C^\infty(\Lambda^o_{n-1} J\mathcal{D}^k(W)) \text{ such that } d((D^k\gamma)^*\beta) = 0$$

for any extremal γ of Ω .

We can state now the "Generalized Noether theorem":

<u>Theorem 2 (Noether theorem for higher order Lagrangians)</u> [2,4,7]:

If X is a symmetry for Ω, then

$$d\left\{(D^k\gamma)^* \tau (\overset{\kappa}{X} \lrcorner \sigma[\Omega])\right\} = 0$$

where:

a) $\sigma[\Omega]$ = Cartan form of Ω,

b) τ is the projection operator $\tau : C^\infty(\Lambda^o_p J\mathcal{D}^k(W)) \to C^\infty(\overset{\bullet}{\Lambda}{}^o_p J\mathcal{D}^{k+1}(W))$ defined by $(D^{k+1}\gamma)^* \tau\omega = (D^k\gamma)^*\omega$ for any $\gamma \in C^\infty(W)$.

We can see [4] that any symmetry of the Lagrangian is also a symmetry of the relating Euler-Lagrange equation (the vice versa is not true!). Then A.Prastaro gets in ref. [2] a geometric method to recognize conservation laws of a general partial differential equation (PDE), $E_k \subset J\mathcal{D}^k(W)$, directly related to the symmetries of E_k.

<u>Theorem 3</u> [2,4,5]:

Let $E_k \subset J\mathcal{D}^k(W)$ be a k-order partial differential equation. Let H be a r-dimensional Lie subgroup of the symmetry group of E_k. Let $\mathbb{E}_H$ be the involutive r-distribution generated by the infinitesimal generators of the restricted action of H on E_k. Let $\{\phi^\tau\}_{\tau=1,\dots s}$, with $s \leq q-r$, $q \equiv \dim E_k$, be a fundamental set of differential invariants of $\mathbb{E}_H$. Then for any function ϕ^{τ} we have some configurations $c: M \to W$ with conservation laws if $Sol(E_k) \cap Ext(\phi^{\tau}) \neq \emptyset$, where $Ext(\phi^{\tau})$ denotes the set of extremals for ϕ^{τ} and $Sol(E_k)$ denotes the set of dynamical configurations for (W, E_k).

4. THE VINOGRADOV APPROACH

In this section we shall consider the fundamental results of A. M. Vinogradov about a generalization of the Noether theorem for linear differential equations [9,10].

Theorem 1 (Vinogradov theorem):

a) Let $\pi : E \to M$ be a vector fiber bundle on a n-dimensional manifold M and $E_k \subset J\mathcal{D}^k(E)$ a k-order linear differential equation obtained as the kernel of a linear operator $\Delta : C^{\infty}(E) \to C^{\infty}(K)$ where K is a vector fiber bundle on M. Let $E_k^* \subset J\mathcal{D}^k(K^* \boxtimes \Lambda_n^o M)$ be the dual of E_k, obtained by means of the formal adjoint $\Delta^*: C^{\infty}(K^* \boxtimes \Lambda_n^o M) \to C^{\infty}(E^* \boxtimes \Lambda_n^o M)$ of Δ.

Then, for any couple (ϕ_1, ϕ_2), where ϕ_1 is a symmetry of E_k and ϕ_2 is a Bäcklund transformation of E_k into E_k^*, we have an associated conservation law, that is it exists a differential $(n-1)$-form β on $J\mathcal{D}^k(E)$ such that for any solution s of E_k it holds:

$$d((D^k s)^* \beta) = 0$$

b) If $\Delta \equiv \varepsilon(L)$ = Euler-Lagrange operator for some h-order Lagrangian $L: J\mathcal{D}^k(E) \to \mathbb{R}$ $(k = 2h)$, then the above conservation law can be calculated directly from the symmetry of E_k.

Proof:

Let $Diff^{(+)}(C^{\infty}(E), C^{\infty}(K)) = \bigcup_{s \geq 0} Diff_s^{(+)}(C^{\infty}(E), C^{\infty}(K))$ be the $C^{\infty}(M)$-bimodule, that is the union of $C^{\infty}(M)$-bimodules of all the differential operators of order $\leq s$. The left (resp. right) $C^{\infty}(M)$-module structure is given by $a\Delta = a \circ \Delta$ where $(a\Delta)p = a\Delta(p)$ (resp. $\Delta a = \Delta \circ a$ where $(\Delta a)p = \Delta(ap)$) for $a \in C^{\infty}(M)$ and $p \in C^{\infty}(E)$.

For simplicity of notations, let us make this position: $P = C^{\infty}(E)$, $Q = C^{\infty}(K)$ and $\Lambda^n = C^{\infty}(\Lambda_n^o M)$. Observe that $Diff_o(P,Q) = Hom(P,Q)$(1)

(1) $Diff_s(P,Q)$ denotes the left $C^{\infty}(M)$-module of differential operators of order $\leq s$ from P to Q; $Diff_s^+(P,Q)$ denotes the right $C^{\infty}(M)$-module.

and $\mathrm{Diff}_o P = \mathrm{Hom}(C^\infty(M),P) = P$.

We have then the following decomposition of the $C^\infty(M)$-module $\mathrm{Diff}_s(C^\infty(M),P) = \mathrm{Diff}_s P$ as the direct sum: $\mathrm{Diff}_s P = P \oplus \widetilde{\mathrm{Diff}}_s P$, where $\widetilde{\mathrm{Diff}}_s P = \{\Delta \in \mathrm{Diff}_s P \mid \Delta(1) = 0\}$ and $\Delta = (\Delta - \Delta(1)) + \Delta(1)$.

In a similiar way we get $\mathrm{Diff}\, P = P \oplus \widetilde{\mathrm{Diff}}\, P$.

Let us denote by $D_s'(P) = \mathrm{Diff}_s^a(\underbrace{C^\infty(M) \times \ldots \times C^\infty(M)}_{s \text{ times}},P)$ the $C^\infty(M)$-module of skew multiderivations in P; it consists of all functions Φ such that for any $(a_1, \ldots, a_s) \in \underbrace{C^\infty(M) \times \ldots \times C^\infty(M)}_{s \text{ times}}$, $\Phi(a_1,\ldots, a_s) \in P$, which are skew symmetric and are derivatives with respect to each variable.

Let us set $D(P) = D_1(P)$ and $P = D_o(P)$. We can also give an iterative definition of $D_s(P)$:

$D_i(P) = D(D_{i-1}(P))$.

Let $\dot{D}_s(\mathrm{Diff}_t^+ Q)$ be the set of the multiderivations $D_s(\mathrm{Diff}_t^+ Q)$ with the $C^\infty(M)$-module structure induced by the left module structure in the bimodule $\mathrm{Diff}_t^{(+)} Q$; i.e. if $\Delta \in D_s(\mathrm{Diff}_t^+ Q)$, then for any $a_i \in C^\infty(M)$ we have: $(a\Delta)(a_1, \ldots, a_s) = a \circ \Delta(a_1, \ldots, a_s)$.

Moreover let us recall that, if $\Delta \in \mathrm{Diff}\,\Lambda^n$ and ω is a volume form on M, the adjoint of Δ ($\Delta^* \in \mathrm{Diff}\,\Lambda^n$) is defined by

$\Delta^*(f) = \Delta_\omega[f\omega]$

where $\Delta_\omega \in \mathrm{Diff}(C^\infty(M), C^\infty(M))$ is such that $\Delta(f) = \Delta_\omega(f)\omega$ and $\Delta_\omega[\bullet]$ denotes the action(2) of Δ_ω on the n-forms.

To any $\Delta \in \mathrm{Diff}(P,Q)$ we can associate a differential operator $\blacktriangle(p,\hat{q}) \in \mathrm{Diff}\,\Lambda^n$ which is defined by

$\blacktriangle(p,\hat{q})(f) = \langle \Delta(fp),\hat{q}\rangle$,

where $f \in C^\infty(M)$, $p \in P$, $\hat{q} \in C^\infty(K^* \boxtimes \Lambda^n)$.

The adjoint Δ^* of Δ is the operator $\Delta^*: C^\infty(K^* \boxtimes \Lambda^n) \to C^\infty(E^* \boxtimes \Lambda^n)$, defined by

---

(2) The action of $\Delta \in \mathrm{Diff}(C^\infty(M), C^\infty(M))$ on $\gamma \in \Lambda^n$ is defined by :
1) $(\Delta_1 + \Delta_2)[\gamma] = \Delta_1[\gamma] + \Delta_2[\gamma]$; $\Delta_1, \Delta_2 \in \mathrm{Diff}(C^\infty(M), C^\infty(M))$
2) $f[\gamma] = f\gamma$; $f \in C^\infty(M)$
3) $X[\gamma] = -\mathcal{L}_X\gamma$; $X \in H(M) = C^\infty(TM)$
4) $(\Delta_1 \circ \Delta_2)[\gamma] = \Delta_2[\Delta_1[\gamma]]$.

Furthermore the action of any Δ on the n-forms is unique and it is possible to write $\Delta[\gamma] = \Delta(1) + \sum (-1)^s \mathcal{L}_{X_s}(\ldots(\mathcal{L}_{X_1}\gamma)\ldots)$, with $X_i \in H(M)$.

$$< \Delta^*(\gamma \boxtimes \omega), p> \ = \ < \Delta(p), \gamma > \omega$$

where $(\gamma \boxtimes \omega) \in C^\infty(K^* \boxtimes \Lambda^n)$ and $p \in P$.

Let us define the following homomorphisms:

$$\square^+ : \mathrm{Diff}^+ P \to P$$
$$\Delta \longmapsto \Delta(1)$$

and

$$i_+ : \mathrm{Diff}_s(P,Q) \to \mathrm{Diff}^+_s(P,Q)$$
$$i^+ : \mathrm{Diff}^+_s(P,Q) \to \mathrm{Diff}_s(P,Q)$$

generated by the identity maps of the corresponding sets.

Let $\Psi : \dot{D}_s(\mathrm{Diff}^+_{k-s}\ \Lambda^n) \to \mathrm{Diff}^+_{k-s}(\Lambda^{n-s})$ denote the following composition of isomorphisms

$$\Psi : \dot{D}_s(\mathrm{Diff}^+_{k-s}\Lambda^n) \xrightarrow{D_s(*)} D_s(\mathrm{Diff}_{k-s}\Lambda^n) \xrightarrow{a} \mathrm{Diff}_{k-s}(D_s\Lambda^n) \xrightarrow{\mathrm{Diff}_{k-s}\Pi} \mathrm{Diff}_{k-s}\Lambda^{n-s} \xrightarrow{i_+}$$
$$\xrightarrow{i_+} \mathrm{Diff}^+_{k-s}\Lambda^{n-s}$$

where:

1) $D_s(*)$ is the functorial extension of $*$.
2) The homomorphism a is given by a permutation of the functors D_s and Diff_{k-s}.
3) $\mathrm{Diff}_{k-s}\ \Pi$ is the functorial extension of the Poincaré isomorphism $\Pi = \Pi_i : D_i(\Lambda^n) \to \Lambda^{n-i}$

given by $Z = X_1 \wedge \ldots \wedge X_i \boxtimes \omega \to \Pi(Z) = X_1 \lrcorner (\ldots X_i \lrcorner \omega))$.

Reminding that Spencer Diff-operators are:

$$S = S_{i,s} = \beta \circ \alpha = \dot{D}_i(\mathrm{Diff}^+_s P) \to D_{i-1}(\mathrm{Diff}^+_{s+1} P)$$

where

$$\dot{D}_i(\mathrm{Diff}^+_s P) \xleftrightarrow{\alpha} \dot{D}_{i-1}(\mathrm{Diff}^+_1(\mathrm{Diff}^+_s P)) \xleftrightarrow{\beta} \dot{D}_{i-1}(\mathrm{Diff}^+_{s+1} P),$$

we can construct the complex $S_k P$

$$0 \longleftarrow P \xleftarrow{\square} \mathrm{Diff}_k P \xleftarrow{S_{1,k-1}} \dot{D}(\mathrm{Diff}^+_{k-1} P) \xleftarrow{S_{2,k-2}} .$$

In particular, the complex $S_k P$ for the Λ^n module becomes the upper line of the following diagram:

$$\begin{array}{ccccccccccc}
0 & \longleftarrow & \Lambda^n & \longleftarrow & \mathrm{Diff}_k \Lambda^n & \xleftarrow{S_{1,k-1}} & \dot{D}(\mathrm{Diff}^+_{k-1}\Lambda^n) & \longleftarrow \dots \longleftarrow & \dot{D}_n(\mathrm{Diff}^+_{k-n}\Lambda^n) & \longleftarrow & 0 \\
 & & \Big\downarrow = & & \Big\downarrow \psi & & \Big\downarrow \psi & & \Big\downarrow \psi & & \\
0 & \longleftarrow & \Lambda^n & \xleftarrow{-\mu} & \mathrm{Diff}^+_k \Lambda^n & \xleftarrow{\mathscr{S}} & \mathrm{Diff}^+_{k-1}\Lambda^{n-1} & \xleftarrow{\mathscr{S}} \dots \longleftarrow & \mathrm{Diff}^+_{k-n} C^\infty(M) & \longleftarrow & 0
\end{array}$$

where $\mu(\Delta) = \Delta^*(1)$, for each $\Delta \in \mathrm{Diff}_k \Lambda^n$, and $\mathscr{S}(\nabla) = d \circ \nabla$ for each $\nabla \in \mathrm{Diff}^+_i \Lambda^s$. It can be shown that each square of the above diagram is skew-commutative.

Let us consider now the skew-commutative diagram

$$\begin{array}{ccc}
\mathrm{Diff}_k \Lambda^n & \xleftarrow{S} & \dot{D}(\mathrm{Diff}^+_{k-1}\Lambda^n) \\
\Big\downarrow * & & \Big\downarrow \psi \\
\mathrm{Diff}^+_k \Lambda^n & \xleftarrow{\mathscr{S}} & \mathrm{Diff}^+_{k-1}\Lambda^{n-1}
\end{array}$$

We notice that $\mathrm{Im}\, S = \widetilde{\mathrm{Diff}_k \Lambda^n}$ and since this is a projective $C^\infty(M)$-module it is possible to find a homomorphism $\lambda\colon \mathrm{Diff}_k \Lambda^n \longrightarrow \dot{D}(\mathrm{Diff}^+_{k-1}\Lambda^n)$ such that $S \circ \lambda = 1$.

Moreover let us construct the homomorphism $\mathfrak{a}_\lambda$ which appears in the Green formula that will be used in what follows: $\mathfrak{a}_\lambda\colon \mathrm{Diff}_k \Lambda^n \longrightarrow \Lambda^{n-1}$, given by:

$$\mathfrak{a}_\lambda = \square^+_{k-1} \circ \psi \circ \lambda \circ \nu$$

with $\nu\colon \mathrm{Diff}_k \Lambda^n \longrightarrow \widetilde{\mathrm{Diff}_k \Lambda^n}$ defined as $\nu(\Delta) = \Delta - \Delta(1)$.

It is now possible to determine the β $(n-1)$-form using the general Green's formula $\langle \Delta(s), \hat{\gamma} \rangle - \langle s, \Delta^*(\hat{\gamma}) \rangle = d\mathfrak{a}_\lambda(\blacktriangle(s,\hat{\gamma}))$ with $s \in C^\infty(E)$ and $\hat{\gamma} \in C^\infty(K^* \boxtimes \Lambda^n)$.

Since indeed ϕ_1 is a symmetry of E_k and hence an endomorphism of $\Delta = 0$, $\Delta(s) = 0$ implies that $\Delta(\phi_1 s) = 0$ and, if $\hat{\gamma} = \phi_2 s$, then $\Delta^*(\hat{\gamma}) = 0$ whenever $\Delta(s) = 0$.

Since a non linear c-density for the equation $\Delta = 0$ is an operator $\nabla\colon C^\infty(E) \to \Lambda^0_{n-1}(M)$ such that $d(\nabla(s)) = 0$ for any local solution s of

$\Delta = 0$, the operator

$$F : s \longrightarrow \mathbf{a}_\lambda(\blacktriangle(\phi_1 s, \phi_2 s)) = \beta_M$$

is a non linear c-density of $\Delta = 0$. Due to the general Green's formula we have indeed $d\mathbf{a}_\lambda(\blacktriangle(\phi_1 s, \phi_2 s)) = 0$.

It is then possible to determine the corresponding non linear conservation law as the equivalence class of the known density, i.e. the totality of the densities which differ from the differential of an operator $F: C^\infty(E) \to \Lambda^{n-2}$. Furthemore $F(s) = \beta_M$ coincides with $(D^k s)^* \beta$ for $\beta \in C^\infty(\Lambda^o_{n-1}(J\mathcal{D}^k(E)))$.

b) If $\Delta = \varepsilon(L)$, i.e. Δ is an Euler-Lagrange operator for some h-order Lagrangian $L: J\mathcal{D}^k(E) \to \mathbb{R}$ $(k = 2h)$, and ϕ_1, ϕ_2 are symmetries of $E_k = \operatorname{Ker} \Delta = \operatorname{Ker} \varepsilon(L)$, then the operator $F: s \to \mathbf{a}_\lambda(\blacktriangle(\phi_1 s, \phi_2 s))$ is actually a c-density for $\Delta = 0$.

Remark:

Since a symmetry of a Lagrangian is also a symmetry of the corresponding Euler-Lagrange equation [4], Noether theorem can be obtained as the direct consequence of the theorem 2.1 (or 1.3).

5. SOME OBSERVATIONS ON THE ABOVE GENERALIZATIONS OF NOETHER THEOREM

In this section we shall develope some critical considerations about previous theorems.

Let us before see under which conditions theorem 3.3 is complete.

Theorem 1:

Let $E_k \subset J\mathcal{D}^k(W)$ be a formally integrable partial differential equation (PDE) and $\phi : J\mathcal{D}^k(W) \to \mathbb{R}$ a differential invariant of E_k.

Iff $F_{2k} \equiv E_{k+k}$ is a submanifold of $(EL)_\phi \subset J\mathcal{D}^{2k}(W)$, then any solution s of E_k brings a dynamic conservation law associated to ϕ.

Proof:

In fact in this case we have $\operatorname{Sol}(E_k) \cap \operatorname{Ext}(\phi) = \operatorname{Sol}(E_k)$.

We obtain a stronger result by the following

Theorem 2:

Let $E_k \subset J\mathcal{D}^k(W)$ be a formally integrable PDE and $\phi: J\mathcal{D}^k(W) \to \mathbb{R}$

a differential invariant associated to 1-dimensional subgroup $\{\phi_t\}_{t\in\mathbb{R}}$ of the symmetry group of E_k. Then E_k admits a conservation law for <u>any</u> dynamic configuration ϕ if the Cartan form $\sigma(\phi)$ is such that $d\sigma(\phi)$ belongs to the ker of τ^+, where τ^+ is the projection operator

$$\tau^+ : C^\infty(\Lambda^0_{n+h} J\mathcal{D}(W)) \to C^\infty(\Lambda^0_{n+h,h}(\pi_1,\pi_{1,0})) \equiv C^\infty(\Lambda^0_{n+h,h}(\pi_1) \cap \Lambda^0_n(\pi_{1,0}))$$

defined by

$$\overset{1}{X}_1 \lrcorner \ldots \overset{1}{X}_n \lrcorner \tau^+\omega = \tau(X_1 \lrcorner \ldots X_n \lrcorner \omega),$$

$$\forall\, \omega \in C^\infty(\Lambda^0_{n+h} J\mathcal{D}(W)), \qquad X_i \in C^\infty(vTJ\mathcal{D}(W)).$$

π_1 and $\pi_{1,0}$ are the canonical projections $\pi_1 : J\mathcal{D}(W) \to M$ and $\pi_{1,0} : J\mathcal{D}(W) \to W$. $\overset{1}{X}_i$ is the first prolongation of X_i on $J\mathcal{D}^2(W)$.

<u>Proof</u>:

From theorem 3.3 we know that to ϕ is associated a conservation law if

$$\mathrm{Ext}(\phi) \cap \mathrm{Sol}(E_k) \neq \emptyset.$$

So we must consider under which conditions it is verified that

$$\mathrm{Ext}(\phi) \cap \mathrm{Sol}(E_k) = \mathrm{Sol}(E_k). \qquad (2)$$

On the other words we will find the condition on ϕ such that any solution s of E_k brings the conservation law

$$\beta = (D^\infty s)^* \tau(\overset{\infty}{X} \lrcorner \sigma(\phi)), \quad X = \partial\phi.$$

As E_k is assumed to be formally integrable, we have that equation (2) is satisfied if the Euler-Lagrange equation $(EL)_\phi \subset J\mathcal{D}^{2k}(W)$ coincides with $J\mathcal{D}^{2k}(W)$. We have indeed that the k-prolongation $E_{2k} \equiv E_{k+k} \subset J\mathcal{D}^{2k}(W)$ of E_k has some solutions of E_k. So, in order to satisfy equation (2), it must be $E_{2k} \cap (EL)_\phi = E_{2k}$. This condition is really satisfied if $(EL)_\phi = J\mathcal{D}^{2k}(W)$, namely $d\sigma(\phi) \in \ker \tau^+$. In fact we have the well known relation: [2,4,7]

$$\varepsilon(\phi) = \tau^+ d\sigma(\phi),$$

where $\varepsilon(\phi)$ is the Euler-Lagrange form associated to ϕ.

<u>Theorem 3 (Noether theorem)</u>:

If $E_k \subset J\mathcal{D}^k(W)$ is a formally integrable PDE and $\phi : J\mathcal{D}^k(W) \to \mathbb{R}$ is a differential invariant of E_k such that $E_{2k} \equiv E_{k+k}$ is an open submanifold of $(EL)_\phi \subset J\mathcal{D}^{2k}(W)$, then any solution s of E_k brings a dynamic conservation law associated to ϕ. Furthermore E_{2k} is locally of va riational type with local Lagrangian given by ϕ.

Finally, if $H^{n+1}(W) = 0, E_{2k}$ is globally variational too ($n=\dim M$).

Corollary 1:

If $E_k \subset J\mathcal{D}^k(W)$ is a linear formally integrable PDE and $\phi : J\mathcal{D}^k(W) \to \mathbb{R}$ is a differential invariant of E_k such that $E_{2k} \equiv E_{k+k}$ is an open submanifold of $(EL)_\phi \subset J\mathcal{D}^{2k}(W)$, then any solution of E_k brings a dynamic conservation law associated to ϕ and E_k is globally of variational type.

Remark:

We want to observe that theorem 4.1 is not a direct generalization of the Noether's theorem as the closed (n-1)-differential form (Vinogradov's law) given by means of the Green's formula

$$\langle\Delta(s), \hat{\gamma}\rangle - \langle s, \Delta^*(\hat{\gamma})\rangle = d\mathfrak{a}_\lambda(\blacktriangle(s,\hat{\gamma}))$$

is not fully characterized by the symmetry properties of the differential equation $E_k(\Delta = 0)$, but it is necessary to add some other "physical entities" $\phi_2 : J\mathcal{D}^k(W) \to K^* \boxtimes \Lambda^o_n M$. In other words the symmetry properties of E_k are not enough to characterize the Vinogradov's laws.

Really a good generalization of Noether's theorem should recognize conservation laws directly studing the symmetries of a differential equation (linear or non-linear) without adding any other structure.

In this sense theorem 3.3 is a generalization of Noether's theorem.

Remark:

Let us now stress some cohomological considerations. Any conservation law of E_k, $\rho = (D^k s)^*\beta$, where s is a suitable solution of E_k and β is a (n-1)-differential form on $J\mathcal{D}^k(W)$ given by theorem 3.3, belongs to some cohomological class of $H^{n-1}(M)$ ($n = \dim M$). In general we can say that any conservation law associated to a differential equation $E_k \subset J\mathcal{D}^k(W)$ is of the type described in theorem 3.3 iff in any cohomological class of $H^{n-1}(M)$ there exists a non trivial (= non exact) conservation law of this type. In fact under this condition we have that any other closed (n-1)-differential form ρ on M can be written as $\rho = (D^k s)^*\beta + d\alpha$, where $(D^k s)^*\beta$ is given by means of theorem 3.3.

We can call such equations as wholly-cohomologic equations.

Of course if the (n-1)-dimensional cohomology of M is trivial ($H^{n-1}(M) = 0$), then any differential equation that admits conservation laws is wholly-cohomologic.

REFERENCES

[1] A. Prastaro, "On the General Structure of Continuum Physics", I. Boll. Unione Mat. Ital., (5) 17-B, (1980), 704-726. II. Boll. Unione Mat. Ital., (5) Suppl.FM, (1981), 69-106. III. Boll. Unione Mat. Ital., (5) Suppl.FM, (1981), 107-129.

[2] A. Prastaro, "Gauge Geometrodynamics", Riv. Nuovo Cimento, 5 (4), (1982), 1-122.

[3] A. Prastaro, "On the Intrinsic Expression of the Euler-Lagrangian Operator", Boll. Unione Mat. Ital., (5) 18-A, (1981), 411-416.

[4] A. Prastaro, "Dynamics Conservation Laws", Geometrodynamics Proceedings, (1985), 302-439.

[5] A. Prastaro, "Geometrodynamics of Non-Relativistic Continuous Media. II: Dynamics and Constitutive Structures", Rend. Semin. Mat. Torino, in press.

[6] H. Goldschmidt and S. Sternberg, "The Hamilton-Cartan Formalism in the Calculus of Variations", Ann. Inst. Fourier, Grenoble, 23-1, (1973), 203-267.

[7] B.A. Kupershmidt, "Geometry of Jet-Bundles and the Structure of Lagrangian and Hamiltonian Formalism", Lect. Notes Math., (775), Springer, (1980), 168-218.

[8] M. Ferraris and M. Francaviglia, "Proceedings IUTAM/ISIMM", Symposium on Modern Development in Analytical Mechanics, Torino, (1982).

[9] A.M. Vinogradov, "On the Algebro-Geometric Foundations of Lagrangian Field Theory", Sov. Math. Dokl., 18 (5), (1977), 1200-1204.

[10] A.M. Vinogradov, "The $\mathcal{C}$-spectral Sequence. Lagrangian Formalism and Conservation Laws", I, II. J. Math. Anal. Appl., 100 (1), (1984), 1-129.

[11] W. Tulczyjew, "The Lagrange Complex", Bull. Soc. Math. Fr., 105, (1977), 419-431.

[12] F. Takens, "A Global Version of the Inverse Problem of the Calculus of Variations", J.Differ.Geom., 14, (1979), 543-562.

[13] V. Aldaya and J.A. de Azcárraga, J. Math. Phys., 19, 1869 (1978).

[14] D. Krupka, Mod. Develop. in Analytical Mechanics, Torino 1982 S. Benenti et al. Eds. Torino (1983).

[15] M. León and P.R. Rodrigues, Generalized Classical Mechanics and Field Theory, North Holland Mathematical Studies, (1985).

THE ISOPERIMETRIC INEQUALITY AND THE GEODESIC SPHERES. SOME GEOMETRIC CONSEQUENCES

A. M. Naveira and S. Segura
Departamento de Geometría y Topología
Facultad de Matemáticas
University of Valencia
BURJASOT (Valencia)-Spain

1. Introduction

It is well-known that among all plane curves of given length the circle of circumference L encloses maximum area.

This statement is expressed in the isoperimetric inequality:

(1) $$L^2 \geq 4\pi A$$

where A is the area enclosed by a curve C of length L, and where equality holds if and only if C is a circle, i.e. a geodesic sphere in R^2.

Although this extremal property of the circle was already known to the Greeks, rigorous proofs of it did not begin to appear till the last century. (See [Os2], [Ba2]).

In 1939 E. Schmidt [Sch1] generalized this result to R^n, obtaining the inequality

(2) $$A^n \geq C_n V^{n-1}$$

where A is the area of the boundary of a domain D, V its volume and C_n the n-dimensional isoperimetric constant given by

(3) $$C_n = n^n \omega_n \quad (1$$

The equality sign in (2) holds if and only if the hypersurface which encloses D is an hypersphere.

During the present century many mathematicians have tried to obtain the isoperimetric inequality for manifolds with curvature.

The first succesful attempt was made by Bernstein [Be] in 1905 who solved this problem for the sphere S^2 proving

(4) $$L^2 \geqq 4\pi A - KA^2$$

where $K=1/R^2$ is the curvature of the sphere of radius R, and L, A are as in (1).

Again equality in (4) holds if the domain on S^2 is a geodesic ball:

---

1) $\omega_n = \frac{\pi^{n/2}}{(n/2)!}$ is the volume of the unit euclidean ball in R^n

---

Work partially supported by CAICYT, 1985-87, nº 120.

(5) $$L_m^2(r) = 4\pi A_m(r) - KA_m^2(r)$$

where $L_m(r)$ is the length of the geodesic circle of radius r centered at m, and $A_m(r)$ is the area of the corresponding geodesic disk.

It is interesting to observe that letting in (4) the radius of the sphere tend to infinity, (1) is obtained.

In the case of arbitrary 2-dimensional Riemannian manifolds, as Osserman [Os2] says, it is convenient to consider separately two parts of the general problem:

Part I. Finding those classes of surfaces and domains on them for which the classical isoperimetric inequality (1) remains valid.

Following this direction it is important the result of Beckenbach and Radó [B-R] which states that a necessary and sufficient condition for a surface S with Gauss curvature K satisfy (1) for all simply-connected domains on it is that $K \leq 0$ everywhere on S.

Part II. In those cases where (1) does not hold finding appropiate analogues such as (4) for the sphere.

A lot of important results have been obtained in this second direction by Fiala [Fi], Bol [Bo], Schmidt [Sch2], Aleksandroff [Al1], [Al2], Huber [Hu], Toponogov [To], Karcher [Ka], Bandle [Ba] and Aubin [Au1], [Au2].

In the case of n-dimensional Riemannian manifolds the problem is very much more difficult.

For instance, although the isoperimetric inequality was proved in S^n by Schmidt [Sch3] -in the sense that among all domains of given volume on S^n the geodesic balls have minimum (n-1)-dimensional area of its boundary-, it is not possible to obtain a proper generalization of (4) because there does not exist a simple expression equivalent to (5) for geodesic balls.

In fact Aubin [Au2] gives a generalization of Bernstein's (5) for the geodesic balls in the sphere S^n, but he does not obtain an equality but only the inequality

(6) $$A_m^{n/(n-1)}(r) \leq (C_n)^{1/(n-1)} V_m(r)\left(1 - \frac{K}{2(n+2)}\, n\omega_n^{-2/n} V_m(r)^{2/n}\right)$$

Our purpose is

1) Using the series expansions of Gray and Vanhecke [G-V1] to study the isoperimetric deficit

(7) $$(A_m(r))^n - C_n(V_m(r))^{n-1}$$

of geodesic spheres for arbitrary n-dimensional Riemannian manifolds and its influence in the geometry of the manifold. Among other results we give a generalization of the necessary condition of Beckenbach and Radó's theorem.

2) Using the area and volume formulae of geodesic spheres and their balls, [Gr], to find expressions analogous to (5) for the complex and quaternionic projective spaces. Several geometric consequences will be obtained from these formulae.

2. The isoperimetric deficit of geodesic balls in arbitrary Riemannian manifolds.

It is a well-known result that, for all surfaces in R^3, the Gauss curvature at a point m can be expressed as

$$(8) \qquad K(m) = -\lim_{r\to 0} \frac{(L_m(r))^2 - 4\pi A_m(r)}{\pi^2 r^4}$$

(See Osserman [Os2])

We have obtained the following generalization of it:

Proposition 1: Let M be an arbitrary n-dimensional Riemannian manifold and $m \in M$. Let $A_m(r)$ be the (n-1)-dimensional area of the geodesic sphere of radius r centered at m, $V_m(r)$ the n-dimensional volume of the corresponding geodesic ball, and τ_m the scalar curvature at the point m. Then

$$(9) \qquad \tau_m = -\lim_{r\to 0} \frac{2(n+2)}{r^2} \cdot \frac{(A_m(r))^n - n^n \omega_n (V_m(r))^{n-1}}{(n\omega_n r^{n-1})^n}$$

Proof: The proof follows from Gray and Vanhecke's formulae [G-V1] by a l'Hospital computation.

The formula (9) shows that the scalar curvature at a point m is a measure of the isoperimetric deficit of a sufficiently small geodesic ball centered at m in a Riemannian manifold.

Corollary 1: The standard n-dimensional isoperimetric inequality (2) can only hold for an arbitrary n-dimensional Riemannian manifold if $\tau_m \leq 0$ everywhere on M.

This Corollary generalizes the necessary condition of Beckenbach and Radó [B-R] for a surface S to satisfy the standard isoperimetric enequality (1); however this condition is not sufficient as it can be deduced from Corollary 2.

If dim M = 4 Croke has recently solved [C] Beckenbach and Radó's conjecture, but it is still open in all dimensions except 2 and 4.

Proposition 2: Let M be a Ricci-flat manifold. Let $||R||^2 = \Sigma R^2_{ijkl}$ where R_{ijkl} are the components of the curvature tensor using the notations of Gray [Gr] . Then

$$(10) \qquad ||R||^2 = -\lim_{r\to 0} \frac{24(n+2)(n+4)}{r^4} \cdot \frac{(A_m(r))^n - n^n \omega_n (V_m(r))^{n-1}}{(n\omega_n r^{n-1})^n}$$

Proof: The proof is analogous to that of Proposition 1.

Corollary 2: The isoperimetric deficit of sufficiently small geodesic balls in Ricci-flat manifolds is negative.

In [G-V1] Gray and Vanhecke studied the geometry of a Riemannian manifold by analyzing the properties of small geodesic balls. In particular they conjectured.

(I) Suppose that $V_m(r) = \omega_n r^n$, for all $m \in M$ and all sufficiently small $r > 0$. Then M is flat.

They gave affirmative answer to this conjecture in some cases (See (G-V1) §4 p.172).

It is easy to see that conjecture (I) can also be written in terms of the isoperimetric deficit of the geodesic balls:

(I)' Suppose that $(A_m(r))^n - n^n\omega_n(V_m(r))^{n-1} = 0$ for all $m \in M$ and all sufficiently small $r > 0$. Then M is flat .

3. The geodesic spheres in the complex projective space.

Proposition 3: Let CP^n be the complex projective space with constant holomorphic sectional curvature $4\alpha^2$. Then for all $m \in CP^n$:

$$(A_m(r))^{2n} = C_{2n}(V_m(r))^{2n-1}\left(1 - \frac{(2n)^2}{C_{2n}{}^{1/n}}\alpha^2 V_m(r)^{1/n}\right)^n \tag{11}$$

Proof: It is well-known (See for instance (Gr)):

$$A_m(r) = \frac{2\pi^n}{(n-1)!\,\alpha^{2n-1}} \sin^{2n-1}(\alpha r)\cos(\alpha r) \tag{12}$$

$$V_m(r) = \frac{\pi^n}{n!\,\alpha^{2n}} \sin^{2n}(\alpha r) \tag{13}$$

The proof follows inmediately from these formulae.

Remark 1: In the particular case $CP^1 \simeq S^2$, it is obtained

$$(L_m(r))^2 = 4\pi A_m(r) - 4\alpha^2(A_m(r))^2 \tag{14}$$

which is (5) because in this case $K = 4\alpha^2$.

Remark 2: (11) seems to be the corresponding result to (6) in S^n, but in CP^n the equality holds for geodesic balls in any dimension.

Remark 3: An expression equivalent to (11) is

$$(A_m(r))^2 = C_{2n}^{1/n}(V_m(r))^{(2n-1)/n} - (2n\alpha)^2(V_m(r))^2 \tag{15}$$

As these formulae seem to be the natural generalization of those in S^2, we state here the following conjecture:

(II) Let CP^n be the complex projective space with constant holomorphic sectional curvature $4\alpha^2$. Then for any domain D on CP^n such that its volume is smaller or equal than that of a geodesic ball with diameter smaller or equal than the injectivity radius in CP^n

(16) $$A^2 \geqq C_{2n}^{1/n} V^{(2n-1)/n} - (2n\alpha)^2 V^2$$

In the particular case n = 1 (16) is Bernstein's result (4).

Remark 4:

(17) $$(A_m(r))^{2n} \leq C_{2n}(V_m(r))^{2n-1}$$

is a result weaker than (11) which can be obtained trivially from it.

Notice that from (6) it can be obtained also

(18) $$(A_m(r))^n \leq C_n(V_m(r))^{n-1}$$

for geodesic spheres in S^n .

Corollary 3: Let $B_1(B_2)$ be a geodesic ball in the complex projective space $CP_1^n(CP_2^n)$ of holomorphic sectional curvature $4\alpha_1^2$ $(4\alpha_2^2)$, so that if V_1 (V_2) is the volume in CP_1^n (CP_2^n), $V_1(B_1)=V_2(B_2)$; then if $\alpha_1 < \alpha_2$

(19) $$A_1(\delta B_1) > A_2(\delta B_2)$$

where $A_i(\delta B_i)$ is the area of the boundary of B_i, i=1,2.

Notice that this Corollary is the corresponding to Corollary 2 of Aubin in [Au2] for spaces of constant curvature.

Proof: The proof follows form (15).

In [G-V1] Gray and Vanhecke establish the following conjecture:

[III] Let M be a Kähler manifold with complex dimension n and suppose that for all $m \in M$ and all sufficiently small $r > 0$, $V_m(r)$ is the same as that of an n-dimensional Kähler manifold with constant holomorphic sectional curvature μ . Then M has constant holomorphic sectional curvature .

They prove this conjecture for Böchner-flat manifolds and for Einstein-Kähler manifolds.

It is easy to see by analysing a differential equation that [III] can also be written as:

[III]' Let M be a Kähler manifold with complex dimension n, and suppose that for all $m \in M$ and for all sufficiently small $r > 0$

(20) $$(A_m(r))^2 = C_{2n}^{1/n} (V_m(r))^{(2n-1)/n} - (2n\alpha)^2(V_m(r))^2$$

Then M has constant holomorphic sectional curvature $4\alpha^2$.

It is interesting to compare [III]' with Gray and Vanhecke's Theorem 3 in [G-V2] .

4. Geodesic spheres in the quaternionic projective space.

Proposition 4: Let QP^n be the quaternionic projective space with maximum sectional curvature $4\alpha^2$. Then for all $m \in QP^n$:

(21) $$(A_m(r))^{4n} = C_{4n}(V_m(r))^{4n-1} \cdot \frac{(2n+1)^{4n-1}\cos^{12n}(\alpha r)}{(2n\cos^2(\alpha r)+1)^{4n-1}}$$

Proof: It follows also inmediately from the area and volume formulae of geodesic spheres and their balls. (See for instance [Gr])

Corollary 4:

(22) $$(A_m(r))^{4n} \leq C_{4n}(V_m(r))^{4n-1}$$

Notice that inequality is the corresponding to (18) for S^n and to (17) for CP^n.

Proof: The result follows from (21) after proving that

(23) $$\frac{(2n+1)^{4n-1}\cos^{12n}(\alpha r)}{(2n\cos^2(\alpha r)+1)^{4n-1}} \leq 1$$

To prove (23) it is enough to check that the function

(24) $$f(r) = \frac{(2n+1)^{4n-1}\cos^{12n}(\alpha r)}{(2n\cos^2(\alpha r)+1)^{4n-1}}$$

is a decreasing function defined on $[0, \pi/2\alpha[$ and such that $f(0)=1$.

Corollary 5: Let $B_1 (B_2)$ be a geodesic ball in the quaternionic projective space $QP^n_1 (QP^n_2)$ of maximum sectional curvature $4\alpha_1^2$ $(4\alpha_2^2)$, so that if V_1 (V_2) is the volume in QP^n_1 (QP^n_2), $V_1(B_1) = V_2(B_2)$; then if $\alpha_1 < \alpha_2$

(25) $$A_1(\delta B_1) > A_2(\delta B_2)$$

where $A_i(\delta B_i)$ is the area of the boundary of B_i, $i=1,2$.

Proof: As $V_1(B_1) = V_2(B_2)$,

(26) $$\int_0^{\delta_1} \frac{\cos^3(\alpha_1 r)\operatorname{sen}^{4n-1}(\alpha_1 r)}{\alpha_1^{4n-1}} dr = \int_0^{\delta_2} \frac{\cos^3(\alpha_2 r)\operatorname{sen}^{4n-1}(\alpha_2 r)}{\alpha_2^{4n-1}} dr$$

The function $g(x) = \dfrac{\cos^3(\rho x)\operatorname{sen}^{4n-1}(\rho x)}{x^{4n-1}}$ is a decreasing function of x for $\rho x < \pi/2$

Then as $\alpha_1 < \alpha_2$,

(27) $$\frac{\cos^3(\alpha_1 r)\operatorname{sen}^{4n-1}(\alpha_1 r)}{\alpha_1^{4n-1}} > \frac{\cos^3(\alpha_2 r)\operatorname{sen}^{4n-1}(\alpha_2 r)}{\alpha_2^{4n-1}}$$

From (26) and (27) we have

(28) $$\delta_1 < \delta_2$$

(25) follows from (28) and from the fact that the function f(r) defined is (24) is decreasing.

We have obtained also the corresponding results for $CayP^2$.

BIBLIOGRAPHY

(Al1) A.D. ALEKSANDROV. Isoperimetric inequalities for curved surfaces. Dokl.Akad.Nauk. USSR 47 (1945), 235-238.

(Al2) A.D. ALEKSANDROV. Die innere Geometrie der Konvexen Flächen. Akademie Verlag, Berlin, 1955.

(Au1) T. AUBIN. Problemes isoperimétriques et spaces de Sobolev, C.R.Acad. Sci. Paris Sér. A 280 (1975) 279-281.

(Au2) T. AUBIN. Problemes isoperimétriques et spaces de Sobolev, J.Differential Geometry 11 (1976) 573-598.

(Ba1) C. BANDLE. On a differential inequality and its applications to geometry. Math. Z. 147 (1976), 253-261.

(Ba2) C. BANDLE. Isoperimetric inequalities, from "Convexity and its applications" Birkhäuser Verlag, Basel-Boston, Mass. 1983.

(B-R) E.F. BECKENBACH and T. RADO. Subharmonic functions and surfaces of negative curvature. Trans. Amer. Math. Soc. 35 (1933) 662-674.

(Be) F. BERNSTEIN. Uber die isoperimetrische Eigenshaft des Kreises auf der Kugeloberfläche und in der Ebene. Math. Ann. 60 (1905), 117-136.

(Bo) G. BOL. Isoperimetrische Ungleichung für Bereiche auf Flächen, Iber Deutsch Math. Verein, 51 (1941), 219-257.

(C) C.B. CROKE. A sharp four dimensional isoperimetric inequality. Comment. Math. Helvetici 59 (1984), 187-192.

(Fi) F. FIALA. Le probleme des isopérimètres sur les surfaces ouvertes à courbure positive, Comment. Math. Helvetici 13 (1940/41), 293-346.

(Gr) A. GRAY. The volume of a small geodesic ball of a Riemannian manifold. Michigan Math. J. 20 (1973), 329-344.

(G-V1) A. GRAY and L. VANHECKE. Riemannian Geometry as determined by the volumes of small geodesic balls. Acta Mathematica 142 (1979), 157-197.

(G-V2) A. GRAY and L. VANHECKE. Oppervlakten van geodetische cirkels op oppervlakten. AWLSK, 1980.

(Hu) A. HUBER. On the isoperimetric inequality on surfaces of variable Gaussian curvature, Ann of Math. (2) 60-(1954), 237-247.

(Ka) H. KARCHER. Anwendungen der Aleksandrowschen Winkelvergleichssätze. Manuscripta Math. 2 (1970) 77-102.

(Os1) R. OSSERMAN. Bonnesen-style isoperimetric inequalities. Amer. Math. Monthly, 86 (1979) 1-29.

(Os2) R. OSSERMAN. The isoperimetric inequality. Bulletin of the American Mathematical Society, 84 Number 6 (1978) 1182-1238

(Sch1) E. SCHMIDT. Uber das isoperimetrische Problem im Raum von n Dimensionen. Math. Z. 44 (1939) 689-788.

(Sch2) E. SCHMIDT . Uber eine neue Methode zur Behandlung einer Klässe isoperimetrisschen Aufgaben im Grossen, Math. Z. 47 (1942), 489-642.

(Sch3) E. SCHMIDT. Beweis der isoperimetrischen Eigenschaft der Kugel im hyperbolischen und sphärischen Raum jeder Dimensionenzahl.Math. Z. 49 (1943/44) 1-109.

(To) V.A. TOPONOGOV. An isoperimetric inequality for surfaces whose Gaussian curvature is bounded above. Siberian Math. J. 10 (1969) 104-113.

ON THE K-DIMENSIONAL RADON-TRANSFORM OF RAPIDLY DECREASING FUNCTIONS

F. RICHTER*
Sektion Mathematik
Humboldt-Universität Berlin

1. INTRODUCTION.-

The present paper deals with the k-dimensional Radon transform which is defined by integrating functions on R^n over k-dimensional planes. So these functions are transformed to functions on $E_{n,k}$ (the space of all k-dimensional planes in the n-dimensional Euclidean space). One of the most interesting questions is the description of the range of certain function spaces, such as $\mathcal{T}(R^n)$ (rapidly decreasing functions) or $\Gamma(R^n)$ (compactly supported smooth functions). For the space $\Gamma(R^n)$ and any k ($1 \leq k \leq n-1$) this was done by Helgason in [5]. There he also described the range of $\mathcal{T}(R^n)$ for $k = n-1$. In all these cases the range is characterized by so-called "moment conditions". In his Thesis ([3]) Gonzales proved that such a moment condition does not suffice to describe the range of $\mathcal{T}(R^n)$ if $k < n-1$.

In this case ($1 \leq k < n-1$) a system of linear partial differential equations of second order characterizes the range of $\mathcal{T}(R^n)$, as stated in [2] by Gel'fand, Gindikin and Graev (chapter 1, § 3, section 7, theorem 5). However, this theorem is not proved there but its analogy to the proof of Theorem 1 in [1] is mentioned.

Theorem 1 of [1] describes the range of rapidly decreasing functions on the n-dimensional complex sapce C^n under the k-dimensional Radon transform if $1 \leq k < n-1$. However, some details are omitted.

In [4] Grinberg stated a theorem wich assumes both the moment condition and the system of partial differential equations; he sketched a proof omitting some details, too.

It is the aim of the present paper to give a complete proof of the result by Gel'fand, Gindikin, Graev [2].

In section 2 we introduce some fundamental notations and define local coordinates on $E_{n,k}$.

---

*The contents of this paper has been communicated by Rolf Sulanke

Section 3 contains the definitions of rapidly decreasing functions on $E_{n,k}$ and of the partial Fourier transform playing an important role for the following investigations. In [3] Gonzales proved some basic properties of this partial Fourier transform listed in section 3.

In section 4 we give a precise definition of the k-dimensional Radon transform and deduce some simple properties. In particular, we explain the connection between Radon and partial Fourier transform, the moment condition and the system of partial differential equations mentioned aboved.

The last section 5 contains the proof of the range theorem.

I am indebted to my tutor Prof. R. Sulanke for his support and very instructive discussions.

2. COORDINATES ON THE SPACE OF K-PLANES $E_{n,k}$.-

The space $E_{n,k}$ of all k-dimensional planes ($1 \leq k \leq n-1$) of the n-dimensional Euclidean space R^n is a fibre bundle over the ordinary Grassmann manifold $G_{n,k}$, the manifold of all k-dimensional subspaces in R^n. Let $\pi: E_{n,k} \longrightarrow G_{n,k}$ be the natural projection and $\sigma = \pi(\xi)$ for $\xi \varepsilon E_{n,k}$. Then the fibres of the bundle are indentified with $R^n/_\sigma = \sigma^\perp$ (the orthogonal subspace according to σ with respect to the usual inner product in R^n).

We fix an origin o in R^n and identify the subspaces of the vector space R^n with the corresponding planes through o. Then

(1) $$v = \sigma^\perp \cap \xi \quad (\xi \varepsilon E_{n,k}, \ \sigma = \pi(\xi))$$

is a uniquely defined vector. The pairs (σ, v), $v \varepsilon \sigma^\perp$, and the planes $\xi \varepsilon E_{n,k}$ correspond to each other bijectively. We write

(2) $$\xi = \xi(\sigma, v)$$

and for a function ρ on $E_{n,k}$

(3) $$\rho(\xi) = \rho(\sigma, v) \quad , \quad v \varepsilon \sigma^\perp$$

The atlas on $G_{n,k}$ is defined as usually. Let $(e_i)_{i=1,\dots,n}$ be a fixed orthonormal basis in R^n and $j = (j_1, \dots, j_k)$ a multiindex $(j_1 < \dots < j_k)$.

Define

$$(4)\qquad U_j := \left\{ \sigma \varepsilon G_{n,k} \;\middle|\; \begin{array}{l}\text{the orthogonal projection of } \sigma \text{ on the}\\ \text{subspace spanned by } e_{j_1},\dots,e_{j_k} \text{ is bijective}\end{array} \right\}$$

Then the open sets U_j satisfy

$$(5)\qquad \bigcup_j U_j = G_{n,k}$$

We abreviate $U_0 := U_{(1,\dots,k)}$ and assume $\sigma \varepsilon U_0$ for simplicity.

For the indices let

$$(6)\qquad \begin{aligned} \alpha,\beta,\dots &= 1,\dots,k \\ \chi,\lambda,\dots &= k+1,\dots,n \end{aligned}$$

If $\sigma \varepsilon U_0$ let $a_\alpha \varepsilon \sigma$ denote the originals of e_α with respect to the orthogonal projection onto the span of $e_1,\dots,e_k$. Then there are uniquely defined real numbers $\sigma_{\chi\alpha}$ fulfilling

$$(7)\qquad a_\alpha = e_\alpha + \sum_{\chi=k+1}^{n} e_\chi \;\sigma_{\chi\alpha}$$

The vectors a_α constitute a basis in σ. By formula (7) the subspaces $\sigma \varepsilon U_0$ correspond bijectively to the n x k - matrices

$$(8)\qquad \Sigma := \begin{pmatrix} I_k \\ (\sigma_{\chi\alpha}) \end{pmatrix},$$

where I_k denotes the unit matrix of order k. Now we define

$$(9)\qquad U_{j,M} := \{\sigma \varepsilon U_j \mid |\sigma_{\chi\alpha}| \leq M\}, \quad M > 0 .$$

Since $G_{n,k}$ is compact, there exists a real number $M > 0$ (sufficiently large) satisfying

$$(10)\qquad \bigcup_j U_{j,M} = G_{n,k} .$$

In this way we obtain an atlas on $G_{n,k}$ consisting of compact coordinate neighborhoods. For $\sigma \varepsilon U_0$ we consider the n x (n-k) - matrix

$$(11)\qquad \bar{\Sigma} := \begin{pmatrix} -(\sigma_{\chi\alpha})^T \\ I_{n-k} \end{pmatrix}.$$

here $(\sigma_{\chi\alpha})^T$ is the transposed matrix of $(\sigma_{\chi\alpha})$. Let $\bar{v}$ be the othogonal projection of $v \varepsilon \sigma^\perp$ on the subspace spanned by $e_{k+1},\dots,e_n$. With respect to our fixed basis (e_i) of

R^n the vectors $v \varepsilon \sigma^{\perp}$ have the following coordinate representation:

(12) $\qquad v = \bar{\Sigma} \cdot \bar{v} \quad ;$

where $\bar{v}$ is regarded as a vector in R^{n-k} (=span $(e_{k+1},\ldots,e_n)$). According to formula (12) we get the following local trivialisations of the bundle $E_{n,k}$

(13) $$\Phi_{j,M} : \pi^{-1}(U_{j,M}) \longrightarrow U_{j,M} \times R^{n-k}$$

$$\Phi_{j,M}(\xi) = (\pi(\xi),\bar{v}),$$

if $\xi = \xi(\sigma,v)$ according to (2).

Remark: If $\pi(\xi) \notin U_0$, the vector $\bar{v}$ is the orthogonal projection of $v \varepsilon \sigma^{\perp}$ onto the subspace spanned by $e_{j_{k+1}},\ldots,e_{j_n}$ and it is regarded as a vector in R^{n-k} again.

The (n-k) coordinates of $\bar{v}$ (with respect to the span of $e_{j_{k+1}},\ldots,e_{j_n}$) and the k(n-k) coordinates $\sigma_{j_\chi j_\alpha}$ of $\sigma = \pi(\xi)$ form a complete coordinate system on $E_{n,k}$.

Now we define other local coordinates for $E_{n,k}$. Let $\xi \varepsilon E_{n,k}$ and $\sigma = \pi(\xi) \varepsilon U_0$. We consider the vector

(14) $\qquad y: = \xi \cap \text{span}\,(e_{k+1},\ldots,e_n).$

Then the pairs (Σ,y) and the planes $\xi \varepsilon \pi^{-1}(U_o)$ correspond to each other bijectively and we write

(15) $\qquad \xi = \xi(\Sigma,y) \quad$ or $\quad \xi = \xi(\sigma_{\chi\alpha},y_\chi).$

Remark: Subsequently we use the vector y either as $y = (0,\ldots,0,y_{k+1},\ldots,y_n)^T \;\varepsilon\; R^n$ or as $y = (y_{k+1},\ldots,y_n)^T \;\varepsilon\; R^{n-k}$.

The following formula holds:

(16) $\qquad y = \bar{\Sigma}^T \bar{\Sigma}.\, \bar{v} = \bar{\Sigma}^T.\, v \qquad\qquad (y \,\varepsilon\, R^{n-k}).$

3. RAPIDLY DECREASING FUNCTIONS ON $E_{n,k}$.-

Calling L_σ the Laplacian on the fibre $\sigma^{\perp}$ we set

(17) $\qquad (\square\rho)|_{\sigma^{\perp}} : = L_\sigma\,(\rho|_{\sigma^{\perp}}) \quad , \; \rho \;\varepsilon\; C^\infty\,(E_{n,k}).$

Then $\square$ is a differential operator on $E_{n,k}$ (see [6]). Using the well-known formula for the Laplace-Beltrami-operator on Riemannian manifolds, from (12) we deduce the following representations of $\square$ in local coordinates:

(18) With $\frac{\partial}{\partial v} := (\frac{\partial}{\partial v_{k+1}}, \dots, \frac{\partial}{\partial v_n})^T$ it holds

(19) $$\square = \langle (\bar{\Sigma}^T \bar{\Sigma})^{-1} \frac{\partial}{\partial v}, \frac{\partial}{\partial v} \rangle .$$

Here $\langle \, , \, \rangle$ denotes the usual inner product (in R^{n-k}).

Let D be any differential operator on $G_{n,k}$ with order N. We represent D in local coordinates by

(20) $$D = \sum_{|s| \leq N} g_s(\sigma_{\chi\alpha}) \frac{\partial^{|s|}}{\partial \sigma_{k+1,1}^{s_{k+1\,1}} \cdots \partial \sigma_{nk}^{s_{nk}}}$$

Here $s = (s_{k+1\,1}, \dots, s_{nk})$ is a multiindex, $|s| := s_{k+1\,1} + \dots + s_{nk}$, $s_{\chi\alpha} \in Z^+$ and g_s are continous functions on $U_{0,M}$.

Remark: (20) and (19) describe $\square$ and D in the neighborhoods $U_{0,M}$ and $U_{0,M} \times R^{n-k}$, respectively. For the representations in the general coordinate neighborhoods $U_{j,M}$ and $U_{j,M} \times R^{n-k}$ we should use double indices.

Now we are able to define the space $T(E_{n,k})$ of rapidly decreasing functions on $E_{n,k}$.

<u>Definition 1</u>: $\rho \varepsilon T(E_{n,k}) :\longleftrightarrow$ a) $\rho \varepsilon C^\infty(E_{n,k})$

b) For all $m, l \in Z^+$, all j and all differential operators D on $G_{n,k}$ the following inequality holds:

(21) $$\sup_{\sigma \in U_{j,M}, \bar{v} \in R^{n-k}} |(1+|v|)^m (\square^l D(\rho \circ \Phi_{j,M}^{-1}))(\sigma, \bar{v})| < \infty .$$

Here $R^{n-k} = \text{span}(e_{j_{k+1}}, \dots, e_{j_n})$ and $v = \bar{\Sigma}\,\bar{v}$ (see section 2). The operators $\square$ and D are given in local coordinates by the formulas (19) and (20).

Remark: The space $T(E_{n,k})$ explained above coincides with the space of rapidly decreasing functions on $E_{n,k}$ defined by Gonzales in [3]. He used another coordinate system on $E_{n,k}$ and his definition differs from our's in a purely technical way.

If $x = (x_1,\dots,x_n)^T \in \sigma$ and $\sigma \in U_o$, then

(22) $$x = \Sigma\cdot(x_1,\dots,x_k)^T$$

with respect to our fixed basis (e_i). According to this parametrisation of σ for the Euclidean measure $d\sigma$ in $\sigma(\sigma \in U_0)$ we obtain

(23) $$d\sigma = (\det \Sigma^T\Sigma)^{1/2}\ dx_1\dots dx_k.$$

Let $x = (x_1,\dots,x_n)^T \in R^n$ be any vector and $\sigma \in U_0$ be a fixed plane. Then there is a plane $\xi \in E_{n,k}$ which is parametrized by $\xi = \xi(\Sigma,y)$ satisfying

(24) $$x = \Sigma\cdot(x_1,\dots,x_k)^T + y \qquad (y \in R^n).$$

From this parametrisation of R^n we conclude

(25) $$dx = dx_1\dots dx_k dy_{k+1}\dots dy_n.$$

According to the decomposition $R^n = \sigma \oplus \sigma^{\perp}$ we split up the Euclidean measure $dx = d\sigma\, d\sigma^{\perp}$, where $d\sigma^{\perp}$ denotes the Euclidean measure in $\sigma^{\perp}$. Now we conclude from (23) and (25) that

(26) $$d\sigma^{\perp} = (\det \Sigma^T\Sigma)^{-1/2}\ dy_{k+1}\dots dy_n.$$

We define the "partial Fourier transform" $\tilde{\rho}$ for a function $\rho \in T(E_{n,k})$ by

(27) $$\tilde{\rho}\,(\sigma,w) := \int_{\sigma^{\perp}} \rho(\sigma,v)\ e^{-i<v,w>} d\sigma^{\perp}(v),\ w \in \sigma^{\perp}.$$

For each function ρ on $E_{n,k}$ we further define a function ρ_0 by

(28) $$\rho|_{\pi^{-1}(U_0)}(\xi) = (\det \Sigma^T\Sigma)^{1/2}\ \rho_0(\Sigma,y).$$

We often write $\rho_0(\sigma_{\chi_\alpha},y_\chi)$ for $\rho_0(\Sigma,y)$. From (26), (27) and (28) we conclude

(29) $$\tilde{\rho}|_{\pi^{-1}(U_0)}(\sigma,w) = \int_{R^{n-k}} \rho_0(\sigma_{\chi_\alpha},y_\chi)\ e^{-i<y,w>} dy_{k+1}\dots dy_n.$$

Equation (29) holds because $(v-y) \in \sigma$ and therefore $<v,w> = <w,y>$ $(y \in R^n)$.

Remark: Equation (29) holds for any $\sigma \in U_0$ only. For $\sigma \in U_j$, $j \neq (1,\dots,k)$ one can define a function ρ_j analogously to (28).

We list some fundamental properties of the partial Fourier transform. For the proofs of the following results we refer to Gonzales [3].

Lemma 1.-: The partial Fourier transform (27) is a linear bijection of $T(E_{n,k})$ onto itself.

A smooth function f on R^n is said to be rapidly decreasing ($f \in T(R^n)$) iff the inequality

(30) $$\sup_{x \in R^n} |(1+|x|)^m (P(\frac{\partial}{\partial x_1},\ldots,\frac{\partial}{\partial x_n})f)(x)| < \infty$$

holds for all $m \in Z^+$ and all polynomials P (or all translation invariant differential operators on R^n). Following Gonzales we define

(31) $$\Psi_f(\sigma,v) := f|_{\sigma^\perp}(v).$$

Lemma 2.-: If $f \in T(R^n)$, then $\Psi_f \in T(E_{n,k})$.

4. THE K-DIMENSIONAL RADON TRANSFORM, DEFINITION AND SIMPLE PROPERTIES.-

Subsequently we suppose that $1 \leq k < n-1$.

The k-dimensional Radon transform for rapidly decreasing functions is defined by

(32) $$\hat{f}(\xi) := \int_{x \in \xi} f(x)\, d\xi(x) \quad , \quad f \in T(R^n),\ \xi \in E_{n,k}.$$

Here $d\xi$ is the Euclidean measure in ξ which is equal to $d\sigma$ for $\sigma = \pi(\xi)$. For $\pi(\xi) \in U_0$ we obtain from (23) and (24):

(33) $$\hat{f}(\xi) = (\det \Sigma^T \Sigma)^{1/2} \int_{R^k} f(\Sigma.\bar{x}+y)\, dx_1 \ldots dx_k.$$

Here $\bar{x}$ denotes the projection of $x = (x_1,\ldots,x_n)^T$ onto span $(e_1,\ldots,e_k)$. Thus, we obtain from (28):

(34) $$\hat{f}_0(\Sigma,y) = \int_{R^k} f(\Sigma.\bar{x}+y)\, dx_1 \ldots dx_k.$$

We consider the usual Fourier transform

(35) $$\tilde{f}(u) = \int_{R^n} f(x)\, e^{-i<x,u>} dx, \quad u \in R^n$$

Now we compute $(\sigma \varepsilon U_0)$:

$$\tilde{f}_{|\sigma^\perp}(w) = \int_{R^n} f(x) \; e^{-i<x,w>} dx$$

$$= \int_{\sigma^\perp} \int_\sigma f(x) \; e^{-i<x,w>} d\sigma \, d\sigma^\perp$$

$$= \int_{\sigma^\perp} \hat{f}(\sigma, v) \; e^{-i<v,w>} d\sigma^\perp(v),$$

where v is the projection of x onto $\sigma^\perp$. Hence

(36) $$\tilde{f}_{|\sigma^\perp}(w) = (\hat{f})^\sim (\sigma, w).$$

Corollary 1.-: The k-dimensional Radon transform is one-to-one on $T(R^n)$.

Proof: If $\hat{f} \equiv 0$, (36) implies $\tilde{f}_{|\sigma^\perp} \equiv 0$ for all $\sigma \varepsilon G_{n,k}$. Hence, $\tilde{f} \equiv 0$ and thus $f \equiv 0$ by Schwartz's theorem.

Corollary 2.-: If $f \varepsilon T(R^n)$, then $\hat{f} \; \varepsilon T(E_{n,k})$.

Proof: Since $\tilde{f} \varepsilon T(R^n)$, t follows from Lemma 2 that $\Psi_{\tilde{f}} \varepsilon T(E_{n,k})$. Now we have $\Psi_{\tilde{f}} = \tilde{f}_{|\sigma^\perp} = (\hat{f})^\sim$. Thus $(\hat{f})^\sim \varepsilon T(E_{n,k})$ and by Lemma 1 $\hat{f} \; \varepsilon T(E_{n,k})$.

Corollary 3.-: If $f \; \varepsilon T(R^n)$, then the function $\hat{f}_0$ defined by (28) satisfies

(37) $$\left(\frac{\partial^2}{\partial y_\chi \partial \sigma_{\lambda\alpha}} - \frac{\partial^2}{\partial y_\lambda \partial \sigma_{\chi\alpha}}\right) \hat{f}_0 = 0 \quad \text{for all} \quad \alpha, \chi, \lambda.$$

Proof: The proposition follows immediately by differentiating (34). Since f is rapidly decreasing, the above mentioned differential operator commutes with the integration sign in (34).

Corollary 4.-: If $f \varepsilon T(R^n)$, then $\hat{f}$ satisfies the so-called moment condition: For each $m \varepsilon Z^+$ and all $\sigma \varepsilon G_{n,k}$ there exists a homogeneous polynomial P_m of degree m such that

(38) $$\int_{\sigma^\perp} \hat{f}(\sigma, v) <v,w>^m d\sigma^\perp(v) = P_{m|\sigma^\perp}(w) \quad , \; w \; \varepsilon \; \sigma^\perp .$$

Proof: We define

(39) $$P_m(u) := \int_{R^n} f(x) <x,u>^m dx \quad , \quad f \; \varepsilon \; T(R^n), \; u \; \varepsilon \; R^n.$$

Since $f \varepsilon T(R^n)$, the integral exists and P_m is clearly homogeneous of degree m.

Assuming that $\sigma \in U_0$ we obtain ($w \in \sigma^\perp$)

$$P_{m|\sigma^\perp}(w) = \int_{R^n} f(x) \; \langle w,x\rangle^m \, dx$$

$$= \int_\sigma \int_{\sigma^\perp} f(x) \langle w,x\rangle^m \, d\sigma \, d\sigma^\perp$$

$$= \int_{\sigma^\perp} \hat{f}(\sigma,v) \langle v,w\rangle^m \, d\sigma^\perp(v),$$

where $v = pr_{\sigma^\perp} x$.

5. THE RANGE OF $T(R^n)$.-

We describe the range of $T(R^n)$ under the k-dimensional Radon transform ($1 \leq k < n-1$). First we give two definitions of the most significant function spaces on $E_{n,k}$.

Definition 2: $\rho \in T_D(E_{n,k}) :\longleftrightarrow$ a) $\rho \in T(E_{n,k})$

b) In each coordinate neighborhood $U_j \times R^{n-k}$ ρ_0 defined by (28) fulfills the following system of partial differential equations:

$$(40) \qquad \left(\frac{\partial^2}{\partial y_\chi \partial \sigma_{\lambda\alpha}} - \frac{\partial^2}{\partial y_\lambda \partial \sigma_{\chi\alpha}}\right) \rho_0(\Sigma,y) \equiv 0$$

$$\chi, \lambda = k+1,\dots,n \;, \quad \alpha = 1,\dots,k.$$

Definition 3: $\rho \in T_H(E_{n,k}) :\longleftrightarrow$ a) $\rho \in T(E_{n,k})$

b) For each $m \in Z^+$ there exists a homogeneous polynomial P_m on R^n of degree m which depends on ρ such that for all $\sigma \in G_{n,k}$ holds

$$(\text{m.c.}) \qquad P_{m|_{\sigma^\perp}}(w) = \int_{\sigma^\perp} \rho(\sigma,v) \langle v,w\rangle^m \, d\sigma^\perp(v) \quad , \quad w \in \sigma^\perp .$$

Remark: Helgason proved (cf. [5]) that in the case of compactly supported functions $f \in \Gamma(R^n)$ the space $\Gamma_H(E_{n,k})$ (compactly supported functions satisfying the moment condition) coincides with the range of $\Gamma(R^n)$ under the k-dimensional Radon transform. Gonzales gave a counterexample (cf. |3|) which shows that the moment condition is not sufficient to describe the range of $T(R^n)$ completely. Grinberg stated (cf. [4]) that the range $T(R^n)^\wedge$ equals $T_D(E_{n,k}) \cap T_H(E_{n,k})$

Now we prove the following

THEOREM: The k-dimensional Radon transform is a linear bijection of $T(R^n)$ onto $T_D(E_{n,k})$.

Proof: If $f \in T(R^n)$, then $f \in T_D(E_{n,k})$ by Corollary 2, 3. By Corollary 1 it remains to show that the Radon transform is surjective. To this goal we show

<u>Proposition 1</u>: If $\rho \in T_D(E_{n,k})$, there exists a function $F \in T(R^n)$ so that for each $\sigma \in G_{n,k}$ and all $w \in \sigma^{\perp}$

(41) $$\tilde{\rho}(\sigma,w) = F|_{\sigma^{\perp}}(w) .$$

Assume that this proposition is proved, Then by Schwartz's theorem there exists an uniquely determined function $f \in T(R^n)$ such that $\tilde{f} = F$. By (36) we have

$$\tilde{f}|_{\sigma^{\perp}}(w) = (\hat{f})^{\sim}(\sigma,w) \quad , w \in \sigma^{\perp}$$

and by (41) it follows $\tilde{f}|_{\sigma^{\perp}}(w) = F|_{\sigma^{\perp}}(w) = \tilde{\rho}(\sigma,w)$. Since the partial Fourier transform is one-to-one, the function f is the original of $\rho \in T_D(E_{n,k})$.

Proposition 1 is a corollary of the following three lemmas we have to prove now.

<u>Lemma 3</u>.-: Suppose, $\rho \in T_D(E_{n,k})$. Then there exists a function F on R^n and smooth away from the origin ($F \in C^{\infty}(R^n \setminus \{0\})$) such that for each $\sigma \in G_{n,k}$ and all $w \neq 0$ ($w \in \sigma^{\perp}$).

(42) $$\tilde{\rho}(\sigma,w) = F|_{\sigma^{\perp}}(w).$$

<u>Lemma 4</u>.-: Let $\rho \in T_D E_{n,k})$ and F be the function according to Lemma 3. Then F is smooth in the origin, too.

<u>Lemma 5</u>.-: Consider $\rho \in T_D(E_{n,k})$ and F by Lemma 4. Then F is rapidly decreasing.

Proof of Lemma 3: According to (29) we write (assume that $\sigma \in U_0$):

(43) $$\Psi(\sigma_{\chi\alpha}, w_{\chi}) := \tilde{\rho}(\sigma,w) = \tilde{\rho}(\sigma(\sigma_{\chi\alpha}), w(w_{\chi})).$$

Now we compute

$$0 = \int_{R^{n-k}} \left(\frac{\partial^2}{\partial y_{\chi}\partial\sigma_{\lambda\alpha}} - \frac{\partial^2}{\partial y_{\lambda}\partial\sigma_{\chi\alpha}}\right) \rho_o(\Sigma,y)\, e^{-i<y,w>} dy_{k+1}\cdots dy_n$$
$$= w_{\chi} \int_{R^{n-k}} \frac{\partial}{\partial\sigma_{\lambda\alpha}} \rho_o(\Sigma,y)\, e^{-i<y,w>} dy_{k+1}\cdots dy_n -$$
$$w_{\lambda} \int_{R^{n-k}} \frac{\partial}{\partial\sigma_{\chi\alpha}} \rho_o(\Sigma,y)\, e^{-i<y,w>} dy_{k+1}\cdots dy_n$$

Hence,

(44) $$w_\chi \frac{\partial\Psi}{\partial\sigma_{\lambda\alpha}} - w_\lambda \frac{\partial\Psi}{\partial\sigma_{\chi\alpha}} = 0 \text{ for all } \chi, \lambda, \alpha.$$

Provided that $w \neq 0$, we assume $w_{k+1} \neq 0$. Now we substitute for each α:

(45) $$\hat{w}_\chi := w_\chi$$
$$\hat{\sigma}_{k+1\,\alpha} := \sum_{\chi=k+1}^{n} \sigma_{\chi\alpha} w_\chi$$
$$\hat{\sigma}_{\lambda\alpha} := -w_\lambda \sigma_{k+1\alpha} + w_{k+1}\sigma_{\lambda\alpha}, \quad \lambda \neq k+1.$$

Thus for each $\alpha = 1,\dots,k$ we have

(46) $$(\hat{\sigma}_{k+1\alpha},\dots,\hat{\sigma}_{n\alpha})^T = A.(\sigma_{k+1\alpha},\dots,\sigma_{n\alpha})^T, \text{ where}$$

(47) $$A=\begin{pmatrix} w_{k+1} & w_{k+2} & \cdot & \cdot & \cdot & w_n \\ -w_{k+2} & w_{k+1} & \cdot & 0 & & \\ \cdot & & \cdot & & & \\ \cdot & 0 & \cdot & & & \\ \cdot & & & \cdot & & \\ -w_n & & & & & w_{k+1} \end{pmatrix}$$

We consider the function $\hat{\Psi}$ defined by

(48) $$\hat{\Psi}(\hat{\sigma}_{\chi\alpha}, \hat{w}_\chi) := \Psi(\sigma_{\chi\alpha}, w_\chi).$$

We show that $\hat{\Psi}$ depends only on $\hat{w}_\chi$ and $\hat{\sigma}_{k+1\,\alpha}$. Then it follows that Ψ is a function that depends only on w_χ and $(\sum_{\chi=k+1}^{n} \sigma_{\chi\alpha} w_\chi)$, $\alpha = 1,\dots,k$. But, $\sum_{\chi=k+1}^{n} \sigma_{\chi\alpha} w_\chi = -w_\alpha$ for $w \in \sigma^{\perp}$. Thus the existence of a function F on $R^n \setminus \{0\}$ satisfying (42) follows. Clearly, F is smooth away from the origin. we have

(49) $$\frac{\partial\hat{\Psi}}{\partial\hat{\sigma}_{\chi\alpha}} = \sum_{\lambda=k+1}^{n} \frac{\partial\Psi}{\partial\sigma_{\lambda\alpha}} \frac{\partial\sigma_{\lambda\alpha}}{\partial\hat{\sigma}_{\chi\alpha}}, \quad \alpha = 1,\dots,k$$

(46) and (47) imply by an easy computation

(50a) If $\chi = k+2,\dots,n$, $$\frac{\partial\sigma_{k+1}}{\partial\hat{\sigma}_{\chi\alpha}} = -w_\chi (w_{k+1}^2+\dots+w_n^2)^{-1}$$

(50b) If $\lambda, \chi = k+2,\dots,n$, $\chi \neq \lambda$, $$\frac{\partial\sigma_{\chi\alpha}}{\partial\hat{\sigma}_{\lambda\alpha}} = -\frac{w_\chi w_\lambda}{w_{k+1}} (w_{k+1}^2+\dots+w_n^2)^{-1}$$

(50c) If $\chi = k+2,\dots,n$, $$\frac{\partial\sigma_{\chi\alpha}}{\partial\hat{\sigma}_{\chi\alpha}} = \frac{1}{w_{k+1}} (w_{k+1}^2+\dots+w_n^2 - w_\chi^2)(w_{k+1}^2+\dots+w_n^2)^{-1}$$

Thus we obtain: if $\chi = k+2,\dots,n$; $\alpha = 1,\dots,k$;

$$\frac{\partial\hat{\Psi}}{\partial\hat{\sigma}_{\chi\alpha}} = \frac{(w_{k+1}^2+\dots+w_n^2)^{-1}}{w_{k+1}} \left(-w_\chi w_{k+1} \frac{\partial\Psi}{\partial\sigma_{k+1\alpha}} + (w_{k+1}^2+\dots+w_n^2 - w_\chi^2)\frac{\partial\Psi}{\partial\sigma_{\chi\alpha}} - \dots\right.$$

$$\left.\dots - \sum_{\lambda=k+2,\lambda\neq\chi}^{n} w_\chi w_\lambda \frac{\partial\Psi}{\partial\sigma_{\lambda\alpha}}\right)$$

$$= \frac{(w_{k+1}^2+\dots+w_n^2)^{-1}}{w_{k+1}} \left(w_{k+1}\left(w_{k+1}\frac{\partial\Psi}{\partial\sigma_{\chi\alpha}} - w_\chi \frac{\partial\Psi}{\partial\sigma_{k+1\alpha}}\right) + \sum_{\substack{\lambda=k+2\\ \lambda\neq\chi}}^{n} w_\lambda\left(w_\lambda \frac{\partial\Psi}{\partial\sigma_{\chi\alpha}} - \dots\right.\right.$$

$$\left.\left.\dots - w_\chi \frac{\partial\Psi}{\partial\sigma_{\lambda\alpha}}\right)\right)$$

Hence, by (44) $\frac{\partial\hat{\Psi}}{\partial\hat{\sigma}_{\chi\alpha}} \equiv 0$ for all $\chi = k+2,\dots,n$; $\alpha = 1,\dots,k$. Thus, the function $\hat{\Psi}$ depends only on $\hat{w}_\chi$ and $\hat{\sigma}_{k+1\alpha}$ and the lemma is proved.

Proof of lemma 4:

Let T be a differential operator on R^n which has the coordinate representation

$$(51) \qquad T: = \frac{\partial^{|\alpha|}}{\partial x_1^{\alpha_1} \quad \partial x_n^{\alpha_n}}$$

Here $\alpha = (\alpha_1,\dots,\alpha_n)$ is any multiindex. We show that for all such operators T there exist smooth functions ψ_T on $E_{n,k}$ such that for each $\sigma \varepsilon G_{n,k}$ and all $w \neq 0$, $w \varepsilon \sigma^\perp$.

$$(52) \qquad (TF)|_{\sigma^\perp}(w) = \psi_T(\sigma,w) .$$

We are going to prove this assertion by induction on the order of the differential operators T. Assume, we would have proved it already. Then

$$(53) \qquad \psi_T(\sigma,w) = (TF)\left(-\sum_{\chi=k+1}^{n} \sigma_{\chi 1} w_\chi,\dots,-\sum_\chi \sigma_{\chi k} w_\chi, w_{k+1},\dots,w_n\right).$$

From (53) we deduce that for $w_{k+1} \neq 0$, $w_{k+2} = \dots = w_n = 0$ the function ψ_T depends only on w_{k+1} and $\sigma_{k+1\alpha}$ ($\alpha = 1,\dots,k$). Hence,

$$(54) \qquad \frac{\partial}{\partial\sigma_{\chi\alpha}}(\psi_T(\sigma,0)) = 0 \text{ if } \chi = k+2,\dots,n;\ \alpha = 1,\dots,k.$$

Analogously from (53) we obtain that for $w_{k+2} \neq 0$, $w_{k+1} = w_{k+3} = \dots = w_n = 0$, the function ψ_T does not depend on $\sigma_{k+1\alpha}$ (for all α). Thus,

$$(55) \qquad \frac{\partial}{\partial\sigma_{k+1\alpha}}(\psi_T(\sigma,0)) = 0 , \ \alpha = 1,\dots,k.$$

By the formulas (54) and (55) we obtain

(56) $\psi_T(\sigma,0) = \text{const}$,

that means the function $\psi_T(\sigma,0)$ does not depend on σ. Hence, we can define

(57) $(TF)(0) := \psi_T(\sigma,0)$.

Now it follows from (57) and the following trivial lemma that F is a smooth function.

Lemma: Let $0 \in U \subset R^n$ be an open set containing the origin. Suppose, that $g: U \to C$ is a complex valued function on U and smooth away from the origin ($g \in C^\infty(U\setminus\{0\})$). Suppose, further, that g and all partial derivations of g can be continuously extended into the origin. That means, in particular, that g is defined on the whole set U.
Then all partial derivations of g in the origin exist and coincide with the corresponding continuous extension (hence, g is a smooth function on U).

The proof of this lemma is obvious.

Now we prove assertion (52) by induction. If the order of T is zero, then $\psi_T = \tilde{\rho}$ and (52) has been shown. Assume, assertion (52) is true for all differential operators T with representation (51) and order $\leq N (N \in Z^+)$. Writing $\psi := \psi_T$ for shortness, we obtain by (53)

(58) $$\left(\frac{\partial}{\partial x_\alpha} TF\right)\Big|_{\sigma^\perp}(w) = -\frac{1}{w_{k+1}} \frac{\partial \Psi}{\partial \sigma_{k+1\alpha}}(\sigma_{\chi_\alpha}, w_\chi),$$

(Since $w \neq 0$, we can assume that $w_{k+1} \neq 0$).

(59) $$\left(\frac{\partial}{\partial x_\chi} TF\right)\Big|_{\sigma^\perp}(w) = \frac{\partial \psi}{\partial w_\chi} + \sum_{\alpha=1}^{k} \sigma_{\chi\alpha}\left(\frac{\partial}{\partial x_\alpha} TF\right)\Big|_{\sigma^\perp}(w).$$

We consider $\psi_{\chi\alpha} := \frac{\partial \psi}{\partial \sigma_{\chi\alpha}}$ and obtain from (53) $\psi_{\chi\alpha} = -w_\chi \left(\frac{\partial}{\partial x_\alpha} TF\right)\Big|_{\sigma^\perp}(w)$ for all χ, α. It follows

(60) $$w_\chi \psi_{\lambda\alpha} = w_\lambda \psi_{\chi\alpha} \qquad \text{for all } \chi, \lambda, \alpha.$$

From (60) it follows setting $w_\chi = 0$ $(\lambda = k+1)$:

(61) $$\psi_{\chi\alpha}\big|_{w_\chi = 0} = 0$$

Using the Taylorformula we can easily prove the following statement:

<u>Lemma</u>: Assume, $f \in C^\infty(R^2)$ and $f(x,o) = 0$ for all $x \in R$.

Then there exist a smooth function g on R^2 such that

$$f(x,y) = y\, g(x,y)$$

According to this lemma and formula (61) there are smooth functions $\hat{\psi}_{\chi\alpha}$ such that $\psi_{\chi\alpha} = w_\chi \hat{\psi}_{\chi\alpha}$. Further we obtain from (60) that $\hat{\psi}_{\lambda\alpha} = \hat{\psi}_{\chi\alpha}$ for all χ, λ, α. Thus there exist smooth functions ψ_α satisfying

$$(62) \qquad \psi_{k+1\alpha} = \frac{\partial\psi}{\partial\sigma_{k+1\alpha}} = w_{k+1}\psi_\alpha \qquad , \quad \alpha = 1,\dots,k.$$

Now it follows from (58) and (59):

$$(63) \qquad \left(\frac{\partial}{\partial x_\alpha} TF\right)_{|\sigma^\perp}(w) = -\psi_\alpha(\sigma,w)$$

$$\left(\frac{\partial}{\partial x_\chi} TF\right)_{|\sigma^\perp}(w) = \frac{\partial\psi}{\partial w_{\chi\alpha}} - \sum_{\alpha=1}^{k} \sigma_{\chi\alpha}\psi_\alpha \quad (\sigma,w),$$

where the right-hand sides are smooth on $E_{n,k}$.

This completes the induction process and the proof of Lemma 4 is finished.

Proof of lemma 5:

Since

$$\sup_{x\in R^n} \left| (1+|x|)^m \left(P\left(\frac{\partial}{\partial x_1},\dots,\frac{\partial}{\partial x_n}\right) F\right)(x)\right| = \dots$$

$$\dots = \max_j \sup_{\sigma\in U_{j,M}} \sup_{w\in\sigma^\perp} \left|(1+|w|)^m \left(P\left(\frac{\partial}{\partial x_1},\dots,\frac{\partial}{\partial x_n}\right)F\right)_{|\sigma^\perp}(w)\right|$$

it suffices to show that

$$(64) \qquad \sup_{\sigma\in U_{O,M}} \sup_{w\in\sigma^\perp} \left|(1+|w|)^m \left(P\left(\frac{\partial}{\partial x_1},\dots,\frac{\partial}{\partial x_n}\right)F\right)_{|\sigma^\perp}(w)\right| < \infty$$

for all $m \in Z^+$ and all polynomials P. Now inequality (64) follows from (65)

$$(65) \qquad \sup_{\sigma\in U_{O,M}} \sup_{w\in\sigma^\perp} (1+|w|)^m (TF)_{|\sigma^\perp}(w)| < \infty$$

for all $m \in Z^+$ and all T having the representation (51).

From assertion (52) we deduce the existence of a smooth function ψ_T satisfying

$$(TF)_{|\sigma^\perp}(w) = \psi_T(\sigma,w), \quad \psi_T \in C^\infty(E_{n,k})$$

for all T. We prove by induction on the order of T that $\psi_T \in T(E_{n,k})$. Then (65) follows.

If the order of T is zero, the $\psi_T = \tilde{\rho}$ and $\tilde{\rho} \in T(E_{n,k})$. Assume, that ψ_T is rapidly decreasing for all T with order $\leq N (N \in Z^+)$. We have to show that $(\frac{\partial}{\partial x_\alpha} TF)_{|\sigma^\perp}(w)$ and $(\frac{\partial}{\partial x_\chi} TF)_{|\sigma^\perp}(w)$ belong to $T(E_{n,k})$. By our assumption $\psi \in T(E_{n,k})$ (writing $\psi := \psi_T$ again) hence $\frac{\partial\psi}{\partial w_\chi}$ belongs to $T(E_{n,k})$. We deduce from (63) that it suffices to show (21) for the smooth functions ψ_α defined in the proof of lemma 4 (formula (62)). Let D be any differential operator on $G_{n,k}$ with local representation (20). We have to prove that

$$\sup_{\sigma \in U_{0,M}} \sup_{w \in \sigma^\perp} ||w|^m \quad (\Box^1 D\psi_\alpha)(\sigma,w)| < \infty$$

for all m, $l \in Z^+$ and all D.

We divide $\sigma^\perp$ into the two subsets

$A_\sigma := \{w \in \sigma^\perp | \, |w_\chi| \leq 1 \quad \text{for all } \chi\}$ and

$B_\sigma := \{w \in \sigma^\perp | \text{ there is an index } \chi \text{ with } |w_\chi| > 1\}$.

Since the function $|W|^m (\Box^1 D \psi_\alpha)(\sigma,w)$ is smooth on $E_{n,k}$ it is bounded on the compact set $U_{0,M} \times A_\sigma$. Let $w \in B_\sigma$ and χ such that $|w_\chi| > 1$. Hence $|\frac{1}{w_\chi}| < 1$ and it follows from (62) by replacing the index k+1 by χ ($\psi_\alpha = \frac{1}{w_\chi} \frac{\partial\psi}{\partial\sigma_{\chi\alpha}}$):

$$||w|^m (\Box^1 D \psi_\alpha) (\sigma,w)| \leq ||w|^m (\Box^1 D \frac{\partial\psi}{\partial\sigma_{\chi\alpha}}) (\sigma, w)| , \quad w \in B_\sigma$$

Since ψ is rapidly decreasing on $E_{n,k}$, the function $|w|^m (\Box^1 D\psi_\alpha)$ is bounded on $U_{0,M} \times B_\sigma$, too. This proves Lemma 5.

REFERENCES

[1] Gel'fand, Graev, Shapiro:
Integral geometry on k-dimensional planes (Russian), Funkcion, analiz i ego prilož. 1-1, 1967, p. 1-31.

[2] Gel'fand, Gindikin, Graev:
Integral geometry in affine and projective spaces (Russian), Itogi nauki, Series: Sovrem. probl. mat. vol. 16, 1980.

[3] Gonzales F.B.:
Radon transforms on Grassmann manifolds,
Ph. D. Thesis, M.I.T., Cambridge, Mass., 1984.

[4] Grinberg E.L.:
Euclidean Radon transforms: Ranges and Restrictions,
Preprint, presented at the AMS summer meeting on Integral Geometry, 1984

[5] Helgason S.:
"The Radon Transform",
Progres in Math., vol. 5, Birkhäuser, Boston 1980

[6] Helgason S.:
The Radon transform on Euclidean spaces, compact two-point homogeneous spaces and Grassmann manifolds,
Acta Math. 113, 1965 , p. 153-180

[7] Schwartz L.:
"Théorie des Distributions",
Hermann, Paris, 1966

KAEHLER SUBMANIFOLDS IN THE COMPLEX PROJECTIVE SPACE

Antonio **Ros**

Departamento de Geometría y Topología
Universidad de Granada
Spain.

In this lecture we will expose some results and problems about complex submanifolds in the complex projective space from the view point of Riemannian geometry. This topic of the theory of submanifolds is rich in examples (even in good-curvature examples), is subjected to great restrictions and admits the compacity as a natural global hypothesis. These properties give us a theory with surprising and nice regularities which, in our opinion, forms an interesting and non sufficiently explored field of study and research.

Firstly, we study the geometry of Kaehler submanifolds with constant holomorphic sectional curvature and specially linear subvarieties. We give some characterizations of these in terms of the first eigenvalue of the Laplacian and the diameter. We also give some Ogiue problems on positively curved Kaehler submanifolds and a generalization of a quantization phenomenon obtained by Lawson.

In the second part we study Kaehler submanifolds of the complex projective space with parallel second fundamental form. These submanifolds are from several view points the simplest complex submanifolds of the complex projective space afther linear subvarieties. We give some characterizations of these by pinching on their curvature and the first eigenvalues of their Laplacian.

Let $CP^m(c)$ be the m-dimensional complex projective space endowed with the Fubini-Study metric of constant holomorphic sectional curvature $c > 0$. Let M^n be a Kaehler manifold of complex dimension n. If

$\psi . M^n \longrightarrow CP^m(1)$ is an isometric and holomorphic immersion we will say that M^n is a Kaehler submanifold of $CP^m(1)$. We will say that ψ is a full immersion when $\psi(M^n)$ is not contained in a proper linear subvariety of CP^m. The fundamental local results for these submanifolds have been obtained by Calabi [C]: an intrinsic characterization of Kaehler manifolds which admit a Kaehler immersion into $CP^m(1)$ and a local rigidity theorem stated as follows

THEOREM 1.(Calabi [C]). Let $\psi_i : M^n \longrightarrow CP^{m_i}(1)$, $i = 1,2$, be two full Kaehler immersions of a Kaehler manifold M^n. Then $m_1 = m_2$ and ψ_1, ψ_2 are congruent immersions, i. e., there exists an holomorphic isometry $P : CP^{m_1}(1) \longrightarrow CP^{m_1}(1)$ such that $\psi_2 = P \circ \psi_1$.

This results is a clear example of the particular behaviour of this type of submanifolds.

I.- KAEHLER SUBMANIFOLDS WITH CONSTANT HOLOMORPHIC SECTIONAL CURVATURE.

For any positive integer k we have a full Kaehler imbedding $\psi_k : CP^n(1/k) \longrightarrow CP^{m(k)}(1)$, $m(k) = \binom{n+k}{k} - 1$, defined in homogeneous coordinates as follows

$$\psi_k : (z_i)_{0 \leq i \leq n} \longmapsto \left(\sqrt{\frac{k!}{k_0! \ldots k_n!}} \; z_0^{k_0} \ldots z_n^{k_n} \right)_{k_0 + \ldots + k_n = k}$$

We call this imbedding the k-th standard imbedding of $CP^n(c)$. These imbedding provide all Kaehler immersions of the complex projective space into itself, even locally. More precisely we have

THEOREM 2.(Calabi[C]). Let M^n be a Kaehler submanifold immersed in $CP^m(1)$. Suppose that M^n has constant holomorphic sectional curvature c. Then there exists a positive integer k such that $c = 1/k$ and M^n is locally congruent to the k-th standard imbedding of $CP^n(1/k)$ into

$CP^{m(k)}(1)$. Moreover if M^n is complete then M^n is an embedded submanifold congruent to the k-th standard imbedding of $CP^n(1/k)$.

So we have a complete classification of Kaehler submanifolds with the simplest intrinsic behaviour. An important extension of this result has been obtained by Nakagawa and Takagi [NT]. They give for an irreducible Hermitian symmetric space of compact type M^n, a Kaehler imbedding $\psi: M^n \longrightarrow CP^m(c)$ and prove that a locally symmetric Kaehler submanifold of $CP^m(1)$ is locally congruent to $M^n \xrightarrow{\psi} CP^m(1/k) \xrightarrow{\psi_k} CP^{m(k)}(1)$ or to a product of submanifolds of the above type via the Segre imbedding. See further results in Takeuchi [T].
The simplest Kaehler submanifolds of $CP^m(1)$ are the totally geodesic ones. They are locally congruent to linear subvarieties (i.e. to the first standard imbedding of $CP^n(1)$ into $CP^m(1)$). Now we give some global characterizations of these submanifolds.
The first characterization involves only complex geometry.

THEOREM 3.(Feder [Fd]). Let M^n be a compact complex submanifold immersed in CP^m. Suppose that M^n is biholomorphic to CP^n. If $m < 2n$ then M^n is a linear subvariety.

This result can be view as the complex version of the spherical Berstein problem for minimal hypersurfaces of the sphere. The proof follows from the computation of the top Chern class of the normal bundle of the immersion.
The second is a characterization in terms of the nullity of the submanifold(i.e. the dimension of the space of Jacobi vector fields of the immersion, since these submanifolds are always minimal). Although the problem initially concerns Riemannian geometry, it is reduced by a result of Simons [S]to complex geometry (Simons proves that Jacobi vector fields are the holomorphic sections in the normal bundle of the

immersion. He also proves that the index is zero for a submanifold in this familly. These properties are the infinitesimal version of another important fact: compact Kaehler submanifolds minimize the volume functional among all compact submanifolds in the same homotopy class [L]).

THEOREM 4 (Kimura [Ki]). *Let* M^n *be a compact Kaehler submanifold immersed in* $CP^m(1)$. *Then*

$$\text{nullity}(M) \geq 2(m-n)(n+1).$$

The equality holds if and only if M^n *is a linear subvariety*.

This is a representative result in this context: It gives basic information about a geometric object and characterizes linear subvarieties as the simplest extremal case. An interesting related problem which has been proposed by Simons in the general situation is to decide whether a Jacobi vector field on a Kaehler submanifold is the normal component of a variational vector field for some deformation of M^n by Kaehler submanifolds. The problem has been solved positively if M is a homogeneous Kaehler submanifold [Ki,T].

Now we give two characterizations in terms of intrinsic Riemannian invariants.

THEOREM 5.(Ejiri [E], Ros[R1]). *Let* M^n *be a compact Kaehler submanifold immersed in* $CP^m(1)$. *Let* $\lambda_1(M)$ *be the first eigenvalue of the Laplacian of* M. *Then*

$$\lambda_1(M) \leq n+1 .$$

The equality holds if and only if M^n *is a linear subvariety*.

The following result has been obtained by Montiel, Urbano and the author

THEOREM 6. Let M^n be a compact Kaehler submanifold immersed in $CP^m(1)$. Then

$$\text{diameter}(M) \geqslant \pi .$$

The equality holds if and only if M^n is a linear subvariety.

These results describe the basic behaviour of some important invariants for Kaehler submanifolds. Theorem 5 follows using the λ_1-eigenfunctions of the Laplacian of $CP^m(1)$, restricted to the submanifold, as test functions for the minimum principle for λ_1 on M^n. The inequality in theorem 6 follows from the fact that a compact Kaehler submanifold immersed in $CP^m(1)$ intersects every linear hyperplane [GK]. So if we take a point p in the submanifold and if we call CP_p^{m-1} the linear hyperplane of points which are to maximum distance from p in $CP^m(1)$, we have that $M_p = M^n \cap CP_p^{m-1}$ is non-void. As the extrinsic distance is dominated by the intrinsic one on M^n any minimal geodesic γ in M^n from p to $q \in M_p$ has length greather than or equal to π. If the equality holds γ is a geodesic in $CP^m(1)$. In particular the holomorphic sectional curvature on M^n corresponding to $\gamma'(0)$ at p is equal to 1 and the intersection of M^n with CP_p^{m-1} is transversal. Hence M_p is a (n-1)-dimensional compact Kaehler submanifold of $CP^m(1)$ contained in a (n-1)-dimensional linear subvariety (the intersection of CP_p^{m-1} with the linear subvariety trough p which has the same tangent space that M^n at this point). So the above (n-1)-dimensional submanifolds are the same and we have produced in this way suficiently many geodesics γ to assure that all the holomorphic sectional curvatures on M^n at p are equal to 1. Hence M^n is a linear subvariety.

A problem related to Theorem 6 is to decide whether the values of the injectivity radius of compact Kaehler submanifolds are accumulated on zero or not. If the holomorphic sectional curvature of M^n is positive we can conclude that the injectivity radius is greather than or equal to π as follows: by an argument of Klingenberg ([CE],p. 98) we have that the injectivity radius is attained by a geodesic with conjugate extremes. Also, as the holomorphic curvature is positive, the maximum of the sectional curvature is attained by the holomorphic curvature [BG] and hence it is less than or equal to 1. So the distance between conjugate points along geodesics in M^n is greater than or equal to π. Theorem 5 and 6 can be used to introduce pinching problems for Kaehler submanifolds. It is known that the linear subvarieties are isolated in this class of submanifolds, i.e., they admit no deformations. This fact implies a discontinuity for the values of some Riemannian invariants. Although some of these invariants take only discrete values (the simplest one is the volume: If M^n is a compact Kaehler submanifold immersed in $CP^m(1)$, then $\text{volume}(M^n) = d\,\text{volume}(CP^n(1))$, where d is a positive integer and $d = 1$ if and only if M^n is a linear subvariety), the most part of the invariants take continuous values afther the first initial discontinuity for linear subvarieties. The study of this discontinuity is known in this context by the generical name of pinching problems. For instance, the expected results for the first eigenvalue of the Laplacian and for the diameter that we propose as problems are the following.

PROBLEM 1. If $\lambda_1(M^n) > n$, then $\lambda_1(M^n) = n+1$ and M^n is a linear subvariety.

PROBLEM 2. If $\text{diameter}(M^n) < \sqrt{2}\,\pi$, then $\text{diameter}(M^n) = \pi$ and M^n is a linear subvariety.

The standard complex quadric Q^n in $CP^{n+1}(1)$ satisfies $\lambda_1(Q^n) = n$ and $\text{diameter}(Q^n) = \sqrt{2}\pi$. The majority of the known pinching results concern the rank of variation of the differents curvatures of the submanifold. The first sharp result of this type for Kaehler submanifolds (with arbitrary dimension and codimension) was obtained by Ogiue using Simons' formula [S, CDK].

THEOREM 7. (Ogiue [O1]). Let M^n be a compact Kaehler submanifold immersed in $CP^m(1)$. Let Ric be the Ricci curvature of M^n. If Ric $> \frac{1}{2}n$, then Ric $= \frac{1}{2}(n+1)$ and M^n is a linear subvariety.

The Ricci curvature of the standard complex quadric Q^n in $CP^{n+1}(1)$ is $\frac{n}{2}$. Let H, K and r be the holomorphic sectional, sectional and scalar curvatures of a compact Kaehler submanifold M^n immersed in $CP^m(1)$. Ogiue conjectured the following facts [O1] :

a) If $H > \frac{1}{2}$, or

b) if $n > 2$ and $K > 1/8$, or

c) if $n-m < \frac{1}{2}n(n+1)$ and $K > 0$, or

d) if $r > n^2$,

then M^n is a linear subvariety. Problem d) has been solved, for embedded submanifolds by Cheng [Ch]. Problems a) and b) have been solved by the author [R3] and Verstraelen and the author [RV] using natural arguments at the minimum of the holomorphic sectional curvature in the unit tangent bundle of M^n. Question c) is of a different nature. It is a simple case of the following problem

PROBLEM 3. Classify all compact Kaehler submanifolds of $CP^m(1)$ with positive bisectional curvature.

Note that combining Theorem 3 with the Frankel conjecture we known that c) is true if $m-n < n$.

Now we give a quantization phenomemon for the curvature of compact Kaehler submanifolds which has been proved for n = 1 by Lawson[L1]:

THEOREM 8. Let M^n be an n-dimensional compact Kaehler submanifold immersed in the complex projective space $CP^m(1)$. Let H and K be the ho-

<u>lomorphic sectional curvature and the sectional curvature of</u> M^n. <u>If</u>

(1) $\frac{1}{k} \geqslant H \geqslant \frac{1}{k+1}$, <u>for some</u> $k = 2,3,\ldots,$ <u>or</u>

(2) $n > 2$ <u>and</u> $\frac{1}{k} \geqslant K \geqslant \frac{1}{4(k+1)}$, <u>for some</u> $k = 1,2,\ldots$

<u>then</u> M^n <u>is congruent to the standard isometric imbedding of</u> $CP^n(1/k)$ <u>or</u> $CP^n(1/(k+1))$ <u>into</u> $CP^m(1)$.

<u>Proof</u>.- If one of the hypothesis (1) or (2) holds then M^n has positive bisectional curvature (recall that if the holomorphic curvature is δ-pinched with $\delta > \frac{1}{2}$, then the bisectional curvature is positive [BG]). Then as the Frankel conjecture is true [SY], M^n is biholomorphic to the complex projective space CP^n. Let $\psi: M^n \longrightarrow CP^m(1)$ be the immersion and let Ω_0 and $\tilde{\Omega}_0$ be the Kaehler classes of M^n and CP^m corresponding to the Fubini-Study metrics of constant holomorphic curvature 1, and $\Omega = \psi^*(\tilde{\Omega}_0)$ the Kaehler class of the induced metric on M^n from the immersion. We know that $(1/4\pi)\Omega_0$ and $(1/4\pi)\tilde{\Omega}_0$ are integer classes. Moreover, any integer class in $H^2(M^n,\mathbb{Z})$ is of the form $(s/4\pi)\Omega_0$ for some $s \in \mathbb{Z}$. As $(1/4\pi)\Omega = \psi^*((1/4\pi)\tilde{\Omega}_0)$ is an integer class we have $\Omega = s\Omega_0$ for some positive integer s. Note that $s\Omega_0$ is the Kaehler class corresponding to the Fubini-Study metric on M^n with constant holomorphic sectional curvature 1/s. As the volume and the integral of the scalar curvature r depend only of the Kaehler class, if dM is the canonical measure associated to the metric induced on M^n from the immersion, we have $\mathrm{vol}(M^n) = \mathrm{vol}(CP^n(1/s))$ and $\int_M r\,dM = \frac{n(n+1)}{s}\mathrm{vol}(CP^n(1/s))$. Hence

(3) $$\int_M r\,dM = \frac{n(n+1)}{s}\mathrm{vol}(M^n).$$

On the othe hand, if UM_p is the unit tangent sphere of M at p and du_p is the canonical measure on UM_p, a direct computation gives

(4) $$\int_{UM_p} H\,du_p = \frac{r(p)}{n(n+1)}\mathrm{vol}(UM_p).$$

Suppose that the hypothesis (1) holds. Then, from (4), we have

$$\frac{n(n+1)}{k} \geqslant r \geqslant \frac{n(n+1)}{k+1}. \tag{5}$$

Integrating (5) on M^n and using (3) we conclude

$$\frac{n(n+1)}{k} \geqslant \frac{n(n+1)}{s} \geqslant \frac{n(n+1)}{k+1} \tag{6}$$

and, so, either $s = k$ or $s = k+1$. If $s = k$ (resp. $s = k+1$), then the first (reps. second) inequality in (1) is everywhere an equality. Hence M^n has constant holomorphic sectional curvature $1/k$ (reps. $1/(k+1)$). By Theorem 2 we conclude the proof in this case.

Suppose now that we have the hypothesis (2). From (4) we obtain the first inequality in (5). Let Ric be the Ricci tensor of M^n and $\{e_i\}_{i=1}^{2n}$ an orthonormal basis of the tangent space of M^n at p with $e_1 = u$, $e_2 = Ju$ J being the complex structure on M^n. Then using (2) and the relation $Ric(u,u) = H(u) + \sum_{i=3}^{2n} K(e_i,u)$ we have $Ric(u,u) \geqslant H(u) + \frac{n-1}{2(k+1)}$. Integrating this inequality on UM_p, taking into account (4) and the identity $\int_{UM_p} Ric(u,u)\, du_p = \frac{r(p)}{2n} vol(UM_p)$ we obtain

$$\frac{r}{2n} \geqslant \frac{r}{n(n+1)} + \frac{n-1}{2(k+1)},$$

and so we have the second inequality in (5). Now we conclude the proof reasoning as above and using, for instance, that the only Kaehler metric on CP^n with constant scalar curvature is the Fubini-Study one.

Note that in the hypothesis (2) with $k = 1$ we do not need the upper bound for K (in the proof we only use this bound for the holomorphic curvature, and for a Kaehler submanifold of $CP^m(1)$ we have always $H \leqslant 1$). In the same way one can prove the problem a) of Ogiue. So the results in [R3] and [RV] follow from the above more general theorem. However the case $k = 1$ in the hypothesis (1) is of different nature and will be studied later.

II.-KAEHLER SUBMANIFOLDS WITH PARALLEL SECOND FUNDAMENTAL FORM.

If in the problem 1 and 2 we change strict inequality by non-strict inequality, what is the appropiate formulation of these problems ?
To solve this question we introduce Kaehler submanifolds with parallel second fundamental form. Let ∇, $\nabla^{\perp}$ and σ be the Riemannian connection of M^n, the normal connection and the second fundamental form of the immersion. The covariant derivative of σ is defined by

$$(\nabla\sigma)(X,Y,Z) = \nabla^{\perp}_X \sigma(Y,Z) - \sigma(\nabla_X Y,Z) - \sigma(Y,\nabla_X Y),$$

for any vector fields X, Y and Z tangent to M^n. From Gauss' equation Kaehler submanifolds with parallel second fundamental form, i.e. with $\nabla\sigma = 0$, are locally symmetric. These submanifolds have been classified by Nakagawa and Takagi

THEOREM 9.(Nakagawa, Takagi[NT]). *Let M^n be a Kaehler submanifold immersed in $CP^m(1)$. Then M^n has parallel second fundamental form if and only if it is locally congruent to the standard embedding of one of the following submanifold:*

Submanifold	n	p	r
$M_1 = CP^n(1)$	n	0	n(n+1)
$M_2 = Q^n$, $n \geq 3$	n	1	n^2
$M_3 = CP^n(\frac{1}{2})$	n	$\frac{1}{2}n(n+1)$	$\frac{1}{2}n(n+1)$
$M_4 = U(s+2)/U(2)xU(s)$, $s \geq 3$	2s	$\frac{1}{2}s(s+1)$	2s(s+2)
$M_5 = SO(10)/U(5)$	10	5	80
$M_6 = E_6/Spin(10)xT$	16	10	192
$M_7 = CP^{n-s}(1)xCP^s(1)$	n	s(n-s)	$s^2 + (n-s)^2 + n$

where n *is the complex dimension,* p *the full complex codimension and* r *the scalar curvature of* M_i, i=1,..,7. *Moreover if* M^n *is complete then it is an embedded submanifold congruent to the standard imbedding of one of the above submanifolds.*

M_1 is a linear subvariety, M_2 the standard complex quadric, M_3 the Veronese imbedding, M_4 the Plücker imbedding of the complex Grassmannian of two planes and M_7 the Segre imbedding. For later use we put $M_7' = CP^s(1) \times CP^s(1)$.

Following Ferus [Fr] and Strübing [St], we have the first nice characterization of the above submanifolds. They are the "extrinsic symmetric" Kaehler submanifolds of $CP^m(1)$.

THEOREM 10. Let $\psi: M^n \longrightarrow CP^m(1)$ be a complete Kaehler submanifold of $CP^m(1)$. Then the following conditions are equivalent:

i) M^n has parallel second fundamental form.

ii) For each point $p \in M^n$ there exists an involutive holomorphic isometry $s_p: M^n \longrightarrow M^n$ and a holomorphic isometry $P: CP^m(1) \longrightarrow CP^m(1)$ such that s_p has p as an isolated fixed point, $\psi \circ s_p = P \circ \psi$ and the differential of P restricted to the normal space of the immersion at p is the identity map.

Now, we return to pinching problems. The expected result in many of these problems is that the Kaehler submanifolds which satisfies a non-strict pinching condition have parallel second fundamental form. So we can propose the following problems:

PROBLEM 1'. Let M^n be a compact Kaehler submanifold immersed in $CP^m(1)$. If $\lambda_1(M^n) \geqslant n$, then either $\lambda_1(M^n) = n+1$ and M^n is a linear subvariety or $\lambda_1(M^n) = n$ and M^n is the standard complex quadric Q^n in some linear subvariety $CP^{n+1}(1)$ of $CP^m(1)$.

PROBLEM 2'. Let M^n be a compact Kaehler submanifold immersed in $CP^m(1)$. Then diameter$(M^n) \leqslant \sqrt{2}\,\pi$ if and only if M^n is congruent to the standard imbedding of some M_i, $i=1,..,7$.

Problem 2' was conjectured in [R4]. We give now two curvature pinching theorems. The first one is proved implicitely in [O1].

THEOREM 11. (Ogiue [O1]). Let M^n be a compact Kaehler submanifold immersed in $CP^m(1)$. Let Ric be the Ricci curvature of M^n. If $Ric \geq \frac{1}{2}n$ then either $Ric = \frac{1}{2}(n+1)$ and M^n is a linear subvariety or $Ric = \frac{1}{2}n$ and M^n is the standard complex quadric in some linear subvariety $CP^{n+1}(1)$ of $CP^m(1)$.

This result is closely related to an important theorem of Kobayashi and Ochiai [KO]. The basic tool to studying curvature pinching problems is Simons' formula [S, CDK]. Using a modified version of Simons' formula in the unit tangent bundle of the submanifold, we have proved the following result which was also conjectured by Ogiue in [O2].

THEOREM 12. (Ros [R4]). Let M^n be a compact Kaehler submanifold immersed in $CP^m(1)$. Let H be the holomorphic sectional curvature of M^n. Then $H \geq \frac{1}{2}$ if and only if M^n is congruent to the standard imbedding of some M_i, $i = 1,..,7$.

Now we give a characterization of the submanifolds M_i in terms of the first and the second eigenvalues of their Laplacian.

THEOREM 13. (Ros [R2], Udagawa [U1,U2]). Let M^n be a compact Kaehler submanifold immersed in $CP^m(1)$. Let λ_1 and λ_2 ($\lambda_1 < \lambda_2$) be the first and second eigenvalues of the Laplacian of M^n and let r be the scalar curvature of M^n. Then

$$n\,\{n+1+(n+1-\lambda_1)(n+1-\lambda_2)\}\,\mathrm{vol}(M^n) \geq \int_M r\,dM$$

where $\mathrm{vol}(M^n)$ denotes the volume of M^n. Equality holds if and only if

M^n <u>is congruent to the standard imbedding of some</u> M_i, $i=1,..,6$ <u>or</u> M_7'.

Note that all the invariants which appear in the above inequality are spectral invariants. So we have an inverse spectral theorem for the submanifolds M_i in the class of compact Kaehler submanifolds of $CP^m(1)$. Inequality was proved by the author in [R2], studying the behaviour of the λ_1-eigenfunctions of $CP^m(1)$, restricted to the submanifold, with respect to the Laplacian of the submanifold. It was also proved in [R2] that if the equality holds then M^n is Einstein and a certain tensor in the normal bundle of the immersion is proportional to the metric. Later Udagawa [U2]proved that these contidions implies that M^n has parallel second fundamental form. Also, Udagawa computed the eigenvalues of the exceptional Hermitian symmetric space M_6. If we suppose that $\lambda_1 = \int_M r\,dM / n\,\mathrm{vol}(M^n)$ (this condition holds for homogeneous Einstein Kaehler submanifolds) we have from theorem 13 that $(n+1-\lambda_1)(n+2-\lambda_2) \geq 0$. So, the following nice eigenvalue regularity seems to be natural:

PROBLEM 4. <u>Let</u> M^n <u>be a compact Kaehler submanifold immersed in</u> $CP^m(1)$. <u>If</u> M^n <u>is not a linear subvariety, then</u> $\lambda_2(M^n) \leq n+2$. <u>Equality holds if and only if</u> M^n <u>is congruent to some</u> M_i, $i=2,..,6$ <u>or</u> M_7'.

For compact hypersurfaces with constant scalar curvature we have the following classification result:

THEOREM 14.(Kobayashi [K], Kon [Ko]). <u>Let</u> M^n <u>be a compact Kaehler hypersurface immersed in</u> $CP^{n+1}(1)$. <u>Then</u> M^n <u>has constant scalar curvature if and only if</u> M^n <u>is a linear subvariety or the complex quadric</u>.

Recently Tsukada [Ts] has proved that a Kaehler submanifold of codimension two in $CP^m(1)$ has parallel Ricci tensor if and only if and only if it is locally congruent to a linear subvariety, the standard complex

quadric or to the Segre imbedding of $CP^1(1)xCP^2(1)$ into $CP^5(1)$.
We finish proposing an interesting complex version (without metric) of the results in Theorems 9 and 10.

PROBLEM 5. Characterize the submanifolds M_i $i=1,..,7$ and their images under holomorphic transformations of CP^m by a suitable version of condition ii) of Theorem 10, where "holomorphic isometry" should be changed to "holomorphic transformation".

These submanifolds (and possibly their projections on linear subvarieties of CP^m) should be characterized as complex submanifolds which contains(in a appropiate sense) many conics of CP^m.

REFERENCES

[BG] R. L. Bishop and S. I. Golberg,"Some implications of the generalized Gauss-Bonnet theorem", Trans. Amer. Math. Soc. 112, 1964, 508-535.

[C] E. Calabi,"Isometric imbedding of complex manifolds", Ann. of Math., 58, 1953,1-23.

[CDK] S.S. Chern, M. Do Carmo, S. Kobayashi,"Minimal submanifolds of a sphere with second fundamental form of constant length", Functional Analisis and related fields, 1970,59-75.

[CE] J. Cheeger, D. G. Ebin,"Comparison theorems in Riemannian geometry", North-Holland Publ. Comp. 1975.

[Ch] J. H. Cheng, "An integral formula on the scalar curvature of algebraic manifolds", Proc. AMS, 81, 1981, 451-454.

[E] N. Ejiri,"The first eigenvalue of Δ for compact minimal submanifolds in a complex projective space" preprint.

[Fd] S. Feder, "Immersions and imbedding in complex projective spaces" Topology 4, 1965, 143-158.

[Fr] D. Ferus, "Symmetric submanifolds of Euclidean space", Math. Ann. 247, 1980, 81-93.

[GK] S. I. Goldberg, S. Kobayashi, "Holomorphic bisectional curvature" J. Diff. Geom. 1, 1967, 225-233.

[Ki] Y. Kimura, "The nullity of compact Kaehler submanifolds in a complex projective space", J. Math. Soc. Japan 29,1977,561-580.

[K] S. Kobayashi, "Hypersurfaces of complex projective space with constant scalar curvature", J. Diff. Geom.,1,1967,369-370.

[KO] S. Kobayashi, T. Ochiai, "Characterizations of comples projective spaces and hyperquadrics", J. Math. Kyoto Univ. 13,1973,31-47.

[Ko] M. Kon, "Complex submanifolds with constant scalar curvature in a Kaehler manifold", J. Math. Soc. Japan, 27,1975,76-81.

[L] H. B. Lawson, "Lectures on Minimal surfaces" IMPA 1973.

[L1] ------------ "The Riemannian Geometry of holomorphic curves", Carolina Conference Proc.,1970,45-62.

[NT] H. Nakagawa, R. Takagi, "On locally symmetric Kaehler submanifolds in a complex projective space", J. Math. Soc. Japan 28,1976, 638-667.

[O1] K. Ogiue, "Differential geometry of Kaehler submanifolds" Adv. Math. 13,1974,73-114.

[O2] -------"Positively curved complex submanifolds in a complex projective space, IV", Tsukuba J. Math. 3, 1979,75-77.

[R1] A. Ros, "Spectral geometry of CR-minimal submanifolds in the complex projective space", Kodai Math. J.,6,1983,88-99.

[R2] -----"On spectral geometry of Kaehler submanifolds",J. Math. Soc. Japan 36, 1984,433-448.

[R3] -----"Positively curved Kaehler submanifolds", Proc. AMS, 1985 329-331.

[R4] -----"A characterization of seven compact Kaehler submanifold by holomorphic pinching", Ann. of Math. 121,1985,377-382.

[RV] A. Ros, L. Verstraelen,"On a conjecture of K. Ogiue", J. Diff. Geom. 19,1984,561-566.

[S] J. Simons,"Minimal varieties in Riemannian manifolds" Ann. of Math. 88, 1968,62-105.

[St] W. Strübing, "Symmetric submanifolds of Riemmanian manifolds" Math. Ann., 245, 1979,37-44.

[SY] Y.T. Siu,S.T. Yau,"Compact Kaehler manifolds of positive bisectional curvature", Invent. Math.,59,1980,189-204.

[T] M. Takeuchi,"Homogeneous Kaehler submanifolds in complex projective spaces", Japan J. Math.,4, 1978,171-219.

[Ts] K. Tsukada,"Einstein Kaehler submanifolds with codimension two in a complex space form", preprint.

[U1] S. Udagawa, "Spectral geometry of Kaehler submanifolds of a complex projective space", J. Math. Soc. Japan (to appear).

[U2] ------"Einstein parallel Kaehler submanifolds in a complex projective space" preprint.

VOLUME-PRESERVING GEODESIC SYMMETRIES ON FOUR-DIMENSIONAL KÄHLER MANIFOLDS

by

K. Sekigawa and L. Vanhecke

Niigata University
Department of Mathematics
Niigata, 950-21, Japan

Katholieke Universiteit Leuven
Department of Mathematics
Celestijnenlaan 200 B
B-3030 Leuven, Belgium

1. INTRODUCTION

In this paper we continue our work on Riemannian manifolds such that all local geodesic symmetries are volume-preserving. The study of this kind of manifolds has been started by D'Atri and Nickerson [4], [5], [6]. Locally symmetric manifolds are the simplest examples but there are a lot of nonsymmetric ones. All naturally reductive spaces and all commutative spaces also have this property. (See [22] for more details.) To our knowledge a nonhomogeneous example is not known and there is some support for an affirmative answer to the following question :

Are Riemannian manifolds such that all local geodesic symmetries are volume-preserving, locally homogeneous manifolds ?

For three-dimensional manifolds this is indeed the case [13] but in the general case, this is still an open problem.

In [17] we considered this problem in a special situation. We proved that all almost Hermitian manifolds (M,g,J) such that all local geodesic symmetries preserve the corresponding Kähler form, are automatically Kählerian and locally symmetric. We also proved that a connected four-dimensional homogeneous Kähler manifold such that all local geodesic symmetries are volume-preserving is locally symmetric. In this paper we shall extend this last result by deleting the condition of homogeneity. We shall prove the following

MAIN THEOREM. *Let* (M,g,J) *be a connected four-dimensional Kähler manifold such that all local geodesic symmetries are volume-preserving. Then* (M,g,J) *is locally symmetric.*

To prove this result we shall use the special features of the geometry on a four-dimensional Einstein space.

2. VOLUME-PRESERVING GEODESIC SYMMETRIES

Let (M,g) be an n-dimensional Riemannian manifold of class C^∞, m a point of M and T_mM the tangent space of M at m. For a unit vector $\xi \in T_mM$ we denote by $\gamma : r \longmapsto \exp_m(r\xi)$ the geodesic of M with arc length r. We always choose r sufficiently small so that $\exp_m$ is a diffeomorphism.

Next, let $\{e_i, i = 1,\ldots,n\}$ be an orthonormal frame at m and let $(x_1,\ldots,x_n)$ be a system of *normal coordinates* centered at m and such that $\frac{\partial}{\partial x_i}(m) = e_i$ for $i = 1,\ldots,n$. The *volume density function* θ_m of $\exp_m$ is defined by

$$\theta_m = (\det(g_{ij}))^{1/2}$$

where $g_{ij} = g\left(\frac{\partial}{\partial x_i},\frac{\partial}{\partial x_j}\right)$. Let $p = \exp_m(r\xi)$. Then we have the following Taylor expansion [3], [9], [10]:

$$\theta_m(p) = 1 + \sum_{k=2}^{7} \alpha_k(m,\xi)r^k + O(r^8),$$

where the α_k are completely determined by the Riemann curvature tensor and its covariant derivatives. For example, we have

$$\alpha_2(m,\xi) = -\frac{1}{6}\,\rho_{\xi\xi}(m), \qquad \alpha_3(m,\xi) = -\frac{1}{12}\,(\nabla_\xi\rho_{\xi\xi})(m),$$

$$\alpha_4(m,\xi) = \frac{1}{24}\left(-\frac{3}{5}\,\nabla^2_{\xi\xi}\rho_{\xi\xi} + \frac{1}{3}\,\rho^2_{\xi\xi} - \frac{2}{15}\sum_{a,b=1}^{n} R^2_{\xi a\xi b}\right)(m),$$

$$\alpha_5(m,\xi) = \frac{1}{120}\left(-\frac{2}{3}\,\nabla^3_{\xi\xi\xi}\rho_{\xi\xi} + \frac{5}{3}\,\rho_{\xi\xi}\nabla_\xi\rho_{\xi\xi} - \frac{2}{3}\sum_{a,b=1}^{n} R_{\xi a\xi b}\nabla_\xi R_{\xi a\xi b}\right)(m),$$

$$\begin{aligned}\alpha_6(m,\xi) = \frac{1}{720}\Bigg(&-\frac{5}{7}\,\nabla^4_{\xi\xi\xi\xi}\rho_{\xi\xi} + 3\,\rho_{\xi\xi}\nabla^2_{\xi\xi}\rho_{\xi\xi} + \frac{5}{2}\left(\nabla_\xi\rho_{\xi\xi}\right)^2 - \frac{5}{9}\,\rho^3_{\xi\xi}\\ &-\frac{8}{7}\sum_{a,b=1}^{n} R_{\xi a\xi b}\nabla^2_{\xi\xi}R_{\xi a\xi b} - \frac{15}{14}\sum_{a,b=1}^{n}\left(\nabla_\xi R_{\xi a\xi b}\right)^2\\ &-\frac{16}{63}\sum_{a,b,c=1}^{n} R_{\xi a\xi b}\,R_{\xi b\xi c}\,R_{\xi c\xi a} + \frac{2}{3}\,\rho_{\xi\xi}\sum_{a,b=1}^{n} R^2_{\xi a\xi b}\Bigg)(m).\end{aligned}$$

Here ∇ denotes the Levi Civita connection, R is the Riemann curvature tensor and ρ the corresponding Ricci tensor. Further, α_7 is given by (see [21])

$$(1) \qquad 2\,\alpha_7(m,\xi) = \sum_{i=2}^{6} \frac{(-1)^i}{(7-i)!}\left(\nabla^{7-i}_{\xi\ldots\xi}\,\alpha_i\right)(m,\xi).$$

The local geodesic symmetry φ_m at m is defined by $\varphi_m : \exp_m(r\xi) \longmapsto \exp_m(-r\xi)$ and it is a local diffeomorphism. It preserves the volume (up to sign) if and only if

$$\Theta_m\left(\exp_m(r\xi)\right) = \Theta_m\left(\exp_m(-r\xi)\right).$$

Hence, using the Taylor expansion, we have

LEMMA 1. *Let* (M,g) *be a Riemannian manifold such that all local geodesic symmetries are volume-preserving (up to sign). Then we have at each* $m \in M$ *and for any* $\xi \in T_mM$:

(2) $$\nabla_\xi \rho_{\xi\xi} = 0,$$

(3) $$\sum_{a,b=1}^{n} R_{\xi a \xi b} \; \nabla_\xi R_{\xi a \xi b} = 0.$$

In the rest of the paper we shall adopt the following notation :

$$F(\xi) = \sum_{a,b=1}^{n} R^2_{\xi a \xi b} \quad , \quad G(\xi) = \sum_{a,b=1}^{n} R_{\xi a \xi b} \; \nabla_\xi R_{\xi a \xi b}$$

for all $\xi \in T_mM$ and all $m \in M$. Then we may regard F and G as differentiable functions on the tangent bundle. Next, we identify, for each $m \in M$, the tangent space T_mM with an n-dimensional Euclidean space $\mathbb{R}^n$ via an orthonormal basis of T_mM. Then we may regard the restrictions of F and G to m as homogeneous polynomials of degree 4 and 5 respectively. Further, we denote by D the Laplacian of $\mathbb{R}^n$. Then we have

LEMMA 2. *Let* (M,g) *be a Ricci-parallel manifold (i.e.* $\nabla\rho = 0$*). Then we have at each point* $m \in M$

(4) $$(D^2G)(\xi) = 12\, \xi \|R\|^2 .$$

We omit the proof which is a straightforward computation. From this and (3) we derive

COROLLARY 3. *Let* (M,g) *be a connected Riemannian Ricci-parallel manifold such that all local geodesic symmetries are volume-preserving. Then* $\|R\|^2$ *is constant on* (M,g).

3. EINSTEIN MANIFOLDS OF DIMENSION FOUR

Now we first suppose that (M,g) is a four-dimensional *Einstein space*. We write down some well-known facts about the special features of the geometry on such manifolds. Following [1], [19] we can connect with any fixed point $m \in M$ a so-called *Singer-Thorpe basis* $\{e_1, e_2, e_3, e_4\}$ of T_mM. This is an orthonormal basis such that all the components of the curvature tensor R are given by the following formulas :

$$(5)\quad \begin{cases} R_{1212} = R_{3434} = a, \quad R_{1313} = R_{2424} = b, \quad R_{1414} = R_{2323} = c, \\ R_{1234} = \alpha, \quad R_{1342} = \beta, \quad R_{1423} = \gamma, \\ R_{ijkl} = 0 \quad \text{whenever just three of the indices } i,j,k,l \text{ are distinct.} \end{cases}$$

Note that $\alpha + \beta + \gamma = 0$ (the first Bianchi identity) and

$$a + b + c = \frac{\tau}{4}$$

where τ denotes the scalar curvature. Moreover it is always possible to choose a Singer-Thorpe basis such that $a = \max \sigma$ and $c = \min \sigma$, where σ denotes the sectional curvature function at m.

In what follows we will also consider the subclass of Einstein spaces satisfying the additional condition

" $\sum_{a,b=1}^{4} R^2_{xaxb}$ is independent of the unit vector $x \in T_mM$ for all $m \in M$."

Such Einstein spaces are called *2-stein spaces* [2]. The following lemma is easy to prove

LEMMA 4. *Let* (M,g) *be a four-dimensional Einstein space. Then* (M,g) *is a 2-stein space if and only if*

$$(6)\qquad \pm\alpha = a - \frac{\tau}{12}, \quad \pm\beta = b - \frac{\tau}{12}, \quad \pm\gamma = c - \frac{\tau}{12}$$

for each $m \in M$.

Note that if we have the + signs for one orientation of T_mM, then we have the - signs for the other orientation.

For the class of 2-stein spaces we have some important and useful freedom in the choice of a Singer-Thorpe basis. Indeed, let I,J,K be a quaternionic structure on T_mM adapted to the given Singer-Thorpe basis. This means, for

$$x = a^1e_1 + a^2e_2 + a^3e_3 + a^4e_4,$$

we put

$$Ix = - a^2e_1 + a^1e_2 - a^4e_3 + a^3e_4,$$

$$Jx = - a^3e_1 + a^4e_2 + a^1e_3 - a^2e_4,$$

$$Kx = - a^4e_1 - a^3e_2 + a^2e_3 + a^1e_4.$$

Then we have

LEMMA 5. *For any unit vector* $x \in T_mM$, $\{x,Ix,Jx,Kx\}$ *is a Singer-Thorpe basis if and only if* (M,g) *is a 2-stein space.*

Proof. The "if" part is proved in [12]. The other part is a straightforward computation which we omit.

Next we give another interpretation of (6), Lemma 4 and Lemma 5. Let $m \in M$, fix an orientation of T_mM and consider the set $\Lambda^2(T_mM)$ of skew-symmetric (0,2)-tensors. Then $\Lambda^2(T_mM)$ splits as $\Lambda^2(T_mM) = \Lambda^+ \oplus \Lambda^-$ where $\Lambda^\pm$ are the eigenspaces with respect to the Hodge star operator $* \in \mathrm{End}\,\Lambda^2(T_mM)$. The *Weyl conformal tensor* $W \in \mathrm{End}\,\Lambda^2(T_mM)$ at m leaves $\Lambda^\pm$ invariant. We denote by $W^\pm$ the restriction of W to $\Lambda^\pm$. Since

$$W^\pm = \frac{1}{2}(W \pm * W),$$

it is easy to see that $W^\pm = 0$ if and only if

$$\pm\alpha = a - \frac{\tau}{12}, \quad \pm\beta = b - \frac{\tau}{12}, \quad \pm\gamma = c - \frac{\tau}{12}$$

where $\{e_1,e_2,e_3,e_4\}$ is a "positive" Singer-Thorpe basis. Note that, because (M,g) is an Einstein manifold, we have

$$W_{XYZU} = R_{XYZU} - \frac{\tau}{12}\{g(X,Z)\,g(Y,U) - g(Y,Z)\,g(X,U)\}.$$

(See for example [19]. See also [7] for a lot of results concerning four-dimensional self-dual and anti-self-dual manifolds.) Hence we have

PROPERTY 6. *Let* (M,g) *be a* 4*-dimensional Einstein space. Then* (M,g) *is a* 2*-stein space if and only if* $W^+ = 0$ *or* $W^- = 0$ *at each* $m \in M$.

Finally we note some useful expressions for some curvature invariants of order 2 and 3 with respect to a Singer-Thorpe basis. We have (see [12])

$$(7)\quad \begin{cases} \|R\|^2 = 8(a^2+b^2+c^2+\alpha^2+\beta^2+\gamma^2), \\ \overset{\vee}{R} = \sum R_{ijkl}R_{klpq}R_{pqij} = 16(a^3+b^3+c^3+3a\alpha^2+3b\beta^2+3c\gamma^2), \\ \underline{\overset{\vee}{R}} = \sum R_{ikjl}R_{kplq}R_{piqj} = 24(abc+a\beta\gamma+b\alpha\gamma+c\alpha\beta). \end{cases}$$

Further we have

$$(8)\qquad \langle \Delta R, R\rangle = \frac{1}{2}\Delta\|R\|^2 - \|\nabla R\|^2$$

and (see [10])

$$(9)\quad \begin{cases} \overset{\vee}{R} = -\dfrac{\tau^3}{12} + \dfrac{2}{3}\tau\|R\|^2 - \dfrac{1}{3}\langle \Delta R, R\rangle, \\ \underline{\overset{\vee}{R}} = \dfrac{\tau^3}{48} - \dfrac{1}{24}\tau\|R\|^2 - \dfrac{1}{6}\langle \Delta R, R\rangle. \end{cases}$$

We will also use the following unpublished result of A.Derdziński :

PROPOSITION 7. *Let* (M,g) *be a four-dimensional Einstein manifold such that* $W \in C^\infty(\mathrm{End}\,\Lambda^2 M)$ *has constant eigenvalues. Then* (M,g) *is locally symmetric.*

It is clear that W has constant eigenvalues if and only if $a,b,c,\alpha,\beta,\gamma$ are constant. Using this, we can give another interesting interpretation of Proposition 7.

A Riemannian manifold (M,g) is said to be *curvature homogenous* [18] if for every $m,p \in M$ there exists an isometry A of the tangent space T_mM onto the tangent space T_pM such that for every $x,y \in T_mM$

$$R_{AxAy} = AR_{xy}A^{-1}.$$

Using Theorem 2.2 in [19] it is clear that the four-dimensional Einstein space is curvature homogeneous if and only if W has constant eigenvalues. Hence we obtain that *a four-dimensional curvature homogeneous Einstein space is locally symmetric.*

This last result includes clearly the result of G.R. Jensen : every four-dimensional locally homogeneous Einstein space is locally symmetric [11]. Jensen proved this result using an algebraic classification procedure but a simple and direct proof of this result has not yet been published for the general case. For the compact case a simple proof is given in [7]. Note that curvature homogeneity does not imply local homogeneity. (See for example [8], [15], [16], [20].)

4. FOUR-DIMENSIONAL KÄHLER MANIFOLDS

Now we turn to the *Kähler manifolds* but before we consider the four-dimensional case, we give some lemmas which are valid for an arbitrary dimension.

LEMMA 8 [17]. *Let* (M,g,J) *be an* n-*dimensional Kähler manifold such that*

$$\nabla_\xi \rho_{\xi\xi} = 0$$

for all $\xi \in T_mM$ *and all* $m \in M$. *Then* ρ *is parallel.*

Note that "$\nabla_\xi \rho_{\xi\xi} = 0$ for all ξ" is equivalent to

$$\mathfrak{S}_{x,y,z} \nabla_x \rho_{yz} = 0$$

for all $x,y,z \in T_mM$. Here $\mathfrak{S}$ denotes the cyclic sum.

Lemma 8 and Lemma 1 imply

COROLLARY 9. *Let* (M,g,J) *be a Kähler manifold such that all local geodesic symmetries are volume-preserving. Then* (M,g,J) *is Ricci-parallel.*

Also we shall use the following characterization of *locally symmetric* Kähler manifolds [17] :

LEMMA 10. *Let* (M,g,J) *be an* n-*dimensional Kähler manifold. Then* (M,g,J) *is locally symmetric if and only if*

$$\nabla R \quad = 0$$

for all $\xi \in T_mM$ *and all* $m \in M$.

Now we consider four-dimensional *Kähler-Einstein spaces*. We call an orthonormal basis $\{e_1,e_2,e_3,e_4\}$ of T_mM an *adapted* basis if

$$e_2 = Je_1 \quad , \quad e_4 = Je_3.$$

We have

LEMMA 11. *Let* (M,g,J) *be a four-dimensional Kähler-Einstein space. Then there always exists an adapted Singer-Thorpe basis at each fixed point* $m \in M$.

Proof. Let $m \in M$ and let $K(x,y)$ denote the sectional curvature of the plane spanned by the orthonormal vectors x,y. Further, let $H(x)$ denote the holomorphic sectional curvature of the holomorphic plane determined by the unit vector $x \in T_mM$.

Next we can choose a unit vector $e \in T_mM$ such that

$$H(e) = \max H(x) \quad \text{for } x \in T_mM \text{ and } \|x\| = 1.$$

Let P be the holomorphic plane determined by e and $P^{\perp}$ the orthogonal complement of P in T_mM. (Note that $P^{\perp}$ is also holomorphic.) Further, choose (e_1,e_3) in $P \times P^{\perp}$ such that

$$K(e_1,e_3) = \max K(x,y) \quad \text{for } x \in P, y \in P^{\perp} \text{ and } \|x\| = \|y\| = 1.$$

An elementary calculation shows that the basis $\{e_1, e_2 = Je_1, e_3, e_4 = Je_3\}$ is indeed a Singer-Thorpe basis.

(See also [23].)

Note that, because of the Kähler identity for R, we have for an adapted Singer-Thorpe basis

$$\text{(10)} \qquad \alpha = b + c = \frac{\tau}{4} - a, \quad \beta = -b, \quad \gamma = -c.$$

Hence, (10) implies with Lemma 4

PROPERTY 12. *Let* (M,g,J) *be a Ricci-flat four-dimensional Kähler manifold. Then* (M,g,J) *is a 2-stein space.*

Now we consider the Ricci-flat Kähler manifolds of dimension 4 more in detail and derive a result which is very useful in some calculations. Let $m \in (M,g,J)$ and let $\{E_1, E_2 = JE_1, E_3, E_4 = JE_3\}$ be an adapted orthonormal frame field on a neighborhood $U(m)$ of m. Further, put

(11) $$\nabla_{E_i} E_j = \sum_{k=1}^{4} B_{ijk} E_k, \quad i,j,k = 1,2,3,4.$$

Then we have

(12) $$B_{ijk} + B_{ikj} = 0.$$

Further, the Kähler condition implies

(13) $$B_{i13} = B_{i24} \quad , \quad B_{i14} = -B_{i23}.$$

Next we consider the following system of partial differential equations :

(14) $$E_i \theta = B_{i12} + B_{i34} \quad , \quad i=1,2,3,4.$$

Using (11),(12) and (14) we obtain

$$[E_i,E_j]\theta - E_i(E_j\theta) + E_j(E_i\theta) = \sum_k \left(B_{ijk} - B_{jik}\right)\left(B_{k12} + B_{k34}\right)$$
$$- E_i\left(B_{j12} + B_{j34}\right) + E_j\left(B_{i12} + B_{i34}\right) = R_{ij12} + R_{ij34}.$$

Using the fact that (M,g,J) is Ricci-flat, we get easily

$$R_{ij12} + R_{ij34} = 0.$$

Hence we may conclude that the system (14) has a unique solution on a neighborhood $V(m) \subset U(m)$ under an initial condition $\theta(m) = \theta_0$.

So, let θ be a solution of (14) on $V(m)$ and define a (1,1)-tensor field J_2 on $V(m)$ by

(15) $$\begin{cases} J_2E_1 = \cos\theta E_3 - \sin\theta E_4, \\ J_2E_2 = -\sin\theta E_3 - \cos\theta E_4, \end{cases}$$

$$\left| \begin{array}{lll} J_2E_3 & = - \cos\Theta E_1 & + \sin\Theta E_2, \\ J_2E_4 & = \sin\Theta E_1 & + \cos\Theta E_2. \end{array} \right.$$

We see at once that J_2 is an almost Hermitian structure on V(m) which satisfies $J_2J = - JJ_2$. Moreover it follows also that J_2 is parallel with respect to the Levi Civita connection ∇.

Further, put $J_3 = JJ_2$. It follows that $(V(m),g,J_1=J,J_2,J_3)$ *is a four-dimensional quaternionic Kähler manifold.*

We finish this section with the derivation of a key formula which we shall need to prove our main result.

PROPERTY 13. *Let* (M,g,J) *be a four-dimensional Kähler-Einstein space. Then, at each point* $m \in M$, *we have*

$$\|R\|^2 = 16\, F(\xi) + \frac{1}{2}\, \tau^2 - 4\, \tau H(\xi) \tag{16}$$

for all unit vectors $\xi \in T_mM$.

Proof. Let $\{e_1 = \xi,\ e_2 = Je_1, e_3, e_4 = Je_3\}$ be an adapted basis at m (not necessarily a Singer-Thorpe basis). Then we have at m :

$$\begin{aligned} \|R\|^2 &= \sum_{i,j,k} R^2_{ijk\ell} \\ &= 4 \sum_{j,\ell} \left\{R^2_{1j1\ell} + R^2_{2j2\ell} + R^2_{3j3\ell} + R^2_{4j4\ell}\right\} \\ &\quad - 4\left\{R^2_{1212} + R^2_{1313} + R^2_{1414} + R^2_{2323} + R^2_{2424} + R^2_{3434}\right\} \\ &\quad + 8\left\{R^2_{1234} + R^2_{1342} + R^2_{1423}\right\}. \end{aligned}$$

Using the fact that (M,g) is Einsteinian and Kählerian we have

$$R_{1212} = \frac{\tau}{4} - R_{1313} - R_{1414} = \frac{\tau}{4} - R_{2424} - R_{1414} = R_{3434}$$

and

$$R_{1313} = R_{2424} = R_{1324} \quad , \quad R_{1414} = R_{2323} = -R_{1423}.$$

Further we have

$$R_{1213} = R_{1224} = -R_{1334} = -R_{2434}$$

and similar relations for other terms. The required result follows then easily using all these relations and the first Bianchi identity.

Note that Property 12 also follows at once from (16) since we have for $\tau = 0$:

$$F(\xi) = \frac{1}{16} \|R\|^2,$$

where ξ is an arbitrary unit vector.

5. PROOF OF THE MAIN THEOREM

Now we are ready to prove the Main Theorem. Let (M,g,J) be a four-dimensional Kähler manifold such that all local geodesic symmetries are volume-preserving. Then (M,g,J) is Ricci-parallel (Corollary 9). Hence, if (M,g,J) is not Einsteinian, it is reducible and hence locally symmetric. So, for our aim, we may suppose that the manifold is Einsteinian.

Next we consider the following two cases :

A. (M,g,J) is not Ricci-flat

Take a unit vector $\xi \in T_mM$ and denote by γ the geodesic tangent to ξ at m with arc length parametrization, i.e. $\gamma(r) = \exp_m(r\xi)$. Since

$$\frac{d}{dr} F(\gamma'(r)) = 2\, G(\gamma'(r)) = 0$$

we obtain that $F(\gamma'(r))$ is constant along γ. This, (16) and the fact that $\|R\|^2$ is constant implies that $H(\gamma'(r))$ is also constant along γ. Hence

$$\nabla_{\gamma'(r)} R_{\gamma'(r)J\gamma'(r)\gamma'J\gamma'(r)} = \frac{d}{dr} H(\gamma'(r)) = 0$$

and so

$$\nabla_{\xi} R_{\xi J\xi\xi J\xi} = 0.$$

Since ξ is arbitrary, the required result follows now from Lemma 10.

B. (M,g,J) is Ricci-flat

In this case, (M,g,J) is a 2-stein space (Property 12). So, following Lemma 5, any arbitrary unit vector ξ can be taken as the first vector $e_1 = \xi$ of a Singer-Thorpe basis $\{e_1,e_2,e_3,e_4\}$. For such a basis, and since $\tau = 0$ and $\|R\|^2$ = constant, we obtain first

$$(17) \qquad a + b + c = 0,$$

and from (7),(8) and (10) we get

$$(18) \qquad \begin{cases} \|R\|^2 = 16(a^2 + b^2 + c^2), \\ \check{R} = 64\,(a^3 + b^3 + c^3) = \frac{1}{3}\|\nabla R\|^2 \\ \underline{\check{R}} = 96\,abc = \frac{1}{6}\,\|\nabla R\|^2. \end{cases}$$

Hence, (17) and (18) imply that a,b,c are constant if $\|\nabla R\|^2$ is constant. Note that in this case α,β and γ are also constant.

Now we will prove that $\|\nabla R\|^2$ is indeed constant on M. To do this we use the condition

$$\alpha_7(m,\xi) = 0$$

for all $m \in M$ and all unit vectors $\xi \in T_mM$. Taking into account that $\alpha_3 = \alpha_5 = 0$ and that α_2 and α_4 are independent of m and ξ we get from (1) that $\alpha_7 = 0$ is equivalent to

$$\nabla_\xi \alpha_6 = 0$$

or

$$8 \sum_{a,b} \nabla_\xi R_{\xi a\xi b}\, \nabla^2_{\xi\xi} R_{\xi a\xi b} + 8 \sum_{a,b} R_{\xi a\xi b}\, \nabla^3_{\xi\xi\xi} R_{\xi a\xi b}$$

$$+ 15 \sum_{a,b} \nabla_\xi R_{\xi a\xi b}\, \nabla^2_{\xi\xi} R_{\xi a\xi b} + \frac{16}{3} \sum_{a,b,c} R_{\xi b\xi c}\, R_{\xi c\xi a}\, \nabla_\xi R_{\xi a\xi b} = 0.$$

Using (3) we see at once that this is equivalent to

$$(19) \qquad 3 \sum_{a,b} \nabla_\xi R_{\xi a\xi b}\, \nabla^2_{\xi\xi} R_{\xi a\xi b} = 16 \sum_{a,b,c} R_{\xi b\xi c}\, R_{\xi c\xi a}\, \nabla_\xi R_{\xi a\xi b}\,.$$

Putting

$$K(\xi) = \sum_{a,b} \nabla_\xi R_{\xi a \xi b} \nabla^2_{\xi\xi} R_{\xi a \xi b},$$

$$L(\xi) = \sum_{a,b,c} R_{\xi b \xi c} R_{\xi c \xi a} \nabla_\xi R_{\xi a \xi b},$$

(19) becomes

$$3\,K(\xi) = 16\,L(\xi). \tag{20}$$

Since (M,g,J) is a 2-stein space, we have

$$\sum_{a,b,c} R_{\xi a \xi b} R_{\xi b \xi c} R_{\xi c \xi a} = A\|\xi\|^6 \tag{21}$$

for any vector $\xi \in T_mM$, where A is independent of ξ (see [1]). To compute A we take $\|\xi\| = 1$ and integrate (21) over the unit sphere in T_mM as is done in [10]. Using (18) we then obtain, by taking the covariant derivative in (21) :

$$L(\xi) = \frac{1}{576}\,\xi\|\nabla R\|^2. \tag{22}$$

Hence, (20) becomes now

$$K(\xi) = \frac{1}{108}\,\xi\|\nabla R\|^2. \tag{23}$$

Next, we write down another expression for $K(\xi)$. Indeed we have

$$K(\xi) = \frac{1}{16} \sum_{a,b,c,d} \nabla_\xi R_{abcd} \nabla^2_{\xi\xi} R_{abcd}.$$

This is easily verified on the Ricci-flat Kähler manifold by using the quaternionic Kähler structures which exist on that manifold (see section 4). Then we get for any vector $\xi \in T_mM$ (not necessarily a unit vector)

$$P(\xi) = \frac{4}{27}\,Q(\xi) \tag{24}$$

where

$$P(\xi) = \sum_{a,b,c,d} \nabla_\xi R_{abcd} \nabla^2_{\xi\xi} R_{abcd},$$

$$Q(\xi) = \|\xi\|^2 \; \xi\|\nabla R\|^2 .$$

To obtain the required result we compute $(DP)(\xi)$ and $(DQ)(\xi)$, where D denotes again, as in section 2, the Laplacian of $T_m M$. We find

$$(DQ)(\xi) = 12 \; \xi\|\nabla R\|^2 \tag{25}$$

and

$$(DP)(\xi) = 2 \sum \Big\{ \nabla^2_{a\xi} R_{kjih} \nabla_a R_{kjih} + \nabla^2_{aa} R_{kjih} \nabla_\xi R_{kjih} + \nabla^2_{\xi a} R_{kjih} \nabla_a R_{kjih} \Big\}.$$

To compute this last expression we note first that, for the Ricci-flat manifold, we have, as is easily verified :

$$\sum R_{acih} R_{acpk} R_{ihqk} = \frac{1}{12} \|\nabla R\|^2 \delta_{pq}. \tag{26}$$

Using this, the Ricci and the Bianchi identities, we then get after a long computation, which we omit :

$$\sum \nabla^2_{\xi a} R_{kjih} \nabla_a R_{kjih} = \frac{1}{2} \; \xi\|\nabla R\|^2 ,$$

$$\sum \nabla^2_{aa} R_{kjih} \nabla_\xi R_{kjih} = - \frac{1}{3} \; \xi\|\nabla R\|^2 ,$$

$$\sum \nabla^2_{a\xi} R_{kjih} \nabla_a R_{kjih} = \frac{11}{18} \; \xi\|\nabla R\|^2 .$$

Hence we have

$$(DP)(\xi) = \frac{14}{9} \; \xi\|\nabla R\|^2 , \tag{27}$$

and so, (24),(25) and (27) lead to

$$\xi\|\nabla R\|^2 = 0,$$

which means that $\|\nabla R\|^2$ is constant.

Finally, this implies that (M,g,J) is curvature homogeneous and hence locally symmetric. (See section 3.)

This finishes the complete proof of the Main Theorem.

Remark.

Note that in the Ricci-flat case, the manifold is actually flat. See for example [14] or consider the classification of symmetric spaces.

6. ANOTHER RESULT

In part A of the proof of the Main Theorem we used only the first two conditions and the key formula (16). Using this formula we also have

PROPOSITION 14. *Let* (M,g,J) *be a connected four-dimensional 2-stein Kähler space which is not Ricci-flat. Then* (M,g,J) *is locally symmetric. (Actually* (M,g,J) *has constant holomorphic sectional curvature.)*

Proof. Since (M,g,J) is a 2-stein space we have, for an arbitrary unit vector $\xi \in T_mM$,

$$F(\xi) = \frac{1}{16} \|R\|^2 + \frac{1}{96} \tau^2 . \tag{28}$$

(See [3], formula 6.3). Then, (28) and (16) imply that $H(\xi) = \frac{1}{6}\tau$ and so the holomorphic sectional curvature is constant. Hence (M,g,J) is locally symmetric.

Remark. In part B of the proof of the Main Theorem we used the fact that (M,g,J) is a 2-stein space in an extensive way. Also we made use of more terms in the power series expansion of the volume density function. Using the same method it seems to be possible to give an affirmative answer to the following

PROBLEM. *Let* (M,g) *be a connected four-dimensional 2-stein space such that all local geodesic symmetries are volume-preserving. Is* (M,g) *locally homogeneous ?*

We hope to come back on this on another occasion.

REFERENCES

[1] A.L. Besse, *Manifolds all of whose geodesics are closed,* Ergebnisse der Mathematik, 93, Springer-Verlag, Berlin, 1978.

[2] P. Carpenter, A. Gray and T.J. Willmore, The curvature of Einstein symmetric spaces, *Quart. J. Math. Oxford* 33 (1982), 45-64.

[3] B.Y. Chen and L. Vanhecke, Differential geometry of geodesic spheres, *J. Reine Angew. Math.* 325 (1981), 28-67.

[4] J.E. D'Atri and H.K. Nickerson, Divergence-preserving geodesic symmetries, *J. Differential Geometry* 3 (1969), 467-476.

[5] J.E. D'Atri and H.K. Nickerson, Geodesic symmetries in spaces with special curvature tensor, *J. Differential Geometry* 9 (1974), 251-262.

[6] J.E. D'Atri, Geodesic spheres and symmetries in naturally reductive homogeneous spaces, *Michigan Math. J.* 22 (1975), 71-76.

[7] A. Derdziński, Self-dual Kähler manifolds and Einstein manifolds of dimension four, *Compositio Math.* 49 (1983), 405-433.

[8] D. Ferus, H. Karcher and H.F. Münzner, Cliffordalgebren und neue isoparametrische Hyperflächen, *Math. Z.* 177 (1981), 479-502.

[9] A. Gray, The volume of a small geodesic ball in a Riemannian manifold, *Michigan Math. J.* 20 (1973), 329-344.

[10] A. Gray and L. Vanhecke, Riemannian geometry as determined by the volumes of small geodesic balls, *Acta Math.* 142 (1979), 157-198.

[11] G.R. Jensen, Homogeneous Einstein spaces of dimension four, *J. Differential Geometry* 3 (1969), 309-349.

[12] O. Kowalski and L. Vanhecke, Ball-homogeneous and disk-homogeneous Riemannian manifods, *Math. Z.* 180 (1982), 429-444.

[13] O. Kowalski, Spaces with volume-preserving symmetries and related classes of Riemannian manifolds, *Rend. Sem. Mat. Univ. e Politec. Torino,* to appear.

[14] K. Sekigawa, On 4-dimensional connected Einstein spaces satisfying the condition $R(X,Y)\cdot R = 0$, *Sem. Rep. Niigata Univ.* A7 (1969), 29-31.

[15] K. Sekigawa, On the Riemannian manifolds of the form $B\times_f F$, *Kōdai Math. Sem. Rep.* 26 (1975), 343-347.

[16] K. Sekigawa, On some 3-dimensional curvature homogeneous spaces, *Tensor* 31 (1977), 87-97.

[17] K. Sekigawa and L. Vanhecke, Symplectic geodesic symmetries on Kähler manifolds, *Quart. J. Math. Oxford,* to appear.

[18] I.M. Singer, Infinitesimally homogeneous spaces, *Comm. Pure. Appl. Math.* 13 (1960), 685-697.

[19] I.M. Singer and J.A. Thorpe, The curvature of 4-dimensional Einstein spaces, in *Global Analysis* (Papers in honor of K. Kodaira), Princeton University Press, Princeton 1969, 355-366.

[20] H. Takagi, On curvature homogeneity of Riemannian manifolds, *Tôhoku Math. J.* 26 (1974), 581-585.

[21] L. Vanhecke, A note on harmonic spaces, *Bull. London Math. Soc.* 13 (1981), 545-546.

[22] L. Vanhecke, Some results about homogeneous structures on Riemannian manifolds, *Differential Geometry*, Proc. Conf. Differential Geometry and its Applications, Nové Město 1983, Univerzita Karlova, Praha, 1984, 147-164.

[23] Y.L. Xin, Remarks on characteristic classes of four-dimensional Einstein manifolds, *J. Math. Phys.* 21 (1980), 343-346.

THE COHOMOLOGY AND GEOMETRY OF HEISENBERG-REITER NILMANIFOLDS

J. F. Torres Lopera (*)

Dep. de Geometría y Topología. Fac. de Matemáticas. Santiago. España.

I.S.E.A., Dep. de Math., 4, rue Frères Lumière. 68093 Mulhouse. France.

1.-INTRODUCTION. H. Reiter, [16], has defined a family of nilpotent topological groups which generalize the well-known Heisenberg groups. He used them to extend some theorems of harmonic analysis. Here our aim is to study the topology and the differential geometry of these Heisenberg-Reiter groups, (HR-groups for short), and their nilmanifolds. More in detail, we show that the universal covering of a topological HR-group is HR too, (Proposition 1). In the Lie group case, a classical theorem of Mal'cev, [13], and the preceding statement allow us to affirm that every compact homogeneous space of a connected HR-Lie group G is a quotient of its universal covering HR-group $\widetilde{G}$ under the natural action of a discrete uniform subgroup Γ of $\widetilde{G}$, (**). The bilinear map that originates the group law of a HR-group G determines its universal covering $\widetilde{G}$, and also the existence of discrete uniform subgroups of $\widetilde{G}$, when G is a Lie group. We give explicitly a family of such subgroups for every HR-Lie group G and study the first integral homology group of their associated compact nilmanifolds, (Proposition 2). Moreover, in §4 we obtain the first Betti number of *every* compact connected HR-nilmanifold, (Prop.3), and a criterium for the existence of non-trivial Massey triple products in their real cohomology rings. It is well-known that these products are obstructions for the existence of Kähler structures on a compact manifold, [5]. These facts are summed up in Proposition 5. Examples are given.

In §5 we present two families of HR-manifolds which can be endowed with several weakly Kähler structures, such as symplectic, semi-Kähler and Hermitian structures, denoted (G,g,J_1), (G,g,J_2). In §6 two results concerning the Riemannian geometry of HR-spaces are included:

i) We give explicitly the geodesics of connected simply connected HR-Lie groups with respect to a suitable left-invariant Riemannian metric, (Proposition 6). ii) It is shown that every conformal transformation on their nilmanifolds must be an isometry.

We are indebted to L.A. Cordero, M. Fernández, M. Goze, A. Gray, T. Hangan, M. de León, R. Lutz and D. Tanré. Acknowledgements to A. Bouyakoub, R. Fernández and P. Piu.

(*) Work supported by a grant of M.E.C., (Spain), and M.R.I., (France).

(**) A quotient $\Gamma\backslash G$ with G nilpotent and Γ discrete will be called nilmanifold even if Γ is not uniform.

2.- HR-GROUPS: DEFINITION AND EXAMPLES. A generalized Heisenberg group, in the sense of Reiter [16], or HR-group, is the product of three Abelian topological groups X,Y,Z, endowed with an operation which arises from a non trivial $\mathbb{Z}$-bilinear continuous map $B: X\times Y \longrightarrow Z$ as follows:

(1) $$(x,y,z)(x',y',z') = (x+x', y+y', z+z'+B(x,y')) \quad .$$

With this law $X\times Y\times Z$ will be denoted G_B.

Examples.- 1) Let A be a topological ring, $X = M_{pq}(A)$, $Y = M_{qr}(A)$, $Z = M_{pr}(A)$ the additive groups of pq, qr and pr matrices with componentes valued in A, and $B: X\times Y \longrightarrow Z$ the usual matrix product. We will denote G_B as $H_A(p,q,r)$. When $A = \mathbb{R}$ and $r = 1$, these are just Haraguchi's groups, [9], [6]. The Heisenberg groups are $H_{\mathbb{R}}(1,q,1)$, $q = 1,2,\ldots$

2) Let $X=Y=\mathbb{R}^k$, $Z=\mathbb{R}^m$ and $B: X\times Y \longrightarrow Z$ an antisymmetric $\mathbb{R}$-bilinear map such that for every linear form ϕ on $\mathbb{R}^m$ the bilinear form ϕB is non degenerated, (such a map B can only exist if and only if there exist an m-frame on the sphere S^{k-1}, [12], [10]). The subgroup $K = \{(x,y,z);\ x = y\}$ of G_B is nothing but a group of type H or of Heisenberg type in the sense of A. Kaplan, [10]. Each group of type H can be seen in this way as a subgroup of a HR-group.

3) Some natural HR-groups can be associated to every pair of real or complex vector spaces X,Y taking $Z = X*Y$, with $*$ the tensor, symmetric or exterior product of X and Y, and B the canonocal projection of $X\times Y$ onto Z.

3.-THE UNIVERSAL COVERING GROUP OF A HR GROUP. For a connected topological space S its universal covering space will be denoted $\tilde{S}$. The covering projection will be written

$$\pi_S : \tilde{S} \longrightarrow S$$

Proposition 1. Let $G_B = X\times Y\times Z$ be a connected HR-group, X,Y,Z being locally simply connected. The universal covering group of G_B is a HR-group $\tilde{G}_B$ which is associated to the $\mathbb{Z}$-bilinear continuous map $\tilde{B}: \tilde{X}\times\tilde{Y} \longrightarrow \tilde{Z}$ obtained as the unique lifting of $B\circ(\pi_X\times\pi_Y)$ sending (0,0) into 0. If B is (anti)symmetric then so is $\tilde{B}$. When G_B is a connected (complex) HR-Lie group then $\tilde{X},\tilde{Y},\tilde{Z}$ are finite dimensional (complex) real vector spaces and $\tilde{B}$ is in fact $\mathbb{R}$-linear.

Proof. The maps $F_1 = (+)_{\tilde{Z}}(\tilde{B}\times\tilde{B})\circ(id_{\tilde{X}}\times P\times id_{\tilde{Y}})\circ(id_{\tilde{X}\times\tilde{X}}\times D)$ and $F_2 = \tilde{B}\circ((+)_{\tilde{X}}\times id_{\tilde{Y}})$, (where $id_{\tilde{X}}$ and $(+)_{\tilde{X}}$ denote identity and sum in $\tilde{X}$, $P(x,y) = (y,x)$, $D(y) = (y,y)$), verify $F_1(0,0,0) = F_2(0,0,0) = 0$ and $\pi_Z\circ F_1 = \pi_Z\circ F_2$, thus $F_1 = F_2$. This is just the $\mathbb{Z}$-linearity of $\tilde{B}$ in its first variable. On the second it is analogous. The last statement follows because $\mathbb{R}^n$, (resp. $\mathbb{C}^n$) are the only Abelian simply connected real (complex) Lie groups and the continuity of $\tilde{B}$.

|||

A theorem of Mal'cev, [13], says that every homogeneous space of a connected nilpotent Lie group N is a quotient $\Gamma\backslash\tilde{N}$ or $\tilde{N}/\Gamma$, with Γ a discrete subgroup of $\tilde{N}$, and that discrete uniform subgroups exist if and only if the Lie algebra of N admits a basis with rational structure coefficients. For HR-groups this can be said in terms of the $\mathbb{R}$-bilinear map B:

Proposition 2.- A connected HR-Lie group G_B admits uniform discrete subgroups if and only if $\tilde{G}_B = \tilde{X}\times\tilde{Y}\times\tilde{Z}$ as a vector space allows a basis $\{e_i, f_u, g_\alpha;\ e_i \varepsilon X, f_u \varepsilon Y, g_\alpha \varepsilon Z\}$ such that $\tilde{B}(e_i, f_u) = \Sigma_\alpha B^\alpha_{iu} g_\alpha$, with B^α_{iu} integer numbers. When such a basis exists, then for every set of strictly positive integer numbers $K = \{k^i, k^u\}$ the subgroup Γ_K of $\tilde{G}_B$

(1) $\Gamma_K = \{(x,y,z)\varepsilon G_B;\ x=\Sigma_i k^i x^i e_i, y=\Sigma_u k^u y^u f_u, z=\Sigma_\alpha z^\alpha g_\alpha, x^i, y^u, z^\alpha \varepsilon \mathbb{Z}\}$

is discrete and uniform. The first integral homology group of $M = \Gamma_K \backslash G_B$ is

(2) $H_1(M, \mathbb{Z}) = \Gamma_K/[\Gamma_K, \Gamma_K] = (\oplus_i \mathbb{Z}e_i) \oplus (\oplus_u \mathbb{Z}f_u) \oplus A$,

where A is the finitely generated Abelian group obtained from the Abelian free group $\oplus_\alpha \mathbb{Z}g_\alpha$ by means of the relations $\Sigma_\alpha r^\alpha_J g_\alpha = 0$, the index J running on the pairs (i,u) and the matrix (r^α_J) being given by $r^\alpha_{(i,u)} = B^\alpha_{iu}\, k^i k^u$,(no sum). The first Betti number $b_1(M)$ is therefore

(3) $b_1(M) = \dim(X) + \dim(Y) + \dim(Z) - \mathrm{rank}(s^\alpha_J)$

where the matrix (s^α_J) is given by $s^\alpha_{(i,u)} = B^\alpha_{iu}$.

Proof. Fix an arbitrary basis $\{e_i, f_u, g_\alpha\}$ of $\tilde{X}\times\tilde{Y}\times\tilde{Z}$ with e_i, f_u, g_α in $\tilde{X}, \tilde{Y}, \tilde{Z}$, respectively. Call x^i, y^u, z^α the coordinate functions associated to it. Then a global moving coframe of left invariant 1-forms on G_B is

(4) $\xi^i = dx^i$; $\eta^u = dy^u$; $\zeta^\alpha = dz^\alpha - \Sigma_{i,u} B^\alpha_{iu} x^i dy^u$.

Thus, the Maurer-Cartan equations are

(5) $d\xi^i = 0$; $d\eta^u = 0$; $d\zeta^\alpha = -\Sigma_{i,u} B^\alpha_{iu} \xi^i \eta^u$,

hence the dual moving frame of left-invariant vector fields on G_B is

(6) $X_i = (\partial/\partial x^i)$; $Y_u = (\partial/\partial y^u) + \Sigma_{i,\alpha} B^\alpha_{iu} x^i (\partial/\partial z^\alpha)$, $Z_\alpha = (\partial/\partial z^\alpha)$

and the only non trivial structure coefficients are

(7) $c^\alpha_{iu} = \zeta^\alpha([X_i, Y_u]) = B^\alpha_{iu} = -c^\alpha_{ui}$.

The first statement follows from the theorem of Mal'cev mentioned above and (7). For the second one, suppose the basis above is chosen in such a way that the coefficients B^α_{iu} are integer numbers. Then the compact subset D of $X\times Y\times Z$ is a fundamental domain

(8) $D = \{(x,y,z);\ 0 \leq x^i \leq k^i,\ 0 \leq y^u \leq k^u,\ 0 \leq z^\alpha \leq 1\}$

is projected onto $\Gamma_K \backslash \tilde{G}_B$ by π_M. Therefore Γ_K is uniform and obviously discrete. On the other hand, $[\Gamma_K, \Gamma_K]$ is the subgroup of Γ_K generated by the elements $(0,0,B(x,y') - B(x',y))$, with $(x,y,0)$ and $(x',y',0)$ in Γ_K. Thus it is a subgroup of the free Abelian group $\oplus_\alpha \mathbb{Z}\, g_\alpha$ with associated quotient group determined by the relations $\Sigma_\alpha r^\alpha_J g_\alpha = 0$. This proves (2). Using the elementary matrix transformations algarithm to classify the Abelian group A, one obtains (3) after observing that both matrices (r^{α}_{J}) and (s^{α}_{J}) have the same rank. |||

Note that the rank of $H_1(M, \mathbb{Z})$ does not depend on the uniform subgroup used to quotient. This is not surprising in view of a theorem of Nomizu, [15], recalled in the next paragraph. But the torsion coefficients depend on Γ_K. They can be obtained from the coefficients of the matrix (r^{α}_{J}) in Proposition 2, via the algorithm to classify finitely generated Abelian groups cited above. At most the number of torsion coefficients can be rank(r^{α}_{J}). In this way one can obtain HR-nilmanifolds having identic first Betti numbers but carrying different torsion coeffients.

4.- THE REAL COHOMOLOGY RING OF COMPACT HR-NILMANIFOLDS. The Lie algebras of HR-Lie groups will be called HR-algebras. They are determined by the system §3.(5). Let us recall that the exterior algebra $\Lambda(g^*)$ of left invariant forms on a Lie group G with Lie algebra g, endowed with the restriction of the exterior derivative has an associated cohomology ring $H^*(g)$, which satisfies Poincaré duality when g is an unimodular Lie algebra. Since HR-algebras are two-step nilpotent, they are in that case.

Proposition 3. If M is a compact and connected HR-nilmanifold then its first Betti number is

(1) $$b_1(M) = \dim(X) + \dim(Y) + \dim(Z) - \mathrm{rank}(B'),$$

where $G_B = X \times Y \times Z$ is the universal covering of M and $B': X \otimes Y \longrightarrow Z$ is the linear map associated to B, (cf. Prop. 1).

Proof. Let g be the Lie algebra of G_B. From the Maurer-Cartan equations §3.(5) of G_B, it follows that the rank of $H^1(g)$ is just the right side of (1). On the other hand, the real comology ring $H^*(M)$ is isomorphic to the cohomology ring $H^*(g)$, (Nomizu's Theorem, [15]; it holds for arbitrary compact connected nilmanifolds). |||

Proposition 4. Let g be a HR-algebra and suppose that there exist an index α and an integer $q > 0$ verifying: i) the 2-form $d\zeta^{\alpha}$, §3(5), is not generated by the other forms $d\zeta^{\beta}$; ii) every cohomology class $h \in H^q(g)$ can be represented by a cocycle θ such that the contraction $(\partial/\partial z^{\alpha}) \lrcorner \theta$ vanishes ; iii) $d\zeta^{\alpha}$ is expressed with exactly q 1-forms $\xi^{i_1},\dots,\xi^{i_q}$ in 3.(5). Then the Massey triple product

(2) $$<[B^{\alpha}_{i_1 u_1}\, \eta^{u_1} \wedge \xi^{i_2} \wedge \dots \wedge \xi^{i_q}], [\xi^{i_1}], [\xi^{i_1}]>$$

is non trivial. It can be represented by the cocycle

(3) $$\zeta^{\alpha} \wedge \xi^{i_1} \wedge \xi^{i_2} \wedge \dots \wedge \xi^{i_q}$$

Proof. We will use two elementary properties:

1) Every monomial $\xi^{j_1} \wedge \dots \wedge \xi^{j_n} \neq 0$ is a non trivial cocycle in the complex $(\Lambda(g^*), d)$.

2) If μ is a 1-form on a vector space V, r a vector in V, and ρ, ϕ multilinear antisymmetric forms on V satisfying

$$r \rfloor \mu \neq 0, \; r \rfloor \rho = 0 = r \rfloor \phi, \; \mu \wedge \rho = \phi ,$$

then both ρ and ϕ must be zero.

Using 1) and 2) it is straightforward to check that (2) is a non-trivial cocycle. It represents the Massey triple product (1),(see [5] or [17] for Massey products and rational homotopy theory) . In order to prove that (1) is not a trivial Massey product one needs to show that

$$[\zeta^\alpha \wedge \xi^{i_1} \wedge \ldots \wedge \xi^{i_q}] \notin [a] H^1(g) + [c] H^q(g)$$

where $a = B^\alpha_{i_1 u_1} \eta^{u_1} \wedge \xi^{i_1} \wedge \ldots \wedge \xi^{i_q}$ and $c = \xi^{i_1}$. This is equivalent to verify that there is no $[\omega]$ in $H^1(g)$ and no $[\theta]$ in $H^q(g)$ such that

(3) $$\zeta^\alpha \wedge \xi^{i_1} \wedge \ldots \wedge \xi^{i_q} - a \wedge \omega - c \wedge \theta = d\lambda$$

for a q-form λ on g. By hypothesis we can suppose that the cocycle θ is a sum of monomials not carrying the 1-form ζ^α. On the other hand, no 1-cocycle can depend on ζ^α, by §3.(5) and the hypothesis on $d\zeta^\alpha$. Thus, writing $\lambda = \zeta^\alpha \wedge \phi + \psi$ and using 1) and 2) in (3), we arrive to

$$\xi^{i_1} \wedge \ldots \wedge \xi^{i_q} + d\phi = 0$$

in contradiction with 1). Permuting ξ's and η's gives analogous products. |||

Example 1. Let X,Y be two real vector spaces and $Z = X \otimes Y$ their tensor product. The hypothesis of Proposition 4 are verified for every index α, (here α is a pair (i,u)), and q = 1, on the Lie algebra of the HR-group associated to $\otimes$.

Example 2. Let $H_{\mathbb{R}}(p,q,r)$ be the HR-group defined in §2. Call $(p) = \{1,\ldots,p\}$, etc., and use i_u, v_α, j_β as indices for X,Y,Z, with respect to their canonical basis of matrices $E_{i_u}, E_{v_\alpha}, E_{j_\beta}$, with 1 in the component (i,u), (v,α), (j,β) respectively, and 0 in the other ones, ($i,j \in (p)$, $u,v \in (q)$, $\alpha,\beta \in (r)$). The bilinear map B, (matrix product), has coefficients

(4) $$B^{j_\beta}_{i_u v_\alpha} = \delta^j_i \delta^v_u \delta^\beta_\alpha ,$$

with respect to these basis. Every form $\zeta^{i_\alpha} = dz^{i_\alpha} - \Sigma_u x^{i_u} dy^{u_\alpha}$ satisfies the hypothesis of Proposition 4 for the integer q. Moreover, in the cohomology of the Lie algebra of $H_{\mathbb{R}}(p,q,r)$ we have matric Massey triple products, [14],

(5) $$< \begin{bmatrix} \xi^{11} \wedge \ldots \wedge \xi^{1q} & 0 & \ldots & 0 \\ 0 & \xi^{21} \wedge \ldots \wedge \xi^{2q} & \ldots & 0 \\ \ldots & \ldots & \ldots & \ldots \\ 0 & 0 & \ldots & \xi^{p1} \wedge \ldots \wedge \xi^{pq} \end{bmatrix}, \begin{bmatrix} \xi^{11} & \ldots & \xi^{1q} \\ \ldots & \ldots & \ldots \\ \ldots & \ldots & \ldots \\ \xi^{p1} & \ldots & \xi^{pq} \end{bmatrix}, \begin{bmatrix} \eta^{11} & \ldots & \eta^{1r} \\ \ldots & \ldots & \ldots \\ \ldots & \ldots & \ldots \\ \eta^{q1} & \ldots & \eta^{qr} \end{bmatrix} > ,$$

which can be represented by the matrix whose components are, up to a sign, the cocycles representing the non-trivial Massey products given by Proposition 4.

Remark. The product of a HR-algebra and an Abelian Lie algebra is HR too. In formula (1) it suffice to add the dimension of the Abelian factor to obtain the rank of the first cohomology group of the product algebra. Massey triple products are non trivial in the product if and only if they are so in the HR-algebra.

Proposition 5. Let $G_B = X\times Y\times Z$ a connected and simply connected HR-Lie group admitting discrete uniform subgroups, Γ any of them, $M = \Gamma\backslash G_B$ the compact nilmanifold associated and g the Lie algebra of G_B. Let T be a compact connected Abelian Lie group,(a torus).
1) If $\dim(G_B\times T)$ is even and the rank of the linear map $B': X\otimes Y \longrightarrow Z$ associated to B is odd then no Kähler structure can exist on $M\times T$.
2) If $\dim(M\times T)$ is even and $H^*(g)$ contains non-trivial Massey products,(in particular: if g fulfils the hypothesis of Prop. 4), then no Kähler structure can exist on $M\times T$.

Proof. Point 1) is a consequence of the Hodge theorem on the Betti numbers of compact Kähler manifolds, Proposition 3 and the Remark above. For 2) notice that Nomizu's theorem, [15], can be seen in terms of Sullivan's theory of minimal models by saying that the complex $(\Lambda(g^*),d)$ of a nilpotent Lie algebra with rational structure coefficients is a minimal model for the de Rham complex of any of its associated compact nilmanifolds. On the other hand, a theorem by Deligne-Griffiths-Morgan-Sullivan, [5], says that Kähler manifolds are *formal*, that implying that every Massey product in their cohomology ring must be zero. 2) follows from this, Prop. 4 and the Remark above. |||

Proposition 5 generalizes results in [4] and [3]. An analogous statement gives obstructions for almost contact Riemannian manifolds to admit cosymplectic structures, (see [1] for definitions). D. Tanré has pointed out to us that the non formality of arbitrary compact nilmanifolds seems plausible.

5.-WEAKLY KÄHLER STRUCTURES ON HR-NILMANIFOLDS. Let $G_B = X\times Y\times Z$ be the HR-group associated to a $\mathbb{R}$-linear map $\neq 0$, $B: X\times Y \longrightarrow Z$, X,Y,Z real vector spaces, (every connected and simply connected HR-Lie group is of this type, by Prop. 1). If T is a fourth real vector space then the product $G = X\times Y\times Z\times T$ is also a HR-group, with law

$$(x,y,z,t)(x',y',z',t') = (x + x',y + y', z + z' + B(x,y'),t + t') \tag{1}$$

For a basis $\{h_\lambda\}$ of T, denote t^λ the associated coordinate functions. The vector fields $T_\lambda = \partial/\partial t^\lambda$ and those defined in §3.(6) constitute a left-invariant moving frame on G. The forms $\tau^\lambda = dt^\lambda$ and those in §3.(4) give the dual moving coframe. Fix the left-invariant Riemannian metric

$$g = \Sigma_i(\xi^i)^2 + \Sigma_u(\eta^u)^2 + \Sigma_\alpha(\zeta^\alpha)^2 + \Sigma_\lambda(\tau^\lambda)^2 \quad . \tag{2}$$

The chosen moving frame is orthonormal with respect to g. Let us define two left-invariant almost Hermitian structures (G,g,J_1),(G,g,J_2), that therefore will pass to the nilmanifolds $\Gamma\backslash G$.

The almost Hermitian structure (G,g,J_1). Suppose that X and Z are of the same dimension, k, and that Y and T have the same dimension, l. We use the same set of indices $\{i\}$ for X,Z, once their basis chosen.Analogously, let us use the same set $\{u\}$ of indices for Y and T. The almost complex structure J_1 is given by,(writing J for short),

(3) $J(X_i) = Z_i$, $J(Z_i) = -X_i$, $J(Y_u) = T_u$, $J(T_u) = -Y_u$.

The Nijenhuis tensor N of J_1,($N(U,V) = [JU,JV] - [U,V] - J[U,JV] - J[JU,V]$, U,V differentiable vector fields), is not zero. For example, since for some i,u,j, $B^j_{iu} \neq 0$,

(4) $N(X_i,Y_u) = -[X_i,Y_u] = -\Sigma_h B^h_{iu} Z_h \neq 0$,

in fact, N = 0 is equivalent to B = 0, yielding the trivial Euclidean case. The metric tensor g,(2), is compatible with J_1, (3). Thus (G,g,J_1) is a left-invariant almost Hermitian structure. Its Kähler form F, ($F(U,V) = g(JU,V)$, for every U,V differentiable vector fields), can be expressed with respect to the coframe $\{\xi^i,\eta^u,\zeta^j,\tau^v\}$ as

(5) $F = \Sigma_i \xi^i \wedge \zeta^i + \Sigma_u \eta^u \wedge \tau^u$

It is straightforward to check that the coderivative of F, δF, is zero. In other words: <u>(G,g,J_1) is always a semi Kähler structure.</u>

On the other hand, the exterior derivative of F is

(6) $dF = \Sigma_{i,j,u} B^i_{ju} \xi^i \wedge \xi^j \wedge \eta^u$,

and therefore: <u>when $B^i_{ju} = B^j_{iu}$ for all indices i,j,u, the structure (G,F) is symplectic, (dF = 0), or equivalently,(G,g,J_1) becomes almost Kähler.</u>

<u>Example 1.</u> Let $G = H_{\mathbb{R}}(p,q,q) \times M_{qq}(\mathbb{R})$ be endowed with the metric (2) and the almost complex structure J_1,(3). The semi Kähler structure (G,g,J_1) is <u>strict</u> in the sense of the classification of Gray-Hervella, [7], if and only if q > 1. The case q = 1 has the index symmetry which makes (G,F) symplectic and is treated in [3], [4].

<u>Example 2.</u> Let $X = Y = Z = T = \{$real symmetric n×n matrices$\}$, B: $X \times Y \longrightarrow Z$ the symmetric matrix product

(7) $B(x,y) = 2(xy + yx)$.

with respect to the basis $\{E_{(i,j)}; i \leq j\}$, (where $E_{(i,j)} = (E_{ij} + E_{ji})/2$, $i,j \in (n)$), we have the symmetry condition

$$B^{(i,j)}_{(k,l)(r,s)} = B^{(k,l)}_{(i,j)(r,s)}$$

sufficient for $G = G_B \times T$ to be symplectic with respect to the metric (2) and the almost complex structure J_1, (3).

<u>The almost Hermitian structure (G,g,J_2).</u> Suppose that X and Y have the same dimension k and that Z and T have the same dimension m. Let us use the same set of indices for each couple of spaces, say $\{i\}$ and $\{\alpha\}$. Define an almost complex structure J_2 on $G = G_B \times T$ as follows

(8) $J_2(X_i) = Y_i$, $J_2(Y_i) = -X_i$, $J_2(Z_\alpha) = T_\alpha$, $J_2(T_\alpha) = -Z_\alpha$.

Fix the metric (2). The Kähler form F of (G,g,J_2) is

(10) $F = \Sigma_i \xi^i \wedge \eta^i + \Sigma_\alpha \zeta^\alpha \wedge \tau^\alpha$,

Since B is not zero it follows from (10) and §3.(5) that dF is not null, in other words, <u>(G,F) is never symplectic.</u> Bearing in mind (8) and §3.(7) one finds that <u>the Nijenhuis tensor of J_2 vanishes if and only if in the given basis the coefficients of B satisfy for every index α the symmetry condition</u>

(11) $$B^{\alpha}_{ij} = B^{\alpha}_{ji} .$$

On the other hand, the coderivative δF of the Kähler form F is given by

(12) $$\delta F(X_i) = \delta F(Y_j) = \delta F(T_\alpha) = 0, \quad \delta F(Z_\alpha) = \Sigma_i B^{\alpha}_{ii} ,$$

hence, <u>(G,g,J_2) is semi Kähler if and only if for every index α</u>

(13) $$\Sigma_i B^{\alpha}_{ii} = 0 .$$

<u>Example 2.</u> Let $X = M_{pq}(\mathbb{R})$, $Y = M_{qp}(\mathbb{R})$, endowed with the basis $\{E_{iu}\}$, $\{E^{t}_{iu}\}$ respectively, where " t " stands for transposing matrices. Let $Z = T = M_{pp}(\mathbb{R})$, and consider

(14) $$B_1(x,y) = xy + (xy)^t , \quad B_2(x,y) = xy - (xy)^t$$

The group $G_a = G_{B_a} \times T$, $(a = 1,2)$, with the metric (2) and the almost complex structure J_2 verifies that (G_1,g,J_2) is integrable, (i.e.: Hermitian) and (G_2,g,J_2) is semi Kähler, (note that g depends on B too).

<u>6. A SHORT ACCOUNT OF THE RIEMANNIAN GEOMETRY OF HR-SPACES.</u> Let $G_B = X\times Y\times Z$ be a connected and simply connected HR-Lie group associated to a bilinear map $B: X\times Y \longrightarrow Z$. Fix the Riemannian metric

(1) $$g = \Sigma_i (\xi^i)^2 + \Sigma_u (\eta^u)^2 + \Sigma_\alpha (\zeta^\alpha)^2$$

where the left-invariant 1-forms ξ^i, η^u, ζ^α are defined in §3.(4). Then:

<u>Proposition 6.</u> <u>The geodesic line of (G_B,g) with initial conditions (x_o,y_o,z_o) and $(\dot{x}_o,\dot{y}_o,\dot{z}_o)$ is given by the real analytic vector valued maps</u>

(2) $$x(t) = \left(\frac{1}{F}(\mathrm{id}-\cos(t\sqrt{F}))\right)M + \left(\frac{1}{\sqrt{F}}\sin(t\sqrt{F})\right)\dot{x}_o + \left(\cos(t\sqrt{F})\right)x_o$$

(3) $$y(t) = \left(\frac{1}{H}(\mathrm{id}-\cos(t\sqrt{H}))\right)N + \left(\frac{1}{\sqrt{H}}\sin(t\sqrt{H})\right)\dot{y}_o + \left(\cos(t\sqrt{H})\right)y_o$$

(4) $$z(t) = z_o + t\Sigma_\alpha C^\alpha g_\alpha + \Sigma_\alpha\left(\Sigma_{i,j} Q^{\alpha}_{ij}\int_0^t x^i(s)x^j(s)\,ds\right)g_\alpha +$$
$$+\Sigma_{i,u,\alpha} B^{\alpha}_{iu} D^u\left(\left[\frac{1}{\sqrt{F}}\sin(t\sqrt{F})\right]x_o + \left[\frac{1}{F}\left(t\,\mathrm{id} - \frac{1}{\sqrt{F}}\sin(t\sqrt{F})\right)\right]M + \left[\frac{1}{F}(\mathrm{id}-\cos(t\sqrt{F}))\right]\dot{x}_o\right)^i g_\alpha$$

<u>where</u> $\dot{x}_o = \Sigma_i \dot{x}^i_o X_i$, $\dot{y}_o = \Sigma_u \dot{y}^u_o Y_u$, $\dot{z}_o = \Sigma_\alpha \dot{z}^\alpha_o Z_\alpha$, $\dot{x}_o = \Sigma_i \dot{x}^i_o e_i$, $\dot{y}_o = \Sigma_u \dot{y}^u_o f_u$, $\{e_i,f_u,g_\alpha\}$ basis of $X\times Y\times Z$, $C^\alpha = \dot{z}^\alpha_o - \Sigma_{i,u} B^{\alpha}_{iu} x^i_o \dot{y}^u_o$, $D^u = \dot{y}_o - \Sigma_{i,\alpha} B^{\alpha}_{iu} x_o C^\alpha$, $Q^{\alpha}_{iu} = \Sigma_{u,\beta} B^{\alpha}_{iu} B^{\beta}_{ju} C^\beta$, $F_{ij} = \Sigma_{\alpha,\beta,u} B^{\alpha}_{iu} B^{\beta}_{ju} C^\alpha C^\beta$, $H_{uv} = \Sigma_{\alpha,\beta,i} B^{\alpha}_{iu} B^{\beta}_{iv} C^\alpha C^\beta$, $E^i = \dot{x}^i_o + \Sigma_{u,\alpha} B^{\alpha}_{iu} y^u_o C^\alpha$, $N^u = \Sigma_{i,\alpha} B^{\alpha}_{iu} E^i C^\alpha$,

$M^i = -\Sigma_{u,\alpha} B^{\alpha}_{iu} D^u C^\alpha$, <u>and where the function</u> $\frac{1}{F}(\mathrm{id} - \cos(t\sqrt{F}))$ <u>is the real analytic matrix-valued map obtained by substitution of the powers of the complex variable z in the power series</u>

(5) $$\frac{1}{z}(z^0-\cos(t\sqrt{z})) = \Sigma_{k=1}^{\infty}(-1)^{k+1}t^{2k}z^{k-1}/(2k)! \quad , \quad t\in\mathbb{R}$$

by the powers of the matrix $F = (F_{ij})$, and analogously for the other matrix-valued functions enclosed between big parentheses $\left(\right)$ in the variables F or $\underline{H} = (H_{uv})$.

Proof. It follows from integration of the Euler-Lagrange equations associated to the Lagrangian $L(r,\dot{r}) = (1/2)g_{ab}\dot{r}^a\dot{r}^b$, where g_{ab} are the coefficients of the Riemannian metric (1) with respect to the coordinate frame. |||

The third term in (4) can be determined in a term by term integration when G_B is the matrix group $H_{\mathbb{R}}(p,q,r)$. Moreover, for p or q equal to 1 that term can be obtained by means of "elementary functions" of $\cos(t\sqrt{A})$ and $\frac{1}{\sqrt{A}}\sin(t\sqrt{A})$ and the constant functions given above. The component z(t) of the geodesics of Kaplan's groups of type H has been fully determined by him, [10].

Proposition 7. Let Γ be an arbitrary discrete subgroup of a HR-Lie group G_B and $\bar{g}$ the Riemannian metric induced on $\Gamma\setminus G_B$ by the metric on the universal covering group of G_B. Then, writing $M = \Gamma\setminus G_B$, we have:
1) $(M,\bar{g})$ is not projectively flat. 2) $(M,\bar{g})$ is not conformally flat. 3) Every global conformal transformation of $(M,\bar{g})$ is necessarily a global isometry.

Proof. 1) is proved by seeing that $(M,\bar{g})$ is not a manifold of constant curvature. Point 3) is a consequence of point 2) and a theorem of Kulkarni, $(\dim(M)>3)$, and Yau, $(\dim(M)=3)$, [11], [19]. Point 2) consists in proving that the conformal Weyl curvature of $(M,\bar{g})$ is zero if and only if $\dim(M) = 3$. In this case the Schouten tensor is not null. |||

We finally point out that $(M,\bar{g})$ is never Einstein. Its scalar curvature is strictly negative. Notice that when Γ is one of the subgroups Γ_K given in Proposition 2 the volume of the compact quotient $\Gamma_K\setminus G_B$ is just the product of the positive integers defining Γ_K.

BIBLIOGRAPHY

[1] D. E. Blair, The theory of quasi-Sasakian structures. J. Diff Geom.1(1967) 331-345.

[2] N. Brotherton, Some parallelizable four manifolds not admitting a complex structure. Bull. London Math. Soc., 10(1978) 303-304.

[3] L. A. Cordero, M. Fernández, M. de León, Examples of compact non-Kähler almost Kähler manifols. Proc. Amer. Math. Soc. 95,2(1985) 280-286.

[4] L. A. Cordero, M. Fernández, A. Gray, Symplectic manifolds with no Kähler structure. To appear in Topology.

[5] P. Deligne, P. Griffiths, J. Morgan, D. Sullivan, Real homotopy theory of Kähler manifolds. Invent. Math. 29 (1975) 245-274.

[6] M. Goze, Y. Haraguchi, Sur les r-systèmes de contact, C.R. Acad. Sc. Paris, 294 Ser. I(1982) 95-97.

[7] A. Gray, L. M. Hervella, The sixteen classes of almost Hermitian manifolds and their linear invariants. Ann. Mat. Pura ed Appl.(IV) 123,(1980) 35-58.

[8] T. Hangan, Sur les transformations géométriques de l'espace de Heisenberg. Rend. Sem. Fac. Sci. Univ. Cagliari (1985).

[9] Y. Haraguchi, Sur une généralisation des structures de contact, Thèse, Univ. de Haute Alsace., Mulhouse (1981).

[10] A. Kaplan, Lie groups of Heisenberg type. Conf. Diff. Geom. Homog. Spaces. Sem Mat. Univ. Polit. Torino, Fasc. Esp. 1983 (1984) 117-130.

[11] R. Kulkarni, Curvature structures and conformal transformations. J. Diff. Geom. 4 (1970) 425-451.

[12] R. Lutz, Structures de contact en codimension quelconque. Lect. Not. Math. 392 (1974)23-29.

[13] A. I. Mal'cev, On a class of homogeneous spaces. Translations A.M.S. 39(1951) 266-307

[14] J. P. May, Matric Massey products. J. of Algebra 12 (1969) 533-568.

[15] K. Nomizu, On the cohomology of compact homogeneous spaces of nilpotent Lie groups. Ann. of Math. 59(1954) 531-538.

[16] H. Reiter, Über den Satz von Wiener und lokalkompakte Gruppen. Comment. Math. Helv. 49 (1974) 333-364.

[17] D. Tanré, Homotopie rationnelle: modèles de Chen, Quillen, Sullivan. Lect. Not. Math. 1025 (1983).

[18] W. P. Thurston, Some simple examples of symplectic manifolds. Proc. Amer. Math. Soc. 55 (1976) 467-468.

[19] S. T. Yau, Remarks on conformal transformations. J. Diff. Geom. 8 (1973) 369-381.

Totally real submanifolds of a complex projective space

Francisco Urbano
Departamento de Geometría y Topología
Universidad de Granada, 18071-Granada, Spain

1. Introduction. Let $CP^n(c)$ be the complex projective space with the Fubini-Study metric of constant holomorphic sectional curvature $c>0$. The totally real submanifolds of $CP^n(c)$ have been studied by several authors and from different view points. So, H. Naitoh and M. Takeuchi ([3],[4]) obtained the complete classification of the parallel totally real submanifolds of $CP^n(c)$, giving important examples of symmetric spaces embeded in $CP^n(c)$ as totally real parallel submanifolds. Among them, we have: $SU(m)/SO(m), m\geqslant 3$; $SU(2m)/Sp(m), m\geqslant 3$; $SU(m)$, $m\geqslant 3$; and E_6/F_4. Also, H. Naitoh ([3]), gave an example of a flat surface embedded in $CP^2(c)$ as a totally real parallel minimal submanifold which will be denote by M_0^2.

Some pinching problems for the curvature of totally real submanifolds have been studied by B.Y. Chen and K. Ogiue ([1]), H. Naitoh and M. Takeuchi ([4]), and S.T. Yau ([8]). An important result for the scalar curvature is the following:

Theorem 1. ([1],[4],[8]). *Let M be an n-dimensional compact totally real minimal submanifold isometrically immersed in $CP^n(c)$. If the scalar curvature ρ of M satisfies:*

$$\rho \geqslant \frac{n^2(n-2)}{2(2n-1)}c$$

then either $\rho = n(n-1)c/4$ and M is totally geodesic or $n=2$, $\rho = 0$ and M is a finite Riemannian covering of M_0^2.

When $n=2$, Theorem 1 was proved by S.T. Yau. The strict inequality was studied by B.Y. Chen and K. Ogiue.

In this paper I expose some pinching theorems for the sectional curvature and the Ricci tensor of a totally real submanifold of $CP^n(c)$

obtained by S. Montiel, A. Ros and the author ([2],[6],[7]) using methods different those of [1],[4], and [8]. Concretely, we solve the pinching problem for the sectional curvature, obtaining Theorem 1 (when $n=2$) as a particular case of our result. Also, we characterize some of the above examples with a pinching for the Ricci tensor.

2. Statement of results. Theorem 1 was proved using Simons' formula. Here, we give a modified version of Simons' formula which is obtained only for totally real submanifolds.

In fact, let M be an n-dimensional compact totally real submanifold of $CP^n(c)$ with parallel mean curvature vector. If σ denotes the second fundamental form of the submanifold, then we can define a 3-covariant symmetric tensor T on M given by

$$T(X,Y,Z) = \langle\sigma(X,Y),JZ\rangle,$$

where J is the complex structure of $CP^n(c)$.

Now, we use a result proved in [5]. If $\mathcal{T}$ is a k-covariant tensor on a Riemannian manifold N, and UN is the unit tangent bundle over N, then

$$(1) \qquad \int_{UN} \{ \sum_{i=1}^{n} (\nabla^2\mathcal{T})(e_i,e_i,v,\ldots,v)\} dv = 0 ,$$

where $e_1,\ldots,e_n$ is an orthonormal basis of T_pN, $p\in N$, $\nabla^2\sigma$ is the second covariant derivative of σ, and dv is the canonical measure on UN.

Then using (1) for the 6-covariant tensor $\mathcal{T}$ on M defined by $\mathcal{T} = T\otimes T$ we obtain the following integral formula (see [7] for details):

$$(2) \qquad \int_{UM} \{|(\nabla\sigma)(v,v,v)|^2 + 3R(A_{Jv}v,v,v,A_{Jv}v)\} dv = 0,$$

where A_{Jv} is the shape operator associated to Jv and R is the curvature operator of M.

Now from (2) and using the classification of the parallel totally real submanifolds of $CP^n(c)$ obtained in [3] and [4], we have:

Theorem 2. Let M be an n-dimensional compact totally real submanifold isometrically immersed in $CP^n(c)$ with parallel mean curvature vector. Then M has nonnegative sectional curvature if and only if M has parallel second fundamental form.

Remark. The integral formula (2) has been obtained in [7] for compact totally real submanifolds with parallel mean curvature vector immersed in any complex space form.

From (2), we can obtain another different proof of the result obtained by the author in [6].

Corollary 1. ([6]). Let M be a compact totally real minimal submanifold isometrically immersed in $CP^n(c)$. If M has positive sectional curvature then M is totally geodesic.

Proof. Since the sectional curvature of M is positive, from (2) we have that $A_{Jv}v$ and v are proportional for all $v \in UM$. Now, as in a totally real submanifold $A_{Jv}v = -J\sigma(v,v)$, we obtain that $\sigma(v,v) = \lambda(v)Jv$ for all $v \in UM$, where $\lambda(v) \in R$. If $e_1, \ldots, e_n$ is an orthonormal basis at any point $p \in M$, we have from the minimality of M:

$$0 = \sum_{i=1}^{n} \sigma(e_i, e_i) = \sum_{i=1}^{n} \lambda(e_i) Je_i,$$

and hence $\lambda(e_i) = 0$ for all $i \in \{e_1, \ldots, e_n\}$. This implies that $\sigma = 0$.

Now, we give a pinching theorem for the Ricci tensor of a totally real submanifold of $CP^n(c)$.

Theorem 3. Let M be an n-dimensional compact totally real minimal submanifold isometrically immersed in $CP^n(c)$. If S is the Ricci tensor of M, then

$$S \geqslant \frac{3(n-2)}{16} c$$

if and only if one of the following conditions is satisfied:

a) $S = (n-1)c/4$ and M is totally geodesic,

b) $S = 0$, $n = 2$ and M is a finite Riemannian covering of M_0^2,

c) $S = 3(n-2)c/16$, $n>2$ and M is an embedded submanifold congruent to the standart embedding of: $SU(3)/SO(3)$, $n = 5$; $SU(6)/Sp(3)$, $n = 14$; $SU(3)$, $n = 8$; or E_6/F_4, $n = 26$.

Sketch of the proof. Using (1) for the 4-covariant tensor T on M defined by

$$T(v_1,v_2,v_3,v_4) = \langle\sigma(v_1,v_2), \sigma(v_3,v_4)\rangle,$$

we obtain the following integral formula (see [2] for details):

$$(3) \qquad 0 = \frac{n+4}{3} \int_{UM} |(\nabla\sigma)(v,v,v)|^2 + \frac{(n+1)c}{2n(n+2)} \int_{UM} |\sigma|^2 dv + $$

$$+ (n+4) \int_{UM} |A_{\sigma(v,v)}v|^2 dv - 4 \int_{UM} \langle Lv, A_{\sigma(v,v)}v\rangle dv -$$

$$- 2 \int_{UM} \langle LA_{Jv}v, A_{Jv}v\rangle dv,$$

where $L : T_pM \longrightarrow T_pM$, $p\varepsilon M$, is a self-adjoint linear map given by

$$Lv = \sum_{i=1}^{n} A_{\sigma(v,e_i)}e_i$$

being $e_1,\ldots,e_n$ an orthonormal basis of T_pM.

Now, if $p\varepsilon M$, let $f : UM_p \longrightarrow T_pM$ be the vectorial function defined by

$$f(v) = A_{\sigma(v,v)}v$$

where UM_p is the fiber of UM over p. As f is odd, we have $\int_{UM_p} f(v)\, dv_p = 0$, where dv_p is the canonical measure on UM_p. Using the minimum principle for the first non-null eigenvalue of the Laplacian Δ of UM_p, we have

$$(4) \qquad -\int_{UM_p} \langle\Delta f, f\rangle dv_p \geqslant (n-1) \int_{UM_p} |f|^2 dv_p.$$

On the other hand, as $S(v,w) = \frac{(n-1)}{4} c \langle v,w\rangle - \langle Lv,w\rangle$, we have

$$(5) \qquad 0 \leqslant \langle Lv,v\rangle \leqslant \frac{n+2}{16} c$$

So, using (4) and (5), (3) gives that $\nabla\sigma = 0$. Now the theorem follows from the classification of the totally real parallel submanifolds of $CP^n(c)$ obtained in [3] and [4].

REFERENCES

[1] B.Y.Chen, K.Ogiue, "On totally real submanifolds" Trans.A.M.S. 193(1974),257-266.

[2] S.Montiel, A.Ros, F.Urbano, "Curvature pinching and eigenvalue rigidity for minimal submanifolds" (Preprint).

[3] H.Naitoh, "Totally real parallel submanifolds of $P^n(C)$" Tokyo J. Math. 4(1981),279-306.

[4] H.Naitoh, M.Takeuchi, "Totally real submanifolds and symmetric bounded domains" Osaka J.Math. 19(1982),717-731.

[5] A.Ros, "A characterization of seven compact Kaehler submanifolds by holomorphic pinching" Ann of Math. 121(1985),377-382.

[6] F.Urbano, "Totally real minimal submanifolds in a complex projective space" Proc.A.M.S. 93(1985),332-334.

[7] F.Urbano, "Nonnegatively curved totally real submanifolds" (Preprint).

[8] S.T.Yau, "Submanifolds with constant mean curvature I" Amer.J. Math. 96(1974),346-366.

Vol. 1062: J. Jost, Harmonic Maps Between Surfaces. X, 133 pages. 1984.

Vol. 1063: Orienting Polymers. Proceedings, 1983. Edited by J.L. Ericksen. VII, 166 pages. 1984.

Vol. 1064: Probability Measures on Groups VII. Proceedings, 1983. Edited by H. Heyer. X, 588 pages. 1984.

Vol. 1065: A. Cuyt, Padé Approximants for Operators: Theory and Applications. IX, 138 pages. 1984.

Vol. 1066: Numerical Analysis. Proceedings, 1983. Edited by D.F. Griffiths. XI, 275 pages. 1984.

Vol. 1067: Yasuo Okuyama, Absolute Summability of Fourier Series and Orthogonal Series. VI, 118 pages. 1984.

Vol. 1068: Number Theory, Noordwijkerhout 1983. Proceedings. Edited by H. Jager. V, 296 pages. 1984.

Vol. 1069: M. Kreck, Bordism of Diffeomorphisms and Related Topics. III, 144 pages. 1984.

Vol. 1070: Interpolation Spaces and Allied Topics in Analysis. Proceedings, 1983. Edited by M. Cwikel and J. Peetre. III, 239 pages. 1984.

Vol. 1071: Padé Approximation and its Applications, Bad Honnef 1983. Prodeedings. Edited by H. Werner and H.J. Bünger. VI, 264 pages. 1984.

Vol. 1072: F. Rothe, Global Solutions of Reaction-Diffusion Systems. V, 216 pages. 1984.

Vol. 1073: Graph Theory, Singapore 1983. Proceedings. Edited by K.M. Koh and H.P. Yap. XIII, 335 pages. 1984.

Vol. 1074: E.W. Stredulinsky, Weighted Inequalities and Degenerate Elliptic Partial Differential Equations. III, 143 pages. 1984.

Vol. 1075: H. Majima, Asymptotic Analysis for Integrable Connections with Irregular Singular Points. IX, 159 pages. 1984.

Vol. 1076: Infinite-Dimensional Systems. Proceedings, 1983. Edited by F. Kappel and W. Schappacher. VII, 278 pages. 1984.

Vol. 1077: Lie Group Representations III. Proceedings, 1982–1983. Edited by R. Herb, R. Johnson, R. Lipsman, J. Rosenberg. XI, 454 pages. 1984.

Vol. 1078: A.J.E.M. Janssen, P. van der Steen, Integration Theory. V, 224 pages. 1984.

Vol. 1079: W. Ruppert. Compact Semitopological Semigroups: An Intrinsic Theory. V, 260 pages. 1984

Vol. 1080: Probability Theory on Vector Spaces III. Proceedings, 1983. Edited by D. Szynal and A. Weron. V, 373 pages. 1984.

Vol. 1081: D. Benson, Modular Representation Theory: New Trends and Methods. XI, 231 pages. 1984.

Vol. 1082: C.-G. Schmidt, Arithmetik Abelscher Varietäten mit komplexer Multiplikation. X, 96 Seiten. 1984.

Vol. 1083: D. Bump, Automorphic Forms on GL (3,IR). XI, 184 pages. 1984.

Vol. 1084: D. Kletzing, Structure and Representations of Q-Groups. VI, 290 pages. 1984.

Vol. 1085: G.K. Immink, Asymptotics of Analytic Difference Equations. V, 134 pages. 1984.

Vol. 1086: Sensitivity of Functionals with Applications to Engineering Sciences. Proceedings, 1983. Edited by V. Komkov. V, 130 pages. 1984

Vol. 1087: W. Narkiewicz, Uniform Distribution of Sequences of Integers in Residue Classes. VIII, 125 pages. 1984.

Vol. 1088: A.V. Kakosyan, L.B. Klebanov, J.A. Melamed, Characterization of Distributions by the Method of Intensively Monotone Operators. X, 175 pages. 1984.

Vol. 1089: Measure Theory, Oberwolfach 1983. Proceedings. Edited by D. Kölzow and D. Maharam-Stone. XIII, 327 pages. 1984.

Vol. 1090: Differential Geometry of Submanifolds. Proceedings, 1984. Edited by K. Kenmotsu. VI, 132 pages. 1984.

Vol. 1091: Multifunctions and Integrands. Proceedings, 1983. Edited by G. Salinetti. V, 234 pages. 1984.

Vol. 1092: Complete Intersections. Seminar, 1983. Edited by S. Greco and R. Strano. VII, 299 pages. 1984.

Vol. 1093: A. Prestel, Lectures on Formally Real Fields. XI, 125 pages. 1984.

Vol. 1094: Analyse Complexe. Proceedings, 1983. Edité par E. Amar, R. Gay et Nguyen Thanh Van. IX, 184 pages. 1984.

Vol. 1095: Stochastic Analysis and Applications. Proceedings, 1983. Edited by A. Truman and D. Williams. V, 199 pages. 1984.

Vol. 1096: Théorie du Potentiel. Proceedings, 1983. Edité par G. Mokobodzki et D. Pinchon. IX, 601 pages. 1984.

Vol. 1097: R.M. Dudley, H. Kunita, F. Ledrappier, École d'Éte de Probabilités de Saint-Flour XII – 1982. Edité par P.L. Hennequin. X, 396 pages. 1984.

Vol. 1098: Groups – Korea 1983. Proceedings. Edited by A.C. Kim and B.H. Neumann. VII, 183 pages. 1984.

Vol. 1099: C.M. Ringel, Tame Algebras and Integral Quadratic Forms. XIII, 376 pages. 1984.

Vol. 1100: V. Ivrii, Precise Spectral Asymptotics for Elliptic Operators Acting in Fiberings over Manifolds with Boundary. V, 237 pages. 1984.

Vol. 1101: V. Cossart, J. Giraud, U. Orbanz, Resolution of Surface Singularities. Seminar. VII, 132 pages. 1984.

Vol. 1102: A. Verona, Stratified Mappings – Structure and Triangulability. IX, 160 pages. 1984.

Vol. 1103: Models and Sets. Proceedings, Logic Colloquium, 1983, Part I. Edited by G.H. Müller and M.M. Richter. VIII, 484 pages. 1984.

Vol. 1104: Computation and Proof Theory. Proceedings, Logic Colloquium, 1983, Part II. Edited by M.M. Richter, E. Börger, W. Oberschelp, B. Schinzel and W. Thomas. VIII, 475 pages. 1984.

Vol. 1105: Rational Approximation and Interpolation. Proceedings, 1983. Edited by P.R. Graves-Morris, E.B. Saff and R.S. Varga. XII, 528 pages. 1984.

Vol. 1106: C.T. Chong, Techniques of Admissible Recursion Theory. IX, 214 pages. 1984.

Vol. 1107: Nonlinear Analysis and Optimization. Proceedings, 1982. Edited by C. Vinti. V, 224 pages. 1984.

Vol. 1108: Global Analysis – Studies and Applications I. Edited by Yu.G. Borisovich and Yu.E. Gliklikh. V, 301 pages. 1984.

Vol. 1109: Stochastic Aspects of Classical and Quantum Systems. Proceedings, 1983. Edited by S. Albeverio, P. Combe and M. Sirugue-Collin. IX, 227 pages. 1985.

Vol. 1110: R. Jajte, Strong Limit Theorems in Non-Commutative Probability. VI, 152 pages. 1985.

Vol. 1111: Arbeitstagung Bonn 1984. Proceedings. Edited by F. Hirzebruch, J. Schwermer and S. Suter. V, 481 pages. 1985.

Vol. 1112: Products of Conjugacy Classes in Groups. Edited by Z. Arad and M. Herzog. V, 244 pages. 1985.

Vol. 1113: P. Antosik, C. Swartz, Matrix Methods in Analysis. IV, 114 pages. 1985.

Vol. 1114: Zahlentheoretische Analysis. Seminar. Herausgegeben von E. Hlawka. V, 157 Seiten. 1985.

Vol. 1115: J. Moulin Ollagnier, Ergodic Theory and Statistical Mechanics. VI, 147 pages. 1985.

Vol. 1116: S. Stolz, Hochzusammenhängende Mannigfaltigkeiten und ihre Ränder. XXIII, 134 Seiten. 1985.

Vol. 1117: D.J. Aldous, J.A. Ibragimov, J. Jacod, Ecole d'Été de Probabilités de Saint-Flour XIII – 1983. Édité par P.L. Hennequin. IX, 409 pages. 1985.

Vol. 1118: Grossissements de filtrations: exemples et applications. Seminaire, 1982/83. Edité par Th. Jeulin et M. Yor. V, 315 pages. 1985.

Vol. 1119: Recent Mathematical Methods in Dynamic Programming. Proceedings, 1984. Edited by I. Capuzzo Dolcetta, W.H. Fleming and T. Zolezzi. VI, 202 pages. 1985.

Vol. 1120: K. Jarosz, Perturbations of Banach Algebras. V, 118 pages. 1985.

Vol. 1121: Singularities and Constructive Methods for Their Treatment. Proceedings, 1983. Edited by P. Grisvard, W. Wendland and J.R. Whiteman. IX, 346 pages. 1985.

Vol. 1122: Number Theory. Proceedings, 1984. Edited by K. Alladi. VII, 217 pages. 1985.

Vol. 1123: Séminaire de Probabilités XIX 1983/84. Proceedings. Edité par J. Azéma et M. Yor. IV, 504 pages. 1985.

Vol. 1124: Algebraic Geometry, Sitges (Barcelona) 1983. Proceedings. Edited by E. Casas-Alvero, G.E. Welters and S. Xambó-Descamps. XI, 416 pages. 1985.

Vol. 1125: Dynamical Systems and Bifurcations. Proceedings, 1984. Edited by B.L.J. Braaksma, H.W. Broer and F. Takens. V, 129 pages. 1985.

Vol. 1126: Algebraic and Geometric Topology. Proceedings, 1983. Edited by A. Ranicki, N. Levitt and F. Quinn. V, 523 pages. 1985.

Vol. 1127: Numerical Methods in Fluid Dynamics. Seminar. Edited by F. Brezzi, VII, 333 pages. 1985.

Vol. 1128: J. Elschner, Singular Ordinary Differential Operators and Pseudodifferential Equations. 200 pages. 1985.

Vol. 1129: Numerical Analysis, Lancaster 1984. Proceedings. Edited by P.R. Turner. XIV, 179 pages. 1985.

Vol. 1130: Methods in Mathematical Logic. Proceedings, 1983. Edited by C.A. Di Prisco. VII, 407 pages. 1985.

Vol. 1131: K. Sundaresan, S. Swaminathan, Geometry and Nonlinear Analysis in Banach Spaces. III, 116 pages. 1985.

Vol. 1132: Operator Algebras and their Connections with Topology and Ergodic Theory. Proceedings, 1983. Edited by H. Araki, C.C. Moore, Ş. Strătilă and C. Voiculescu. VI, 594 pages. 1985.

Vol. 1133: K.C. Kiwiel, Methods of Descent for Nondifferentiable Optimization. VI, 362 pages. 1985.

Vol. 1134: G.P. Galdi, S. Rionero, Weighted Energy Methods in Fluid Dynamics and Elasticity. VII, 126 pages. 1985.

Vol. 1135: Number Theory, New York 1983–84. Seminar. Edited by D.V. Chudnovsky, G.V. Chudnovsky, H. Cohn and M.B. Nathanson. V, 283 pages. 1985.

Vol. 1136: Quantum Probability and Applications II. Proceedings, 1984. Edited by L. Accardi and W. von Waldenfels. VI, 534 pages. 1985.

Vol. 1137: Xiao G., Surfaces fibrées en courbes de genre deux. IX, 103 pages. 1985.

Vol. 1138: A. Ocneanu, Actions of Discrete Amenable Groups on von Neumann Algebras. V, 115 pages. 1985.

Vol. 1139: Differential Geometric Methods in Mathematical Physics. Proceedings, 1983. Edited by H. D. Doebner and J. D. Hennig. VI, 337 pages. 1985.

Vol. 1140: S. Donkin, Rational Representations of Algebraic Groups. VII, 254 pages. 1985.

Vol. 1141: Recursion Theory Week. Proceedings, 1984. Edited by H.-D. Ebbinghaus, G.H. Müller and G.E. Sacks. IX, 418 pages. 1985.

Vol. 1142: Orders and their Applications. Proceedings, 1984. Edited by I. Reiner and K. W. Roggenkamp. X, 306 pages. 1985.

Vol. 1143: A. Krieg, Modular Forms on Half-Spaces of Quaternions. XIII, 203 pages. 1985.

Vol. 1144: Knot Theory and Manifolds. Proceedings, 1983. Edited by D. Rolfsen. V, 163 pages. 1985.

Vol. 1145: G. Winkler, Choquet Order and Simplices. VI, 143 pages. 1985.

Vol. 1146: Séminaire d'Algèbre Paul Dubreil et Marie-Paule Malliavin. Proceedings, 1983–1984. Edité par M.-P. Malliavin. IV, 420 pages. 1985.

Vol. 1147: M. Wschebor, Surfaces Aléatoires. VII, 111 pages. 1985.

Vol. 1148: Mark A. Kon, Probability Distributions in Quantum Statistical Mechanics. V, 121 pages. 1985.

Vol. 1149: Universal Algebra and Lattice Theory. Proceedings, 1984. Edited by S. D. Comer. VI, 282 pages. 1985.

Vol. 1150: B. Kawohl, Rearrangements and Convexity of Level Sets in PDE. V, 136 pages. 1985.

Vol 1151: Ordinary and Partial Differential Equations. Proceedings, 1984. Edited by B.D. Sleeman and R.J. Jarvis. XIV, 357 pages. 1985.

Vol. 1152: H. Widom, Asymptotic Expansions for Pseudodifferential Operators on Bounded Domains. V, 150 pages. 1985.

Vol. 1153: Probability in Banach Spaces V. Proceedings, 1984. Edited by A. Beck, R. Dudley, M. Hahn, J. Kuelbs and M. Marcus. VI, 457 pages. 1985.

Vol. 1154: D.S. Naidu, A.K. Rao, Singular Pertubation Analysis of Discrete Control Systems. IX, 195 pages. 1985.

Vol. 1155: Stability Problems for Stochastic Models. Proceedings, 1984. Edited by V.V. Kalashnikov and V.M. Zolotarev. VI, 447 pages. 1985.

Vol. 1156: Global Differential Geometry and Global Analysis 1984. Proceedings, 1984. Edited by D. Ferus, R.B. Gardner, S. Helgason and U. Simon. V, 339 pages. 1985.

Vol. 1157: H. Levine, Classifying Immersions into $\mathbb{R}^4$ over Stable Maps of 3-Manifolds into $\mathbb{R}^2$. V, 163 pages. 1985.

Vol. 1158: Stochastic Processes – Mathematics and Physics. Proceedings, 1984. Edited by S. Albeverio, Ph. Blanchard and L. Streit. VI, 230 pages. 1986.

Vol. 1159: Schrödinger Operators, Como 1984. Seminar. Edited by S. Graffi. VIII, 272 pages. 1986.

Vol. 1160: J.-C. van der Meer, The Hamiltonian Hopf Bifurcation. VI, 115 pages. 1985.

Vol. 1161: Harmonic Mappings and Minimal Immersions, Montecatini 1984. Seminar. Edited by E. Giusti. VII, 285 pages. 1985.

Vol. 1162: S.J.L. van Eijndhoven, J. de Graaf, Trajectory Spaces, Generalized Functions and Unbounded Operators. IV, 272 pages. 1985.

Vol. 1163: Iteration Theory and its Functional Equations. Proceedings, 1984. Edited by R. Liedl, L. Reich and Gy. Targonski. VIII, 231 pages. 1985.

Vol. 1164: M. Meschiari, J.H. Rawnsley, S. Salamon, Geometry Seminar "Luigi Bianchi" II – 1984. Edited by E. Vesentini. VI, 224 pages. 1985.

Vol. 1165: Seminar on Deformations. Proceedings, 1982/84. Edited by J. Ławrynowicz. IX, 331 pages. 1985.

Vol. 1166: Banach Spaces. Proceedings, 1984. Edited by N. Kalton and E. Saab. VI, 199 pages. 1985.

Vol. 1167: Geometry and Topology. Proceedings, 1985. Edited by J. Alexander and J. Harer. VI, 292 pages. 1985.

Vol. 1168: S.S. Agaian, Hadamard Matrices and their Applications. III, 227 pages. 1985.

Vol. 1169: W.A. Light, E.W. Cheney, Approximation Theory in Tensor Product Spaces. VII, 157 pages. 1985.

Vol. 1170: B.S. Thomson, Real Functions. VII, 229 pages. 1985.

Vol. 1171: Polynômes Orthogonaux et Applications. Proceedings, 1984. Edité par C. Brezinski, A. Draux, A.P. Magnus, P. Maroni et A. Ronveaux. XXXVII, 584 pages. 1985.

Vol. 1172: Algebraic Topology, Göttingen 1984. Proceedings. Edited by L. Smith. VI, 209 pages. 1985.